D0848148

GEOLOGICAL ASPECTS OF
Hazardous Waste
Management

STEPHEN M. TESTA

WITHDRAWN

LEWIS PUBLISHERS
Boca Raton   Ann Arbor   London   Tokyo

Library of Congress Cataloging-in-Publication Data

Testa, Stephen M.
    Geological aspects of hazardous waste management / by Stephen M. Testa.
      p. cm.
    Includes bibliographical references and index.
    ISBN 0-87371-630-2
    1. Hazardous wastes.   2. Environmental geotechnology.   I. Title.
TD1064.T47   1993
628.4'2—dc20   93-4821
                                                             CIP

© 1994 by CRC Press, Inc.
Lewis Publishers is an imprint of CRC Press

No claim to original U.S. Government works
International Standard Book Number 0-87371-630-2
Library of Congress Card Number 93-4821
Printed in the United States of America   2 3 4 5 6 7 8 9 0
Printed on acid-free paper

"You must teach your children that the ground beneath their feet is the ashes of our grandfathers. So that they will respect the land, tell your children that the earth is rich with the lives of our kin. Teach your children what we have taught our children — that the earth is our mother. Whatever befalls the earth, befalls the sons of the earth.

"This we know. The earth does not belong to man; man belongs to the earth. All things are connected like the blood which unites one family.

"Whatever befalls the earth befalls the sons of the earth. Man did not weave the web of life; he is merely a strand in it. Whatever he does to the web, he does to himself . . ."

(Chief Seattle, 1882)

ABOUT THE AUTHOR

Stephen M. Testa is president and founder of Applied Environmental Services, Inc., located in San Juan Capistrano, California. He received his B.S. and M.S. in Geology from California State University at Northridge. For the past 17 years, he has worked in the areas of geology, hydrogeology, engineering geology, and hazardous waste management with firms such as Bechtel, Inc., Converse Consultants, Dames and Moore, and Engineering Enterprises, Inc.

He has participated in numerous subsurface hydrogeologic site characterization projects associated with nuclear hydroelectric power and petrochemical plants, hazardous waste disposal facilities, and other industrial and commercial complexes. For the past five years, his main emphasis has been in the area of nonaqueous phase liquid hydrocarbon recovery and aquifer restoration. Maintaining overall management and technical responsibilities in engineering geology, hydrogeology, and hazardous waste-related projects, he has participated in numerous projects involving geologic and hydrogeologic site assessments, soil and water quality assessments, soil remediation, design and development of groundwater monitoring and aquifer remediation programs, waste minimization and recycling, and expert testimony.

Mr. Testa is the author of three books and more than 60 technical papers. He is an active member of numerous organizations including the American Association for the Advancement of Science, Association of Engineering Geologists, American Association of Petroleum Geologists, Association of Groundwater Scientists and Engineers, Hazardous Materials Control Research Institute, Geological Society of America, California Groundwater Association, and Sigma Xi. He is also a member of the American Institute of Professional Geologists, where he has served on the Executive Committee and on various committees, including the National Committee for Professional Development, Continuing Education and Screening. He also conducts workshops on various environmental topics and teaches part-time at the University of Southern California and California State University at Fullerton.

Lovingly dedicated to
my wife,
Lydia Payne Testa

PREFACE AND ACKNOWLEDGMENTS

Today's environmental specialist has to be half geologist, half chemist, half attorney, and half engineer. This is certainly mathematically absurd, but the fact of the matter is that multidisciplinary skills are critically important to those individuals interested in participating in and solving the challenging environmental problems we all face since the commencement of the industrial revolution in the early 1900s. Geological aspects of hazardous waste management comprise a multifaceted and broad field that includes geology, geophysics, hydrology, hydrogeology, soil science, risk assessment, environmental engineering, environmental geochemistry, and environmental law, among others. The rapid expansion of the environmental industry as applied to the subsurface has made it very difficult not only to keep pace with ambitious regulatory programs mandating that certain geologic and hydrogeologic issues be addressed, but to maintain a mechanism for technology transfer and enhanced communication among subfields. In addition, traditional geology departments have been very slow in preparing students with the necessary skills required to fulfill the immediate and future needs of this industry.

Successful subsurface characterization, detection monitoring, and ultimate remediation are predicated upon a solid conceptual understanding of the geologic and hydrogeologic conditions beneath the site in question. The mere presence of constituents considered hazardous or toxic at certain quantities does not necessarily equate to an adverse risk to public health, safety, and welfare, or to the degradation of groundwater resources. Nor does their presence warrant that remediation be performed. The risks associated with the subsurface presence of these constituents will vary depending upon geologic and hydrogeologic conditions, the potential for adverse impact on groundwater resources considered of beneficial use, existing and planned site usage, magnitude of the problem, and time and financial constraints. Determination as to whether a risk exists requires adequate characterization of subsurface conditions regarding the occurrence and detection of these constituents, assessment of fate and transport mechanisms and processes, and adequate conceptualization of preferential migration pathways, followed by development of an appropriate remediation strategy. Having determined that corrective action is deemed necessary, a remediation strategy must be developed that is environmentally sound, cost-effective, and time-efficient. With numerous strategies and innovative techniques, or combination of techniques, being employed in the assessment of subsurface conditions and subsequent remediation, the geologist plays a very important role in the decision-making process.

This book focuses upon the waste problem and the role of the geologist in addressing this problem. The main features of the subsurface waste disposal problem are discussed in sufficient depth for this book to be of practical use to the professional consultant, regulator, project manager, and attorney as a useful tool and reference in dealing with the varied and complex problems associated with subsurface hazardous waste-related issues. This book also maintains adequate technical depth and mathematical treatment to serve as a textbook suitable for undergraduate and graduate courses in environmental science, environmental geology with an emphasis in waste disposal, and hazardous waste management certification programs. Traditionally, the undergraduate curriculum was and still is rarely adequate for graduate studies in what the author considers as modern environmental geology and hydrogeology, nor have these programs adequately prepared the student to participate in this expanding and growing field which is rapidly employing the largest segment of geology professionals. College geology curricula have also traditionally emphasized geologic assessment and analysis on a regional scale, whereas today's hazardous waste management industry demands focused studies on a site-specific scale. And with the continued decline in employment within the petroleum, mining, and geotechnical industries, more and more geologists are becoming employed in environmental-related fields. Thus, this book will be useful in cross-training geologists displaced by downturned markets in other industries, and may, in fact, enhance their continued contribution to their respective industries, since all industries generate waste and contribute to the overall waste problem.

This book stresses the importance of understanding the regional geologic and hydrogeologic framework or setting, while concurrently focusing on the site-specific subsurface environment. Chapter 1 introduces the reader to the waste problem and the role of geology and the geologist in addressing this problem. Chapter 2 discusses the major regulatory programs where geologic input and application is required, including discussion of environmental site assessments, with illustrative discussions of oil field and geothermal properties. Despite our good intentions, it is the regulations that drive the environmental machine. Geologic principles are presented in Chapter 3, including discussion of the concepts of porosity, permeability, and diagenesis, with an emphasis on depositional environments, facies architecture, and fractured media. Hydrogeologic principles within the vadose and saturated zones, hydrodynamics, and water quality are discussed in Chapter 4. Conventional techniques for conduct of subsurface assessment and site characterization-related activities are covered in Chapter 5, and geophysical applications are presented in Chapter 6.

Discussion of the complex concept of what makes a material or substance a hazardous or toxic waste, and how these materials are classified, is presented in Chapter 7. The predominant physical, biological, and chemical subsurface processes which affect these constituents and contaminants are discussed in Chapter 8. Light and dense nonaqueous phase liquid occurrence, detection, and behavior in the subsurface are discussed in Chapter 9.

Siting of new hazardous waste disposal facilities initiate at the regional level and then are followed by localized and site-specific geologic and hydrogeologic studies. A hazardous waste disposal facility must be designed to protect all aspects of the environment during design, construction, operations, and post-closure for some undefined period of time into the future. To minimize or prevent the degradation of the land and waters from the dissolution and transport of hazardous and toxic constituents, the regional and site-specific geologic and hydrogeologic conditions of the proposed disposal facility must be meticulously studied and designed appropriately. The remaining four chapters address these issues and discuss how the preceding concepts and principles are applied in respect to our four major subsurface disposal alternatives: landfills, underground injection, underground geologic repositories, and ocean disposal. Issues pertaining to siting considerations, geologic and hydrogeologic considerations on a regional and site-specific basis, geochemical aspects, and design criteria are presented. Chapter 10 covers landfill disposal and includes discussion of detection monitoring within both the vadose and saturated zone. Underground injection of hazardous and toxic wastes is discussed in Chapter 11. The use of underground repositories of varying rock types is presented in Chapter 12. The disposal of hazardous and toxic waste in the ocean is discussed in Chapter 13.

This book could not have been completed without the gracious assistance and contributions of numerous individuals. My thanks and appreciation go out to Dr. Jim Conca (Washington State University Tri-Cities, Richland, Washington), Duane Winegardner (American Environmental Consultants, Norman, Oklahoma), and my technical staff at Applied Environmental Services, Inc., notably, Dennis Patton and Mike Patton. Ms. Beverly Dowdy assisted in literature searches and retrieval of documents. The bulk of the administrative tasks, word processing, and editing was performed by my wife Lydia, who dedicated her time in lieu of horseback-riding, movies, vacations, etc., and devotedly complained but still prepared drafts and provided support throughout this project. Word processing support was also provided by Regina Moore, Ellen Patton (a whiz at tables!), Brenda Saltiban, and Elissa Palmer. To all of you, my deepest gratitude and appreciation.

Finally, I wish to take this opportunity to express my appreciation to those who during my career have knowingly or unknowingly strongly influenced me professionally, and served as role models. Included in this special group of individuals are Bob Bean, Bill Bliton, Richard Frankian, Dr. Richard Harlan, Dr. Allan Hatheway, Jon Lovegreen, Cole McClure, Alan O'Niell, Richard Proctor, Dr. James Slosson, and Glen Tarbox.

Stephen M. Testa
January 1993

GEOLOGICAL ASPECTS OF
Hazardous Waste Management

CONTENTS

1 INTRODUCTION

"Man is everywhere a disturbing agent. Wherever he plants his foot, the harmonies of nature are turned to discords."
(George Perkins Marsh, 1874)

THE WASTE PROBLEM

Early environmental concerns in the United States focused on conservation and preservation, beginning with the publication of *The Earth as Modified by Human Action* in 1874 (previously published in 1864 as *Man and Nature*) by George Perkins Marsh. Marsh wrote about the irrevocable damage to what we now call the ecosystem as a result of man's actions such as deforestation and the drainage of lakes and marshes. With the publication of *Silent Spring* by Rachel Carson in 1962, it was clear to the American public that pollution, not preservation, was to be the principal environmental concern in the future.

The problems associated with providing a sanitary method of waste disposal and a safe water supply are as old as civilization. Historically, a camp was set up near a water body, and the primary strategy for disposing of the waste generated was simply to scatter the waste over the adjacent countryside. If the waste was not dumped onto the land, it was discharged into surface water bodies or buried. When sanitary conditions became undesirable, the camp was moved. With the establishment of permanent towns and cities, more sophisticated and adequate disposal facilities were required as increasing urbanization increased the complexities of this problem. Numerous documented cases illustrate the adverse effects of poor waste management practices — including health risks due to soil, surface water, and groundwater contamination, reduced property values, aesthetics, and physical harm.

Waste disposal today is an important social, economic, and political issue in these environmentally conscious times. About 200 million tons of hazardous waste is produced each year in the U.S., of which about 150 billion gallons is

1

liquid waste. The problem is such that, like the national debt or distances in space, the amount of waste produced has become unfathomable. By the 1970s, developed countries started to initiate legislation and the legacy of an "out-of-sight, out-of-mind" mentality was now beginning to catch up with us. In addition to the ever-increasing volumes of domestic and industrial waste being generated, about 40 years ago radioactive waste was granted the title of "man's ultimate problem," although generally restricted to government-controlled lands or certain mining communities, mostly remote from urbanized areas. Although much research has focused on the safe disposal of radioactive waste during this period, we have only recently started to apply the wealth of information generated to the safe disposal of hazardous waste, which generally occurs in close proximity to population centers and environmentally sensitive areas of beneficial value.

The waste disposal problem that we have inherited today reflects upon several factors, notably population growth, urban growth, the generation of literally thousands of new synthetic materials, exploration and development of strategic minerals, and the development of an affluent society with a prevailing attitude that it is cheaper to produce something in quantity than to repair it.

Population growth and technological change are the chief causes of environmental stress. The population growth in the U.S. has been tremendous with an increasing trend since the 1850s, causing a heavy draft upon this country's resources. In the 1870s, the U.S. had fewer than 40 million people. With more than six times that amount today, all else being equal, five times as many people produces five times as much waste with no concurrent increase in environment. About 1 billion people are added to the world's population every 11 years, with the world's population essentially doubling in the past 40 years. The current population of about 5.4 billion is anticipated to double in 50 years. Of this total, about 1.2 billion inhabit the industrial countries, with the remainder inhabiting developing countries. By the year 2000, it is estimated that 50% of the population will reside in urban areas.

Before the Industrial Revolution, most wastes were basically food. Now there are about 50,000 chemical products, with at least 1,000 new chemical compounds added annually. The U.S. average per capita production of waste is on the order of about 4.5 kg (10 lb) generated daily. Of the 4.5 billion tons produced per year, about 90% is derived from agricultural and mining activities, and about 3% from industrial processes, some of which is recycled. In addition, 34 billion gallons of water is disposed of daily and 35 million tons of hazardous waste produced annually.

The industrial regions of Central Europe are alarmingly poor from an environmental perspective. Raw sewage, industrial effluents laced with heavy metal, and toxic chemicals are the principal contributors to the deterioration of water quality in Central Europe's rivers. Pre-industrial growth of the environmental problem in Europe, however, differed from that in the U.S. because of the relatively smaller dimensions of space and magnitude. The worst-case example is the extensive Silesian coal belt across southwestern Poland, north-western Czechoslovakia, and southeastern (formerly East) Germany, where much of Central Europe's heavy industry lies (i.e., steel, cement, chemical, petrochemical, and coal-burning power plants). The cause of regional environmental degradation of Central Europe is very complex and reflects many political and socioeconomic factors related to initiation of socialist rule and the imposition of the Stalinist model of industrial growth in the late 1940s. However, in certain areas such as Katowice, Poland, and parts of Czechoslovakia, industrial growth was developed prior to World War II. After almost a century of environmental abuse, the same industrialized nations are now becoming aware of the damage that has been inflicted.

Another aspect of the waste problem is finding space for it. In Europe, about 10% of the roughly 25 million tons of waste generated each year is passed from one country to another for disposal. The real problem, however, lies in developing countries, where disposal problems that were comparatively small (since everything has value) are now escalating. This escalation reflects the receiving of wastes from other countries exacerbated by few restrictions on how these wastes are disposed of.

In the U.S. and a few other countries, the problems faced today are inherited in that they reflect antiquated facilities and poor operational practices and procedures for doing business which are no longer considered acceptable. Generally, the U.S. is progressing in a favorable manner toward environmental awareness with more and more emphasis on waste minimization, recycling, and reuse. However, we are just beginning to get a glimpse, with the opening of eastern Europe and the demise of the former Soviet Union, of the environmental problems that exist in those countries as well as within many developing countries (Figure 1-1).

The most profound global environmental change in the brief history of the human species is currently being witnessed with human activity as the major agent of these changes. All human activity produces waste which pollutes all environmental media (i.e., air, water, and soil) and eventually enters the biosphere via one or several of these pathways (Figure 1-2). To adapt Descartes: I am, therefore I pollute. Reduction of waste or emissions in one medium may lead to an increase in another. From production to ultimate disposal, many opportunities exist for its escape into the environment and human exposure, and the regulatory authorities to deal with the overall cycle is fragmented. In some cases as with heavy metals, such waste material is permanent and cannot be destroyed, but only stabilized or fixated.

WASTE DISPOSAL ALTERNATIVES

There will always be a need for waste storage and disposal facilities, especially where the waste cannot be recycled or reused. The adverse effects

Figure 1-1   Aerial view of Smoky Mountain, an above-ground landfill situated immediately north of Manila, Republic of the Philippines. This landfill smokes continuously due to spontaneous combustion. Note the living accommodations (shanties) that abut the landfill.

Figure 1-2   Al-Rawdhatayn oil field shows the environmental impact of the Gulf War on the desert environment as of March 5, 1991 (Photo courtesy of Dennis Patton).

of waste disposal facilities in general are well known and include reduced property values, aesthetics, and physical harm via gases, microorganisms and toxins through primary pathways such as air, land, and water (Figure 1-3). There also exists a large degree of public distrust (Figures 1-3 and 1-4). The EPA estimates about 750,000 sources of hazardous waste, of which 90% is disposed of improperly. How do we dispose of waste in a way that does not damage the environment or pose health risks to people living near disposal sites? Many states now are reluctant to allow new sites and in recent years have curbed imports of out-of-state wastes. Disposal of waste anywhere beneath the land or ocean surface typically means that the waste will be located in relatively close proximity to the generators (i.e., densely populated or industrialized areas), and that public opinion will be negative. The public's negative reaction to waste disposal "in their backyard" reflects a lack of appreciation for the role geology plays in ensuring the safety of any underground disposal site.

Geologically, waste is commonly disposed of via landfills. Within the last few decades with the advent of drilling technology, waste was and continues to be reinjected into subsurface rock formations, or water-bearing zones. Within the past decade, waste has been and continues to be considered for storage in underground geologic repositories. Waste is also disposed of off-shore.

These various methods of waste disposal in the U.S. cost an estimated $4.5 billion a year, or more than $10 per ton. About half the cost is for the collection and transportation of the waste to the processing or disposal site; the remaining cost is incurred in the processing and ultimate disposal of the residue at some permanent storage site. Texas, for example, hosts a large oil and chemical industry and is the biggest producer of hazardous and toxic waste, generating 20% (270 million tons) of the nation's annual production. No recent estimate of the volume of waste disposed of per selected method of disposal exists; however, a survey (referred to as the Eckhardt Survey) of 53 of the largest chemical manufacturing companies in the U.S. was conducted for the period between 1950 and 1979 (Table 1-1). The survey showed that of the 822 million tons of waste disposed of, 99% was disposed of in the state of origin, 86.6% went to sites still in use, and 13% to sites now closed. Not shown or broken out is the volume of contaminated soil generated which in some areas of the country, such as Oklahoma, accounts for about 98% of the waste generated on a one-time occurrence.

THE ROLE OF THE GEOLOGIST

The geologist's role in society has changed dramatically over the years. Historically, an emphasis in natural history existed, followed by energy development, notably, as man moved from or evolved from an agricultural society to an industrial society at the turn of the century. Since the 1950s, the role of

Figure 1-3   View to north shows partial surface clean-up of Western Processing, a defunct waste-recycling Superfund site located in Kent, WA.

Figure 1-4   Excavated drums containing paint wastes from a farm site in southwestern Washington that were illegally buried. Such actions, regardless of quantity, can have significant long-term consequences by reducing adjacent property values for years.

Table 1-1　Amount of Waste Disposed of per Disposal Method[a,b]

SIC[c] Division	Untreated				Treated				Unknown	Others[d]	Total
	Land-fills	Pits, Ponds, Lagoons	Injection Wells	(Sub-total)	Incin-erated	Recycled or Reused	Evapo-rated	(Sub-total)			
Organics	2,200	5,237	3,225	10,662	485	539	48	1,072	42	5,067	16,843
Inorganics	2,568	4,702	2,584	9,854	357	77	5,856	6,290	239	3,798	20,181
Plastics	1,112	254	322	1,688	115	385	7	507	11	559	2,765
Agricultural	1,145	351	708	2,204	912	39	2,328	3,279	179	10,686	16,348
Misc. Chemicals	132	122	10	264	19	27	10	56	1	8	329
Other	986	4,240	170	5,396	60	60	332	432	0.65	12	5,861
Medicinal & Soaps	122	9	0.96	132	85	0.65	0	86	0	135	353
Unknown	33	4	5	42	3	0	179	182	0.90	120	345
Total	8,298	14,919	7,025	30,242	2,036	1,128	8,760	11,924	474	20,385	63,025

a Modified after PEDCO (1979).
b Data is presented in units of thousands of tons.
c Standard Industrial Classification.
d Includes storage, land applications, and ocean disposal among other methods.

the geologist has transformed mainly to the application of the earth sciences to geotechnical problems, and during the past two decades to environmental problems.

The geologist's participation in the waste problem-solving effort is significant. Geology encompasses several subdisciplines including hydrogeology, hydrology, structural geology, stratigraphy, geophysics, and geochemistry, among others. The multifaceted role of the geologist is to specify the type of information required for a certain project, evaluate the best way to obtain and analyze the data, interpret the data, provide input and participate in the decision-making and planning process, and translate and interpret information for the general public and policy user.

A thorough understanding and knowledge of the subsurface environment and its various characteristics is essential to assess the likely hazard posed by a particular contaminant or waste, and its behavior and fate over time. The geologist must be relied upon for evaluation of the significance of the hazard and whether soil and/or notably beneficial-use groundwater is jeopardized. The geologist must also contribute to educating the public.

The geologist not only provides guidance in addressing these issues, but also participates in subsequent mitigation activities, playing a significant role in the remediation of hazardous waste sites. Projected costs for assessing the need for and conduct of remediation have significantly increased over the past decade. This increase has commonly been attributed to the increase in the number of sites being discovered and associated treatment costs, but also reflects premature design of a remediation strategy before the site has been properly characterized and/or lack of proper understanding of the subsurface environment. Unfortunately, in dealing with adversely affected sites, there is a strong tendency toward emphasizing site remediation at the expense of the site investigation. Much too often this syndrome of "analysis paralysis" is reflected in a diminished quality and effectiveness of the remedial system. The end result of this ineffectiveness is costs being significantly exceeded beyond what was anticipated, and creating a significant potential for exposure to liabilities which add to the cost and could result in holding up the remediation process for years, let alone the possibility of certain entities going out of business.

The geologist is also relied upon for the evaluation of suitable subsurface disposal alternatives, meticulous study of existing and potential disposal sites, and the subsequent engineering, monitoring, and eventual closure of such sites. Whether the intended disposal site is suitable for the type of wastes being disposed of, via looking at natural analogs or case histories, must also be evaluated.

The role of the geologist today not only requires a level of credibility with the client, regulator, or public, but also with other geologists. With the public expectations being unrealistically high, there is a pervasive problem of promising more certainty regarding the exposure to short- and long-term risk than can actually be provided. The geologist must thus predict future processes and events over the course of thousands of years without understating the risks involved — a most difficult task when the more sensitive sites such as Yucca Mountain, Nevada, are politically selected rather than being selected upon their technical merits relative to other candidate sites. It is necessary to differentiate between risk, which is an event with a known probability, and true uncertainty, which is an event with an unknown probability. The geologist must thus be prepared to provide: (1) conservative (vs optimistic) accounting of uncertainty, (2) accountability as to what and why certain studies and investigations are needed, and (3) the courage to give both the good and the bad news to the appropriate decision makers.

BIBLIOGRAPHY

Carson, R., 1962, *Silent Spring:* Houghton Mifflin, Boston.

Coates, D.R., 1985, *Geology and Society:* Chapman and Hall, New York, 406 p.

Cole, J.A. (Editor), 1974, *Groundwater Pollution in Europe:* Water Information Center, Inc., Port Washington, NY, 546 p.

Commoner, B., Corr, M. and Stamier, P.J., 1971, The Causes of Pollution: *Environment,* April.

Cavallero, A., Corradi, C., DeFelice, G. and Grassi, P., 1985, Underground Water Pollution in Milan by Industrial Chlorinated Organic Compounds: In *Effects of Land Use Upon Fresh Waters,* Ellis Harwood, Chichester.

Dummer, M. and van Straaten, L., 1988, The Influence of Industrialisation and Hydrogeology on the Quality of Ground and Surface Water in Bielefeld (case studies): In Proceedings of the Symposium on Hydrological Processes and Water Management in Urban Areas, Duisburg, FRG, April 24-28, pp. 423-428.

Ehrlich, P.R. and Haldren, J.P., 1971, Impact of Population Growth: Science, March 26.

Fleetwood, A., 1969, Urbant Paverkat Gundvatten: *Vatten,* Vol. 2, No. 69, pp. 107-112.

Foster, S.S.D., 1988, Impacts of Urbanisation on Groundwater: In Proceedings of the Symposium on Hydrological Processes and Water Management in Urban Areas, Duisburg, Germany, pp. D1-D24.

Gosk, E., Bishop, P.K., Lerner, D.N. and Burston, M., 1990, Field Investigation of Solvent Pollution in the Groundwaters of Coventry, UK: In *Proceedings of the IAH Conference on Subsurface Contamination by Immiscible Fluids* (Edited by K.U. Weyer), Balkema Publications, Rotterdam.

Handa, B.K., Goel, D.K., Kumar, A. and Sondhi, T.N., 1983, Pollution of Groundwater by Chromium in Uttar, Pradesh: Technical Annual X, *Journal of Indian Association Water Pollution Control,* pp. 167-176.

Jackson, P.E. (Editor), 1980, Aquifer Contamination and Protection: UNESCO Studies and Reports in Hydrology, Vol. 30.

Johnson, K.S. (Editor), 1990, Hazardous-Waste Disposal in Oklahoma, A Symposium: Oklahoma Geological Survey Special Publication 90-3, 87 p.

Kakar, Y.P. and Bhatnagor, N.C., 1981, Groundwater Pollution Due to Industrial Effluents in Ludhiana, India: In *Quality of Groundwater* (Edited by W. vanDuijvenbooden, Glasbergen and van Lelvveld) Study in Environmental Sciences Vol. 17, pp. 265-275.

Katz, B.G., Lindner, J.B. and Ragone, S.E., 1980, A Comparison of Nitrogen in Shallow Groundwater From Sewered and Unsewered Areas, Nassau County, New York, From 1952 Through 1976: *Ground Water,* Vol. 18, pp. 607-618.

Kowalski, J. and Janink, Z., 1983, Groundwater Changes in the Urban Area of Wroclaw in the Period 1874-1974: In *Proceedings of the IAHS Conference on Relation of Groundwater Quantity and Quality,* IAHS Publication No. 146, pp. 45-56.

Krauskopf, K.B., 1988, *Radioactive Waste Disposal and Geology: Topics in the Earth Sciences, Vol. 1,* Chapman and Hall, New York, 145 p.

Ku, H.F.H., 1980, Ground-water Contamination by Metal-plating Wastes, Long Island, New York, USA: In *Aquifer Contamination and Protection* (Edited by Jackson, R.E.), UNESCO Studies and Reports in Hydrology, Vol. 30, pp. 310-317.

Lloyd, J.W., Lerner, D.N., Rivett, M.O. and Ford, M., 1988, Quantity and Quality of Groundwater Beneath an Industrial Conurbation in Birmingham, UK: In Proceedings of the Symposium on Hydrological Processes and Water Management in Urban Areas, Duisburg, Germany, pp. 445-452.

Marsh, G.P., 1874, *The Earth as Modified by Human Activity:* Scribner, Armstrong & Company, New York, 656 p.

Melosi, M.V., 1987, *Garbage in the Cities - Refuse, Reform, and the Environment,* 1880-1980: Dorsey Press, Chicago.

Olania, M.S. and Saxena, K.L., 1977, Groundwater Pollution by Open Refuse Dumps at Jaipur: *Journal of Environmental Health (India),* Vol. 19, pp. 176-188.

Patrick, R., Ford, E. and Quarles, J., 1988, *Groundwater Contamination in the United States:* University of Pennsylvania Press, Second Edition, Philadelphia, 513 p.

PEDCO, 1979, PEDCO Analysis of Eckhardt Committee Survey for Chemical Manufacturer's Association: PEDCO Environmental, Inc., Washington, D.C.

Razack, M., Drogue, C. and Baitelem, M., 1988, Impact of an Urban Area in the Hydrochemistry of a Shallow Groundwater (alluvial reservoir), Town of Narbonne, France: In Proceedings of the Symposium on Hydrological Processes and Water Management in Urban Areas, Duisburg, Germany, pp. 487-494.

Roxburgh, I.S., 1987, *Geology of High-Level Nuclear Waste Disposal - An Introduction:* Chapman and Hall, New York, 229 p.

Sahgal, U.K., Sahgal, R.K. and Kakar, Y.P., 1989, Nitrate Pollution of Ground Water in Lucknow Area: In Proceedings of the International Workshop on Appropriate Methodologies for Development and Management of Groundwater Resources in Developing Countries, National Geophysical Research Institute, Hyderabad, India, pp. 871-892.

Savini, J. and Kammerer, J.C., 1961, Urban Growth and the Water Regime: U.S. Geological Survey Water-Supply Paper, No. 1591-A, 43 p.

Shahin, M.M.A., 1988, Impacts of Urbanisation of the Greater Cairo Area on the Groundwater in the Underlying Aquifer: In Proceedings of the Symposium on Hydrological Processes and Water Management in Urban Areas, Duisburg, Germany, pp. 517-524.

Sharma, V.P., 1988, Groundwater and Surface Water Quality in and around Bhopal City in India: In Proceedings of the Symposium on Hydrological Processes and Water Management in Urban Areas, Duisburg, Germany, pp. 525-532.

Somasundarum, M.V., Ravindran, G. and Tellam, J.H., 1993, Ground-Water Pollution of the Madras Urban Aquifer, India: *Ground Water,* Vol. 31, No. 1, pp. 4-11.

The World Resources Institute, 1992, *World Resources 1992-93:* Oxford University Press, New York, 385 p.

Thomson, J.A.M. and Foster, S.S.D., 1986, Effects of Urbanisation on Groundwater of Limestone Islands: An Analysis of the Bermuda Case: *Journal of the Institute of Water Engineers and Scientists,* Vol. 40, pp. 527-540.

2 REGULATORY FRAMEWORK

"Despite our good intentions, it is the regulations that drive the environmental machine."

(S. M. Testa, 1993)

INTRODUCTION

Regulations and statutes of geological and hydrogeological interest are rarely highlighted in the literature. In a few instances, familiar geological terms are evident; however, in many cases the geological terms are disguised under terms such as "environmental criteria." The absence of such terms does not obviate the need for geologists' responsible input and role in the decision-making process, over and above their contributions made to the profession and to society in general. Many cases exist where geological input is indirectly evident and coveted under tax codes, real estate transactions, property condemnation codes, and waste classification or declassification regulations, to identify a few. Information has been compiled and provided in convenient handbooks published by the U.S. Environmental Protection Agency (EPA), the Government Institutes, Inc., and other federal and state agencies. However, it is prudent for all serious practitioners to dedicate time to reading the actual statutes and reviewing case histories. Certain professional organizations such as the American Institute of Professional Geologists provide a summary of federal and state regulations affecting the geological community on a periodic basis. In addition, regulatory terms are not necessarily defined or interpreted in other sources the same as in definitions provided by the geologic profession (i.e., American Geological Institute). A glossary of geologic terms as used in myriad federal regulations is provided in Appendix A.

Government regulations pertaining to the potential presence, assessment, monitoring, and remediation of hazardous constituents in the subsurface environment have been promulgated since the early 1970s. The regulatory programs in place under the major federal environmental statutes of concern to

15

American industry and the geologic profession are summarized in Table 2-1 and are discussed below. These federal statutes define most of the substantive compliance obligations; however, the statutes do not operate alone — they are supplemented and complemented by other environmental programs. Many of the federal statutes (i.e., Clean Water Act) allow for the establishment of federal-state regulatory programs in which the states are given the opportunity to enact and enforce laws to achieve the regulatory objectives that Congress has established, providing they meet minimum federal criteria.

States have usually taken over certain regulatory programs in their jurisdiction when the opportunity has presented itself, thus becoming the primary permitting and enforcement authorities. The states are, however, subject to federal intervention if they do not enforce the statutes effectively or rigorously enough. Since enforcement interpretation may vary considerably from state to state, separate state environmental law handbooks are readily available from such sources as Government Institutes, Inc. Certain states, such as California and New Jersey, have taken the initiative to provide additional environmental protection beyond that generally available through the federal statutes.

Discussion of the more pertinent federal regulations that impact the environmental community is presented in this chapter. Included are the National Environmental Policy Act (NEPA); Clean Water Act (CWA); Safe Drinking Water Act (SDWA); Toxic Substance Control Act (TSCA); Resource Conservation and Recovery Act (RCRA); and the Comprehensive Environmental Response, Compensation, and Liability Act (CERCLA). Environmental site assessments as exemplified by oil-field and geothermal properties are also discussed.

NATIONAL ENVIRONMENTAL POLICY ACT (NEPA)

NEPA was signed into law by President Richard M. Nixon on January 1, 1970, resulting in a decade of legislative activity aimed at environmental protection that ended on December 11, 1980 with the passing of CERCLA by President Jimmy Carter. Under NEPA, every project that has the potential to affect the environment requires an environmental impact statement (EIS) and, hence, needs a federal permit. The EIS in turn serves to quantify the anticipated environmental impact, identify potential benefits, and describe and compare alternatives.

CLEAN WATER ACT (CWA)

The basic framework for federal water pollution and ocean-dumping control regulation was promulgated in 1972 by the enactment of the Federal Water Pollution Control Act (FWPCA) and the Marine Protection, Research, and

Sanctuaries Act (MPRSA), respectively. The FWPCA was renamed the CWA in 1977 and focused upon a rigorous control of toxic water pollutants. Extensive amendments to improve water quality in areas where compliance with nationwide minimum discharge standards was insufficient for attainment of water quality goals were enacted in 1987. Then, in 1990 the Oil Pollution Act (OPA) was enacted, affecting oil and hazardous substance discharges; it included a $1 billion cleanup fund. The MPRSA was also amended on several occasions, most recently in 1988, prohibiting offshore dumping of industrial waste and sewage sludge.

The CWA provides rigorous control of toxic water pollutants by imposing effluent limitations and preventing discharges of pollutants into any waters from any point source. This system includes six basic elements:

- Technology-based effluent limitations
- Water quality-related effluent limitations
- A nationwide permit program (National Pollutant Discharge Elimination System)
- Specific deadlines for compliance or noncompliance
- Specific provisions applicable to certain toxic and other pollutant discharges, oil spills, or hazardous chemical spills
- A loan program to help fund publicly owned treatment works' attainment of the applicable requirements.

The regulation of discharges in general, however, provided incentive to dispose of waste materials on or below the ground surface, thus increasing the potential threat to overall groundwater quality. Underground discharges are regulated under SDWA and RCRA, as discussed later.

The CWA does substantially affect development in areas adjacent to navigable waters, controlling dredging activities and the disposal of dredged or fill material into navigable waters. This program is overseen by the Corps of Engineers, which maintains the authority to designate disposal areas and issue permits to discharge dredged and fill material.

The MPRSA governs all discharges of wastes to ocean waters within U.S. jurisdiction or by U.S. vessels or citizens to the oceans anywhere. Due to numerous amendments, no ocean dumping of industrial waste or sewage sludge is permitted, with the exception of exigent circumstances such as in the New York City area, where certain acid and alkaline industrial wastes and sewage sludge are disposed of. New York has the only remaining deep ocean site in the U.S., Deepwater Dumpsite No. 106 (a 106-mile ocean waste dump site). Located in the New York Bight just off the continental slope of New Jersey, this dump site has been in operation since 1961 and is anticipated to be terminated.

Ocean disposal of chemical and biological warfare agents and radiological and high-level radioactive wastes is also prohibited. The only permitted ocean dumping activity of any significance is the disposal of dredged spoil.

Table 2-1   Summary of Major Federal Regulations

Regulation Title	Year	Purpose	Regulatory Reference
National Environmental Policy Act (NEPA)	1969 (1975a, 1975b)	Provides a national environmental policy and promotes consideration of environmental concerns by federal agencies	Pub. L. No. 91-190, 42 U.S.C.; Pub. L. Nos. 95-52 and 94-83
Clean Water Act (CWA)	1972 (1977, 1987, 1988, 1990)	Provides regulatory focus to rigorous control of toxic water pollutants protecting surface and groundwater quality to maintain "beneficial uses" of water; also includes provisions for ocean dumping and solid waste land disposal facilities	40 CFR Parts 112, 123, 125, 130, 131, and 136; 220-225, 227-233; 240-241, 255, and 257
Safe Drinking Water Act (SDWA)	1974 (1986)	Provides national standards for levels of contamination in drinking water, underground injection of hazardous waste, and for the protection of sole source aquifers	40 CFR Parts 141-142 and 143; 144-148 and 149
Toxic Substance Control Act (TSCA)	1976	Provides testing of chemical substances, old and new, entering the environment and regulate them if necessary	
Resource Conservation and Recovery Act (RCRA)	1976 (1984)	Provides "cradle-to-grave" control of hazardous waste by imposing management requirements on generators and transporters of hazardous wastes, and upon owners and operators of treatment, storage, and disposal facilities; also provides a comprehensive regulatory program for underground storage tank systems; mainly applies to active facilities	40 CFR Parts 260-268, 270-272, and 280-281

Act	Year	Description	CFR Reference
Comprehensive Environmental Response, Compensation, and Liability Act (CERCLA or Superfund)	1980 (1986)	Provides funding and enforcement authority for assessment and remediation of hazardous waste sites, and requires reporting of releases of hazardous chemicals; mainly applies to past, abandoned, and inactive sites; provides protocol for preparing and responding to discharges of oil and releases of hazardous substances, pollutants, and contaminants into or upon navigable waters and adjoining shorelines under the National Oil and Hazardous Substances Pollution Contingency Plan (NCP)	40 CFR Part 300
Nuclear Waste Policy Act (NWPA)		Provides standards for management and disposal of spent nuclear fuel, high-level, and transuranic radioactive wastes, and uranium and thorium mill tailings	40 CFR Parts 190-192; 50 CFR

The term "open ocean" refers to that area of the ocean beyond the continental shelf and includes what is called the deep ocean. The MPRSA extends this definition to those waters over the outer continental shelf that are measurably affected by freshwater input from rivers (Figure 2-1).

Overall, the MPRSA has been relatively successful in managing dumping and providing some degree of protection for the open ocean. Over the years, dumping of industrial wastes has declined dramatically, whereas dumping of sewage sludge has increased. Dumping and dredged material has varied significantly, with an overall decrease since 1974 despite increases from 1982 to 1984 (Office of Technology Assessment, 1987).

SAFE DRINKING WATER ACT (SDWA)

The SDWA in 1974 expanded the federal role of strictly enforcing standards used on interstate carriers to preventing communicable disease, promulgating regulations, establishing national standards for levels of contaminants in drinking water, creating state programs to regulate injection wells, and protecting sole source aquifers. This dramatically expanded role reflected information suggesting that (1) chlorinated organic chemicals were contaminating major surface and underground drinking water supplies, (2) widespread underground injection operations were threatening major aquifers, notably the Edwards aquifer supplying much of Texas, and (3) existing public water supply systems were antiquated, underfunded, understaffed, and an increasing threat to public health.

The 1986 amendments were aimed to speed up EPA's pace in issuing standards and implementing other provisions of the act by:

- Mandating the issuance of standards for 83 specific contaminants by June 1989, with standards for 25 additional contaminants to be issued every 3 years thereafter;
- Increasing EPA's enforcement powers;
- Providing increased protection for sole source aquifers and well head areas; and
- Regulating the presence of lead in drinking water systems.

The SDWA is structured into 11 parts; the most pertinent parts are summarized in Table 2-2.

Drinking Water Standards

The SDWA requires EPA to identify contaminants in drinking water which may have an adverse effect on public health and to specify a maximum

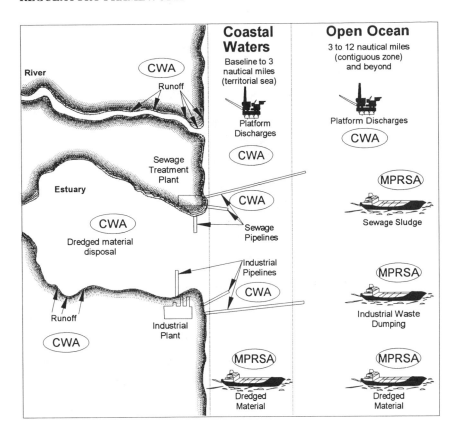

Figure 2-1 Jurisdictional boundaries of environmental laws affecting marine disposal (modified after the Office of Technical Assessment, 1987).

contaminant level (MCL) for each contaminant, where feasible. The MCL is the maximum permissible level of a contaminant in water which is delivered to any user of a public water system. In addition, MCLs established under the SDWA have been used as groundwater cleanup standards under CERCLA and have the potential to be used as standards not only for drinking water but also any groundwater regardless of its end use, if used at all.

A total of 83 contaminants are currently required to be regulated under the SDWA of 1986, and 77 substances or classes of substances comprise a priority list of contaminants for regulation after EPA completes standards for the initial list of 83. Although EPA was required to promulgate MCLs and maximum contaminant level goals (MCLGs) for 25 of the contaminants by January 1991 and every 3 years thereafter, EPA has not kept up with that pace.

Table 2-2   Hydrogeologic Issues Addressed Under the SDWA

Title	Description	Part
Maximum contaminant levels	Provides primary and secondary MCLs for inorganic, organic, turbidity, and radioactive constituents	141, 142, and 143
Underground injection control program	Provides criteria and standards for the federal and state UIC programs	144, 145, 146, 147
Sole source aquifers	Provides criteria for identifying critical aquifer protection areas and review of projects affecting the Edwards Underground Reservoir	149

Underground Injection Control (UIC) Program

The SDWA allows regulation of underground injection to protect usable aquifers from contamination with additional restrictions and control under the 1984 amendments to RCRA. Underground injection is the subsurface emplacement of fluid through a well or hole whose depth is greater than its width. Fluids can include hazardous waste, brine from oil and gas production, and certain mining processes. A well can include a septic tank, cesspool, or dry well. States that do not have full UIC authority, which includes about 23 jurisdictions, are regulated by EPA, including Indian lands for all states. EPA is also restricted from issuing regulations that interfere with the reinjection of brine from oil and gas production, gas storage requirements, or secondary or tertiary oil or gas recovery, unless public health is directly threatened. The UIC program must, however, protect aquifers that are, or may reasonably be expected to be, sources of drinking water (i.e., exceeds an MCL).

Sole Source Aquifers (SSA)

Special protection may be given to areas where (1) an aquifer is the sole or principal drinking water source for an area, supplying drinking water to 50% or more of an area's population, or (2) if contaminated, an aquifer would create a significant hazard or risk to public health. In determining whether an aquifer supplies drinking water to more than 50% of an area's population, EPA must determine whether readily available alternative supplies are available. Since some alternative water supplies are always available, an economic evaluation is performed to determine if the supply may be economically feasible. Availability thus equates to a cost to the typical household of not more than 0.4 to

0.6% of the household's average annual income. Thus, the mere construction of costly treatment and distribution systems will not be economically feasible.

As of January 1991, 51 SSA have been designated, most after the 1986 amendments reflecting the possibility of federal funding for local governments; delegation of authority was not decided until 1987. The legal effect of SSA designation is to bar federal financial assistance to projects that pose a threat to the aquifer from any source, public or private, and qualify these areas for special federal planning grants. A listing of SSA including their state locations and statutory references is presented in Table 2-3. The locations of SSA are shown in Figure 2-2.

SSA are situated in some of the most populated areas, with the majority situated in the northeastern part of the country. Also included are less-populated but well-known resort areas including Cape Cod, Nantucket, and Block Island, reflecting the acute vulnerability of coastal groundwater and the relative ease of aquifer identification.

SSA boundaries are typically geologic or hydrologic in nature in lieu of political or geographical boundaries. This task is not always easy to determine. SSA beneath islands are relatively easier to designate due to their ease in delineation, localized extent, high dependency, and high degree of vulnerability. Inland aquifers are more difficult to designate, reflecting a poor correlation between their overall lateral extent and population center, presence of multiaquifer systems, and whether they are hydraulically connected.

Included as part of the SSA program is the protection of wellhead areas from contaminants. The program defines a wellhead protection area as "the surface and subsurface area surrounding a water well or wellfields, supplying a public water system, through which contaminants are reasonably likely to move toward and reach such well or wellfield." States have had to adopt and submit wellhead programs to EPA, the financial incentive being the availability of federal grants up to 90% of the cost of developing such programs. The importance of the SSA and wellhead protection program cannot be overstated due to the potentially large populations that can be affected and the ability of such programs to set the stage for more elaborate federal and state groundwater protection legislation.

TOXIC SUBSTANCE CONTROL ACT (TSCA)

Enacted in 1976, TSCA provides EPA with the authority to require testing of both old and new chemical substances and mixtures of substances entering the environment and regulate them, where necessary, prior to human or environmental exposure. Although off to a slow start and considered to date as mainly an information-generating act, TSCA has two main regulatory components: (1) acquisition of sufficient information by EPA to identify and evaluate potential hazards from chemical substances and (2) regulation of the produc-

Table 2-3   Summary of Designated Sole Source Aquifers

Map No.[a]	Aquifer and/or Location	State	Statutory Ref.	Date Implemented
1	Edwards Aquifer	TX	40 FR 58344	02/16/75
2	Spokane Valley Rathdrum Prairie Aquifer	WA, ID	43 FR 5566	02/09/78
3	Northern Guam	Guam	43 FR 17868	04/26/78
4	Nassau/Suffolk counties, Long Island	NY	43 FR 26611	06/21/78
5	Fresno County	CA	44 FR 52751	09/10/79
6	Biscayne Aquifer	FL	44 FR 58797	10/11/79
7	Buried Valley Aquifer System	NJ	45 FR 30537	05/08/80
8	Maryland Piedmont Aquifer Montgomery, Frederick, Howard, Carroll counties	MD	45 FR 57165	08/27/80
9	Camano Island Aquifer	WA	47 FR 14779	04/06/82
10	Whidbey Island Aquifer	WA	47 FR 14779	04/06/82
11	Cape Cod Aquifer	MA	47 FR 30282	07/13/82
12	Kings/Queens counties	NY	49 FR 2950	01/24/84
13	Ridgewood Area	NY, NJ	49 FR 2943	01/24/84
14	Upper Rockaway River Basin Area	NJ	49 FR 2946	01/24/84
15	Upper Santa Cruz-Ayra Altar Basin Aquifer	AZ	49 FR 2948	01/24/84
16	Nantucket Island Aquifer	MA	49 FR 2952	01/24/84
17	Block Island Aquifer	RI	49 FR 2952	01/24/84
18	Schenectady-Niskayuna Schenectady, Saratoga, Albany counties	NY	50 FR 2022	01/14/85
19	Santa Margarita Aquifer Scotts Valley, Santa Cruz County	CA	50 FR 2023	01/14/85
20	Clinton Street - Ballpark Valley Aquifer System Broome and Tioga counties	NY	50 FR 2025	01/14/85
21	Seven Valleys Aquifer York County	PA	50 FR 9126	03/06/85
22	Cross Valley Aquifer	WA	52 FR 18606	05/18/87
23	Prospect Hill Aquifer Clark County	VA	52 FR 21733	06/09/87
24	Pleasant City Aquifer	OH	52 FR 32342	08/27/87
25	Cattaraugus Creek - Sardinia	NY	52 FR 36100	09/25/87
26	Bass Island Aquifer Catawba Island	OH	52 FR 37009	10/02/87

Table 2-3 Summary of Designated Sole Source Aquifers (continued)

Map No.[a]	Aquifer and/or Location	State	Statutory Ref.	Date Implemented
27	Newberg Area Aquifer	WA	52 FR 37215	10/05/87
28	Highlands Aquifer System	NY, NJ	52 FR 37213	10/05/87
29	North Florance Dunal Aquifer	OR	52 FR 37519	10/07/87
30	Volusia-Floridan Aquifer	FL	52 FR 44221	11/18/87
31	Southern Oahu Basal Aquifer	HI	52 FR 45496	11/30/87
32	Martha's Vineyard Regional Aquifer	MA	53 FR 3451	02/05/88
33	Buried Valley Aquifer System	OH	53 FR 15876	05/04/88
34	Pawcatuck Basin Aquifer System	RI, CT	53 FR 17108	05/13/88
35	Hunt-Annaquatucket-Pettaquamscutt Aquifer System	RI	53 FR 19026	05/26/88
36	Chicot Aquifer	LA	53 FR 20893	06/09/88
37	Edwards Aquifer-Austin Area	TX	53 FR 20897	06/07/88
38	Missoula Valley Aquifer	MT	53 FR 20895	06/07/88
39	Cortland-Homer-Preble Aquifer System	NY	53 FR 22045	06/13/88
40	St. Joseph Aquifer System Elkhart County	IN	53 FR 23682	06/23/88
41	NJ Fifteen Basin Aquifer System	NJ/NY	53 FR 23685	06/23/88
42	NJ Coastal Plain Aquifer System	NJ	53 FR 23791	06/24/88
43	Monhegan Island	ME	53 FR 24496	06/29/88
44	OKI - Miami Buried Valley Aquifer	OH	53 FR 25670	07/08/88
45	Southern Hills Aquifer System	LA, MS	53 FR 25538	07/07/88
46	Cedar Valley Aquifer	WA	53 FR 39779	10/03/88
47	Lewiston Basin Aquifer	WA, ID	53 FR 38782	10/03/88
48	North Haven Island	ME	54 FR 29934	07/17/89
49	Arbuckle-Simpson Aquifer	OK	55 FR 39236	09/25/89
50	Pootatuck Aquifer	CT	55 FR 11055	03/26/90
51	Plymouth-Carver Aquifer	MA	55 FR 32137	08/07/90

[a] See Figure 2-2 for approximate locations.

tion, use, distribution, and disposal of such substances where necessary. TSCA accomplishes these two objectives by:

- Premanufacture notification wherein a manufacturer must notify EPA 90 days before producing a new chemical substance, or for older chemicals if there is significant new use which increases human or environmental exposure

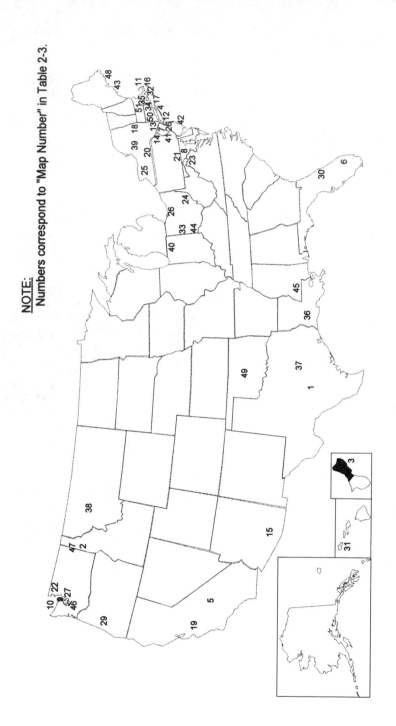

NOTE:
Numbers correspond to "Map Number" in Table 2-3.

Figure 2-2   Location of designated sole source aquifers.

- EPA compilation of an inventory of chemicals manufactured or processed in the U.S.
- Reporting requirements for manufacturers, including miners, importers, and some processors regarding chemical substances information, including production and exposure-related data
- Testing of old and new chemicals, including some mixtures under certain circumstances, if an unreasonable risk to health or the environment is suspected
- EPA authority to take whatever regulatory measures are deemed necessary to restrict chemicals suspected of posing harm to man or the environment
- Ultimate prohibition of certain chemical substances, notably, polychlorinated biphenyl (PCB), chlorofluorocarbons (CFC), and asbestos
- EPA protection of trade secrets while assuring public access to necessary information.

TSCA also allows EPA certain powers. EPA has the authority to refer regulation of a given chemical substance to another agency, or rely on the provisions of TSCA. TSCA provides EPA a mechanism for funding the monitoring of the occurrence, migration, and transformation of toxic substances in groundwater. The control of toxic substances discharged to groundwater via household waste-disposal systems could also be overseen by EPA under TSCA.

RESOURCE CONSERVATION AND RECOVERY ACT (RCRA)

Standards for owners and operators of hazardous waste treatment, storage, and disposal facilities are regulated under RCRA. Prior to 1976, disposal of hazardous waste into the atmosphere or bodies of water was restricted. Economics in conjunction with the prevailing regulations favored land disposal and deep-well injection of hazardous waste. In 1976, RCRA was promulgated to provide cradle-to-grave management of hazardous waste, thus closing the environmental loop by providing the first comprehensive federal regulatory program for protection of the environment including air, surface water, groundwater, and the land.

Hazardous and Toxic Wastes

Under RCRA, interim status and permitted transport, disposal, and storage (TDS) facilities must fulfill certain obligations for (1) leachate minimization and control through design and operating requirements and (2) groundwater monitoring and response requirements should leachate reach the groundwater.

Prior to issuance of a permit, geologic and hydrogeologic characterization of the facility to the extent necessary to proceed with development of a detection groundwater monitoring well system must be completed. Criteria established under 40 CFR Part 264, Subpart F, include identification of the uppermost aquifer and aquifers hydraulically interconnected thereto beneath the waste management area, evaluation of their respective groundwater flow rates and direction, and the basis for such identification. Under interim status, this requirement may all or in part be waived if a demonstration can be made that there is a low potential for migration of hazardous waste or hazardous waste constituents from the facility via the uppermost aquifer to water supply wells or to surface water. Since this demonstration is very difficult to prepare in such a manner that is scientifically sound, a subsurface investigation is usually conducted.

The development of a groundwater detection monitoring system to detect the immediate potential release of contaminants from a specific waste management unit or facility is conducted in two phases. Phase I is the subsurface characterization and demonstration to EPA that the site has been adequately characterized. Phase II is the design of the detection groundwater monitoring system and the demonstration that the system will serve for the immediate detection of a release. Specific technical elements of system design is discussed in Chapter 9 (Landfill Disposal). Specific requirements are outlined in Table 2-4.

Subsurface geologic characterization will at minimum include a review of existing geologic information, soil borings and/or rock corings, material testing (i.e., grain size, mineralogy, etc.), and interpretation of the structure and stratigraphy beneath the facilities. For geologically complicated sites, surface and downhole geophysical techniques may be employed along with aerial photography (i.e., lineament analysis). Hydrogeologic characterization will at minimum include the installation of groundwater monitoring wells and piezometers, hydraulic conductivity testing via aquifer pumping tests, *in situ* rising or falling head tests, and laboratory testing of undisturbed samples. Tracer tests may also be performed. Identification of water-bearing zones and interpretation of horizontal and vertical flow components are evaluated. The investigator bears the burden of properly defining the uppermost aquifer and the presence or absence of hydraulic continuity between aquifers. For major hazardous waste disposal facilities, costs associated with the characterization process can run into millions of dollars.

Upon approval from EPA that the site has been adequately characterized, the groundwater detection monitoring system is designed. It must be capable of determining a facility's impact on the groundwater quality in the uppermost aquifer underlying the facility. Background wells must be selected to produce representative samples of groundwater unaffected by the facility. The

**Table 2-4   Summary of Leachate Containment Strategy per Waste
Management Unit Under RCRA**

40 CFR 264 and 265 Subpart	Waste Management Unit	Containment Strategy
J	Tank system	Secondary containment system (i.e., liner, vault, double walled, or equivalent device)
K	Surface impoundments	Double liner (at minimum) and leachate collection system
L	Waste piles	Liner and leachate collection and removal system
M	Land treatment	Unsaturated zone monitoring
N	Landfills	Liner and leachate collection and removal system
O	Incinerator	Monitoring of combustion temperature, waste feed rate, indicator of combustion gas velocity, CO, and waste and exhaust emissions
P[a]	Thermal treatment	Temperature and emission control
Q[a]	Chemical, physical and biological treatment	Discharge control and confinement structures (i.e., dikes)
R[a]	Underground injection	Excluded from RCRA
W	Drip pads	Impermeable pad (i.e., sealed or coated concrete as asphalt run-off control, bermed, and of sufficient structural strength
Y	Miscellaneous units	In accordance with other units

[a] Pertains to 40 CFR Part 265 only.

downgradient wells must be selected to "immediately detect statistically significant increases in contaminants migrating from a particular waste management unit or facility into the uppermost aquifer should a release occur." Data from upgradient and downgradient wells are statistically compared to determine if such increases have occurred. A detection monitoring system that meets interim status standards for design and location should generally meet permitting standards. Although the interim status standards require placement of the downgradient wells "at the limit of the waste management area" and permitting standards require well placement at the "point of compliance," the definitions of these two terms mean essentially the same thing and are discussed further in Chapter 9.

Under interim status, groundwater sampling is conducted quarterly for 1 year to assess background water quality. Sampling frequently for certain parameters is subsequently performed semiannually or annually, along with measurement of groundwater level elevations at least annually. The major difference between interim and permitted status reflects the units covered by the regulations, and the parameters monitored to assess whether contamination has occurred.

To fulfill the requirements of groundwater monitoring and response under RCRA, the program is divided into three response modes: detection monitoring, compliance monitoring, and corrective action program. Detection monitoring is implemented when no release or contamination has been identified to date. Quarterly sampling is routinely performed, with subsequent testing for the constituents listed in Appendix VIII of CFR Part 261, and establishes a background value for each constituent. The determination as to whether contamination, if reported, exceeds acceptable levels is addressed under the compliance monitoring program. Acceptable levels could reflect the National Interim Primary Drinking Water Standards (NIPDWS) or background levels of other constituents, whichever is higher. The latter would apply for most constituents since the NIPDWS are mostly metals. An alternate concentration limit (ACL) could also be pursued providing that each constituent detected above acceptable levels could be shown not to pose a substantial presence or potential hazard as long as the ACL is not exceeded. In addition, mechanisms exist for demonstration of sampling and analytical errors, documentation of a non-RCRA-regulated unit as the source within certain time constraints, and reporting requirements. The corrective action program requires employment of remedial technologies. Constituents are thus removed or treated in place. A monitoring program is also implemented to monitor the effectiveness of the corrective action program. This monitoring program continues until the standard is no longer exceeded for at least 3 consecutive years.

Under interim and permitting statutes, unsaturated zone monitoring is also required for land treatment units. A leachate collection and removal system and/or liner(s) are required for certain waste management units, as shown in Table 2-4.

Mining Wastes

Mining in the U.S. on a grand scale commenced in the mid 1800s with such mining districts as Leadville in Colorado, the Tri-State Mining Area (Oklahoma, Kansas, and Missouri), and Anaconda in Montana, among others. In addition, the exploration and production of uranium and other strategic minerals since the 1940s have left a legacy of uranium mill tailing piles, asbestos-

and metal-contaminated soils, and numerous cases of groundwater contamination. Mining sites can vary in size from less than 1 acre to several square miles, thus producing large volumes of waste. Virtually all of the mining waste generated is disposed of on-site. Overburden and waste rock are typically placed in piles adjacent to the mine or used as backfill. Tailings are commonly disposed of as a slurry in unlined tailing impoundments, covering some 1500 ha within 24 sites in the western states. Primarily derived from extensive mining activities of the 1950s and 1960s, the material is composed of finely comminuted uranium ore in which all the uranium has been removed, but the radioactive daughter elements produced by radioactive decay remain. The most important of the radioactive daughter elements is radium (^{226}Ra). If ingested, Ra can cause damage due to its intense alpha radiation. Ra also decays to produce radon (^{222}Ra), a radioactive gas which can further decay into suspended solid radioactive products such as lead (^{210}Pb) which, if inhaled, can lead to cancer. No longer covered by rock and soil to absorb the radiation and prevent appreciable migration, unconsolidated tailing piles provide an environment for the migration of decay products by dissolving in water and diffusing into air.

EPA has estimated that about 31 trillion kg of mine waste and 13 trillion kg of tailings were accumulated between the years 1910 and 1981, with over 80% of the overburden and tailing wastes generated by the phosphate rock, uranium, copper, and iron ore mining segments of the industry. Uranium mill tailings currently total more than 230 million tons at mill sites throughout the U.S. Out of 1174 sites listed on the National Priority List under CERCLA, which is discussed later in this chapter, a minimum of 109 of these sites contain mining wastes as a result of past or current mining activity, or the disposal of mining waste on the site. Although some form of mining is conducted in every state, most metal mines are currently located in the western U.S. with similar commodities usually restricted to certain geologic environments. Twenty-five states are chief producers of copper, gold, iron, lead, phosphate, silver, and zinc. Major producing states, the quantity of waste produced, and environmental concerns are presented in Table 2-5.

In 1976, RCRA required EPA to conduct an investigation of all solid waste management practices pertaining to the mining industry, including analysis of the sources and volume of mining wastes, disposal practices, potential danger, alternative disposal methods and cost, and potential uses of the waste. When RCRA was amended in 1980, the Bevill Amendment directed EPA to conduct a comprehensive study to prohibit from regulation but assess the adverse environmental effects, if any, of the disposal and utilization of solid wastes from the extraction, beneficiation, and processing of ores and minerals. Furthermore, EPA was required to state whether regulation of mining wastes was warranted and to express the need for regulations to be promulgated if deemed necessary. In 1986, EPA determined that RCRA's hazardous waste management standards ". . . are likely to be environmentally unnecessary, technically

Table 2-5 Summary of Major Mining Areas of Environmental Concern

Ore	No. of Sites	Quantity of Wastes (Million Metric Tons per Year)	Environmental Concern	Principal Producing States[a]
Copper	19	502	Elevated metals, low pH	Arizona, New Mexico, Utah
Gold and Silver	117	100	Radioactivity	Nevada, Montana, California
Iron	26	177	Elevated metals, low pH	Minnesota, Michigan, Missouri
Lead and zinc	23	18	Residual cyanide	Missouri, Tennessee, Idaho
Phosphate	27	403	Elevated metals, low pH	Florida, North Carolina, Idaho
Other metals	24	62	Elevated metals, low pH	Colorado (molybdenum)
Total	226	1262		

[a] Source: *Bureau of Mines Yearbook* (1985).

infeasible, or economically impractical when applied to mining wastes." Mining wastes were excluded from regulation under RCRA Subtitle C since these wastes were considered raw materials used in the production process or product, and only leach solutions that escape from the production process and abandoned leach piles were considered wastes. In 1988, 6 processing wastes were classified as hazardous and excluded from the Bevill Amendment, and 15 high-volume mineral processing wastes that would come under the Bevill Amendment were identified. All other processing wastes would be excluded from the Bevill Amendment.

Mining waste is defined as those wastes generated from the mining, milling, smelting, and refining of ores and minerals. Mining specifically refers to the extraction or removal of ores, whereas milling refers to the concentration of extracted ores via crushing, screening, or chemical treatment of the ore rock. Primary smelting or refining reflects to further processing of the ore product. Mining wastes include soil and rock that is removed to expose an ore body in a surface mine (i.e., overburden). Rock that is excavated during underground mining operations (i.e., development waste rock) includes tailings, slag piles, and leach residue derived from dump and heap leach operations. Typical mining waste types and associated mining activity are presented in Table 2-6.

The majority of the contaminants are metals (arsenic, cadmium, copper, mercury, lead, and zinc) and nonmetals (asbestos and radium). A by-product of metals mining is acid-mine drainage. Characterized by low pH, acids are produced when rain water or groundwater migrates through the ore zone or waste material, subsequently leaching out high concentrations of metals that ultimately discharge to surface water bodies or groundwater. Contaminants are also derived from the washing of crushed rock and leaching from waste piles and tailing ponds (Figure 2-3).

COMPREHENSIVE ENVIRONMENTAL RESPONSE, COMPENSATION, AND LIABILITY ACT (CERCLA)

CERCLA, or Superfund, was enacted in 1980 and amended in 1986 for the basic purpose of providing funding and enforcement authority for cleaning up the thousands of past hazardous waste sites and for responding to hazardous substance spills. In comparison to RCRA, which establishes a cradle-to-grave regulatory program for present hazardous waste sites, CERCLA primarily establishes a comprehensive response program for past hazardous waste activities. CERCLA also subjects businesses to reporting requirements for spills and other kinds of environmental releases. Unlike other media-specific regulatory programs, CERCLA incorporates all environmental media: air, soil, surface water, and groundwater. CERCLA also applies to any type of industrial, commercial, or noncommercial facility despite the waste stream and how the environment might be impacted.

Table 2-6   Typical Mining Wastes and Associated Mining Activities

Waste Type	Mining Activity
Acid	Oxidation of naturally occurring sulfides in mining waste, notably copper, gold, and silver
Asbestos	Asbestos mining and milling operations
Cyanide	Precious metals heap-leaching operation
Leach liquors	Copper-dump leaching operations
Metals (Pb, Cd, Ar, Cu, Zn, and Hg)	Mining and milling operations
Radionuclides (radium)	Uranium and phosphate mining operations

A site comes under CERCLA if there is a release or threat of release of a hazardous substance, pollutant, or contaminant into the environment. A hazardous substance under CERCLA is any substance designated for special consideration under the Clean Air Act (CAA), CWA, or TSCA, and any hazardous waste under RCRA. Currently, there are a total of 724 hazardous substances and 1500 radionuclides on the list. A pollutant or contaminant is any other substance not on the hazardous substance list which will or may reasonably be anticipated to cause any type of adverse effects in organisms and/or their offspring. Not included as hazardous substances, contaminants, or pollutants are petroleum, natural gas, and synthetic gas usable for fuel, providing the petroleum substance does not contain other substances that are not normally found in refined petroleum fractions or crude, or are present at concentrations which exceed those normally found in such fractions. Although CERCLA can respond to either type of substance, private parties are liable solely to the extent hazardous substances are involved.

When a release has occurred such that an imminent and substantial danger may exist, EPA can undertake removal and/or remedial action, depending whether the activity requires a limited short-term response or a more permanent and long-term approach. A short-term response could simply be the removal of a few drums, underground storage tanks, or significant contaminated shallow soil. If, for example, groundwater was adversely affected, then a long-term response would be required such as pump-and-treat.

As of October 1990, there were 1187 sites on the National Priority List (NPL). From over 33,000 potential sites, a preliminary assessment (i.e., file review) is initially performed and those sites where no further attention is required are discarded. A site inspection is then performed, followed by evaluation and scoring under the hazard ranking system (HRS, or MITRE model). Pertinent data such as waste type and volume, toxicity, distance to population, distance to underground drinking water, etc. is evaluated and scored. Any score at or above 28.50 is included on the NPL. Since EPA does

Figure 2-3 Anaconda Company Yerington Copper Mine, Lyon County, Nevada, showing oxide tailings dump ponds (1976). Oxide ore is treated by leaching with sulfuric acid in a plant adjacent to the mine. The copper in the leach solution is recovered in the form of metallic precipitate by passing the solution through vats filled with iron scrap. Mined pit is in lower right foreground (Courtesy of Bill Dubois, Steve Friberg, and the Nevada Bureau of Mines).

not have the financial or human resources to attend to all sites at once, the Superfund Comprehensive Accomplishments Plan (SCAP) was developed to outline which phase of the cleanup process, if any, is scheduled for each NPL site per fiscal quarter.

Remedial Investigation/Feasibility Study

Of geological concern is the remedial investigation/feasibility study (RI/FS). A wide variety of remediation options are typically available for a particular site. EPA then determines "how clean is clean" and "how expensive is expensive" by conducting an RI/FS and subsequently selecting the remedial option, or record of decision (ROD). The RI and FS are usually performed together, with the overall intent of the RI being to characterize subsurface geologic and hydrogeologic conditions to the extent that remedial alternatives can be developed and evaluated. In addition, source(s) and extent of contamination, migration pathways, and potential for human or other environmental exposures are also evaluated. The FS, on the other hand, utilizes the information developed by the RI and provides a series of specific engineering or construction alternatives for the site accompanied by a detailed analysis of cost, engineering feasibility, and environmental impact per remedial alternative.

To maintain control of costs for the RI/FS, a workplan is sometimes required prior to the RI/FS. Addressed in the workplan will be the soil and groundwater sampling plan (i.e., number, location, depth, etc.), schedule, and overall level of effort. This in turn may ultimately affect the scope of work for the RI/FS and remedial options presented.

Environmental Site Assessments

It is now customary that an environmental assessment be performed prior to most sales of real estate in order to evaluate the financial risk of the transfer regarding potential environmental liabilities from an actual or threatened release of a hazardous or toxic substance. Prior to the sale or redevelopment of commercial and industrial property, CERCLA requires an owner of nonresidential property who knows, or has reasonable cause to believe, that a release of a hazardous substance is located on or beneath that real property, to notify any buyer of that property. As a result of this legislation, an environmental site assessment (ESA) is typically conducted when a property ownership transfer, or a significant change in site usage, is being considered. Additionally, the lead regulatory agency involved in overseeing redevelopment of the property will require detailed information as to the environmental condition of the property before redevelopment commences. To that end, an ESA is conducted which provides sufficient detail to allow remediation levels to be determined prior to actual redevelopment; thus, both the property owner and lead agency will have ample time to develop a remediation strategy. Several tasks comprise an ESA. An ESA of a known oil-field property requires specific tasks not usually required for other industrial property. An ESA typically involves a phased approach incorporating the tasks outlined in Table 2-7. Phase I activities assess the potential for adverse subsurface impact; Phase II activities are performed to determine the actual presence of subsurface constituents in soil or groundwater that are considered hazardous; Phase III activities are conducted to assess the lateral and vertical extent of the contaminant plume(s) and to generate data sufficient to develop a cost-effective and technically efficient remediation strategy. Pilot studies may be conducted to assess the feasibility and overall effectiveness of a particular remediation strategy. Phase IV activities incorporate remediation of adversely affected soil and groundwater.

Oil-Field Properties

From a regional industry perspective, consider the petroleum exploration and production industry. Documented cases of water well and stream pollution in the U.S. resulting from petroleum production activities extend back to the turn of the century. In Marion, IN, many of the major local streams were polluted, and the single source of groundwater supply seriously threatened, when 200 to 300 surface and rock wells were found to be contaminated due to adjacent petroleum production activities. More recent discussion of extensive soil and groundwater contamination at the Martha Oil Field in Kentucky has been reported. Approximately 1578 wells (including oil and gas production wells, injection wells, and industrial water supply wells) have been designated for plugging and abandonment, reflecting their adverse impact on adjacent groundwater resources.

Unlined surface petroleum impoundments historically have also been identified for excessive seepage losses as far back as the early 1900s. In the Kern River field in California, fluid loss on the order of 500,000 bbl from an unlined surface oil reservoir was documented. In this instance, excavated pits disclosed that the oil had penetrated to depths exceeding 20 ft. Another loss of 1 million bbl over a period of 6 years occurred from an unlined reservoir in the same field, although in this instance some of the loss resulted from evaporation. This surface reservoir was ultimately abandoned due to excessive seepage losses.

As oilfields within urbanized areas reach the end of their productive lives, they are rapidly taken out of production and redeveloped. Nowhere is this more evident than in southern California. California has a rich history of oil and gas exploration and exploitation dating back to the 1860s, the first years of commercial production. California comprises 56 counties, of which 30 are known to produce petroleum; 18 of these 30 counties produce chiefly oil, while the remaining 12 counties produce mostly gas. Generally, hydrocarbon occurrence is found in oil south of approximately $37°$ North latitude. Specifically, most oil production in southern California occurs within four basins: the Los Angeles Basin, Ventura Basin, San Joaquin Basin, and Santa Maria Basin (Figure 2-4). The Los Angeles Basin, for example, is known to contain more than 10 giant oil fields (a giant field is defined as one yielding an ultimate recovery in excess of 100 million bbl). Within the City of Long Beach area alone, approximately 12 mi^2 or 3840 acres are currently, or historically have been, petroleum-producing properties.

As oil fields within the urbanized Los Angeles Basin area reach maturity and the end of their productive lives, the property associated with these production facilities faces high demand for more profitable land usage. Therefore, such properties are being redeveloped at a rapid rate due to the burgeoning Los Angeles real estate business. In addition to the oil fields proper, numerous high-volume petroleum-handling facilities exist in close proximity to production fields, which can significantly contribute to subsurface degrada-

**Table 2-7 Tasks To Be Performed as Part of an
Environmental Site Assessment**

Phase		
No.	Task Name	Task Description
I	Due diligence records review	Historical records review Regulatory agency file review Historical aerial photograph review DOG oil field maps and records Site reconnaissance Documentation of findings and recommendations
II	Initial delineation	Preparation of workplan Soil borings and sampling Groundwater monitoring well(s) construction and sampling Analytical program Abandonment of oil/gas wells Documentation of findings and recommendations
III	Detailed delineation	Preparation of phase III workplan Soil borings and sampling Groundwater monitoring well(s) construction and sampling Analytical program Selection of remediation options Pilot study to determine remediation effectiveness Documentation of findings and recommendations
IV	Remediation	Preparation of workplan Field implementation of remediation technique Soil/water sampling to ensure effectiveness during remediation Confirmatory soil/water sampling for agency sign-off Documentation finds

tion of soil and local groundwater quality. Major oil fields and associated structural features situated on the Los Angeles Basin coastal plain are shown in Figure 2-5. Impeding urban congestion associated with oil-field development within the City of Signal Hill along the regional Newport-Inglewood Structural Zone in the late 1920s, late 1930s, and late 1980s is shown in Figure 2-6.

Several potential sources of chemicals or constituents, which when placed in contact with soil and groundwater may result in their being classified as a hazardous material, are typically present on or adjacent to petroleum production sites. Some of these potential sources are outlined in Table 2-8.

The primary and most common compounds and constituents which are associated with oil-field properties and may be considered hazardous include:

- Methane
- Crude oil and constituents
- Drilling mud and constituents
- Refined petroleum products and constituents including volatile organic compounds.

These compounds and constituents can be considered as either hazardous wastes or materials, or designated wastes as further discussed in Chapter 7. Crude oil is not considered a hazardous waste, since it typically does not contain significant concentrations of aromatic hydrocarbons (i.e., benzene, toluene, ethylbenzene, and xylene isomers) or other toxic constituents. However, crude oil may, as in the case in California, be considered a designated waste. Thus, crude oil is excluded as a hazardous waste or material, but can be classified as a "designated" waste on the basis of analytical testing if it can be shown that the substance in question meets the criteria outlined in Table 2-9. In cases where both refined petroleum products and crude oil are present at a specific site, the differentiation between the two can have a significant economic impact reflecting the level of remediation warranted.

Drilling muds present special concerns. Drilling muds are admixtures of water and numerous other chemical additives used to lubricate the drill bit and remove cuttings from the well bore, in addition to conditioning the well bore prior to installation of production casing or prior to abandonment. Such muds can be classified as either water- or oil-based. Oil-based muds are generally defined as muds with a continuous liquid phase of oil. True oil-based muds contain 5% water or less by volume and use crude oil as a major constituent. Oil-based muds therefore feature little environmental difference from crude oil, and may thus be considered a designated waste within California. Water-based muds utilize both organic and inorganic additives to achieve their desired "properties". Additives consist primarily of clays (bentonite, attapulgite, natu-

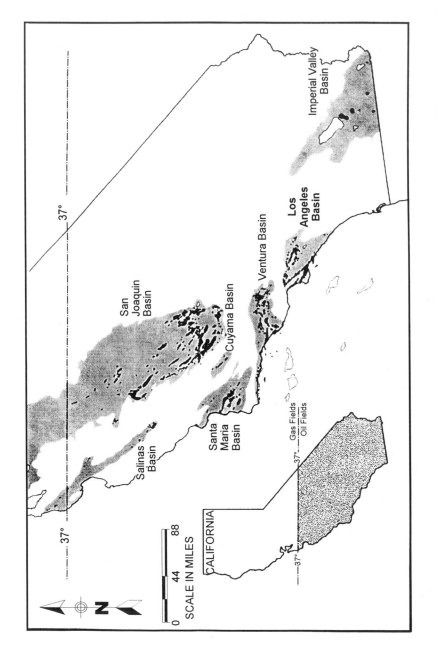

Figure 2-4   Location of major basins and oil fields in southern California (after Testa and Winegardner, 1991).

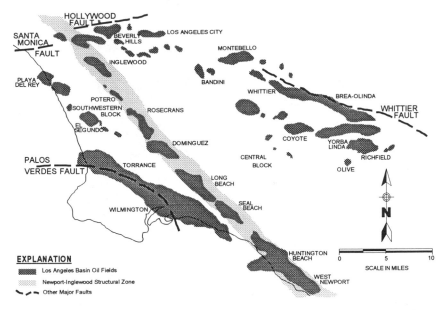

Figure 2-5   Location of major oil fields in the Los Angeles Basin (after Testa and Winegardner, 1991).

ral formation clays); barite; dispersants (tannins, quebracho phosphates, lignites, and lignosulphonates); starch; sodium carboxy-methyl cellulose; polymers (cypan, drispac); detergents; and defoamers. Some of these constituents by themselves can be of environmental concern. Different additives are mixed with water to yield the desired mud "properties" when drilling through different subsurface lithologies. Typically, a pit or sump excavated adjacent to the drill rig is used for the mixing of drilling mud and as a settling pond. Consequently, drill-site mud sumps can have numerous additives, including heavy metals, which may exceed state and/or local regulatory contaminant levels. In addition, historical disposal practices for spent oil-field chemicals included dumping into the mud pit (sump) or spreading onto the ground adjacent to the drill site.

Whether abandoned as a dry hole or at the end of a useful (productive) life, improperly abandoned oil and gas wells present significant concerns, since they can serve as a potential source for leakage of methane gas and/or crude oil, or act as potential migration pathways of petroleum. Redevelopment of an oil-field property must take into account both the cost of remediation and potential liability of building above or adjacent to old oil wells. Maps showing the location of oil and wildcat wells (and sumps) need to be reviewed (Figure 2-7). Wells abandoned prior to the most current standards must be reabandoned in accordance with current standards before building permits will be issued for a site with documented wells. This includes oil wells within 100 ft, or wildcat wells within 500 ft, of a proposed construction or redevelopment site.

Figure 2-6   The Signal Hill, CA area during (A) the late 1920s, (B) the late 1930s (showing encroachment of urbanization along the Newport-Inglewood Structural Zone), and (C) the late 1980s (showing the coexistence of urbanization and an active oil-producing area).

Figure 2-6B

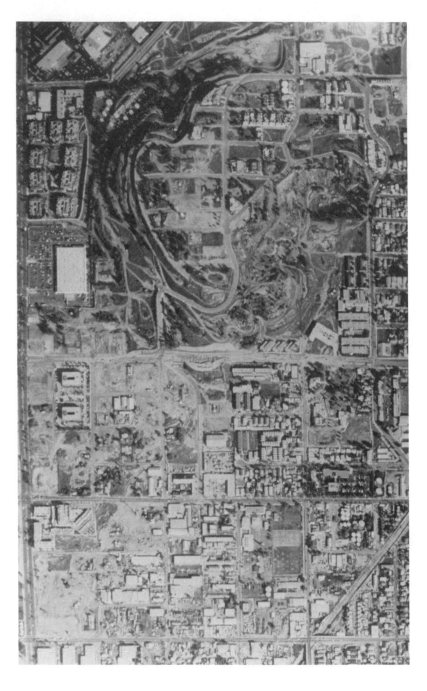

Figure 2-6C

Table 2-8   Potential Sources for Oil-Field Contaminants

Primary Sources	Secondary Sources
Oil wells	Pumping station
Sump, pits, and dumps	Piping ratholes
Above-ground storage tanks and containers	Transformers
Random spillage or leakage	Underground storage tanks
Reservoirs	Well cellars

Table 2-9   Criteria for Determining Whether Crude Oil Is Classified
as a Hazardous Waste

Parameter	Maximum Contaminant Level (parts per million)
Arsenic	5
Chloride	3,000
Chromium	10
Flash Point	Less than minimum standard[a]
Lead	50
Polychlorinated biphenols (PCBs)	5

[a] Minimum standard as set by the American Society for Testing and Materials for recycled products.

When determining whether improperly abandoned oil and/or gas wells are onsite, if well locations are documented, an exploration pit 10 ft square by 10 ft deep is usually excavated centered on the documented well location. Once the pit is excavated, and no visible signs of the well (i.e., casing, cellar boards, etc.) are found, a magnetometer is issued to scan the pit for signs of metal well casing. If a casing is located, proper abandonment of the well must be conducted in accordance with certain standards prior to redevelopment. This process involves the redrilling of the well, beginning inside the casing, down to a depth equal to the known historic production depth. Once the well bore has been reopened, the well is abandoned using a cement grout which is pumped into the well bore to a level approximately 100 ft below the ground surface. The well is then "squeezed-off" or pinched closed, and additional cement is pumped down into the well bore. Abandonment at the surface requires proper venting to avoid methane accumulation. Once all known wells have been properly abandoned, sign-offs on building permits for the site are obtained. Most city building departments require a sign-off from the lead agency before building permits can be issued for oil field properties.

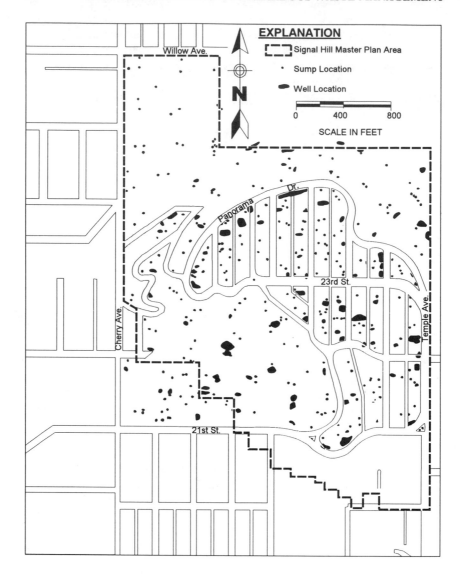

Figure 2-7   Location of abandoned and active wells and sumps within the City of Signal Hill, CA.

In some regions of the country, volatile emissions may be required to be monitored during excavation- and remediation-related activities to insure that the organic vapors released do not exceed certain levels (i.e., 50 ppm). However, this usually does not apply to:

- Soils containing organic compounds having an initial boiling point of 302°F or greater, providing the soil is not heated

- Soils that contain organic compounds resulting from natural seepage of volatile organic compounds from oil- and gas-producing properties
- Soils containing organic compounds having a Reid vapor pressure (RVP) less than 80 mm or 1.55 psi of Hg or an absolute vapor pressure of less than 36 mm or 0.7 psi at 20°C.

Crude oil is thus usually not of concern with regard to air quality, since it typically has a boiling point greater than 302°F, is of a natural source, and maintains a RVP usually less than 80 mm Hg at 20°C.

Should it be determined, as a result of an ESA, that regulated material has impacted onsite soil or groundwater, remediation must be undertaken if redevelopment is to proceed. Soil remediation is commonly conducted during activities related to foundation preparation, which has proven to be cost effective since some grading and subsurface excavation has to be performed during redevelopment of most properties. Different remediation options which achieve the regulated clean-up levels must thus be evaluated with respect to time constraints, accessibility, economics, potential impact, and financial and legal liability.

Geothermal Properties

Geothermal operations as with oil-field properties provide another example of the close relationship between site usage and the potential for adverse environmental impact. Geothermal plants typically comprise a power plant, brine storage pond(s), drill sump(s), and possibly a leach field (Figure 2-8). Constituents associated with geothermal operations which may be considered potentially hazardous fall into two categories: brine and line-mine scale. Brines are geothermal mineralizing fluids denoting warm to hot saline waters containing calcium, sodium, potassium, chloride, and minor amounts of other elements. Line-mine scale is the calcified material that forms within the interior of the process piping system associated with geothermal steam generation operations. As part of these operations, geothermal wells are drilled into a superheated water-bearing strata. Superheated water and steam is recovered from these wells and directed through process piping and condensers as a power source to drive the turbines of the electrical generators. A variety of mineral and elemental metal compounds exist naturally in the superheated water and steam which causes calcification to occur in the form of a scale which builds up on the interior of the process piping. Unrestricted buildup of this scale inhibits the operation of the units by inhibiting the flow of water and steam to the turbines, and necessitating costly shutdown and line-cleaning programs. To overcome this situation, UGD has perfected a method to manage this scale buildup via a process referred to as "line mining". Line mining is the insertion of wire mesh rolls within the process piping and is conducted in

Figure 2-8   Former geothermal generating plant, Salton Sea area, California.

conjunction with alteration of the water's pH levels, whereby the scale buildup is directed to the mesh rolls and away from the piping walls. The process water flow is thus literally "mined" of the minerals and metal compounds by allowing the scale to build up on the wire mesh rolls. These rolls are periodically removed and replaced with new rolls.

Chemical constituents of primary concern, certainly during abandonment of geothermal operations or a change in site usage, include arsenic, copper, lead, and zinc. General representative chemical characteristics for both brine and line-mine scale are presented in Table 2-10.

Table 2-10   General Chemical Characteristics for Brine and Scale

| | | Concentration | |
| | | Scale (in wt %)[a] | |
Parameter	Brine (mg/L)	Well Head Separator Scale	Injection Line Scale
Aluminum	ND (<2)	0.3–0.72	ND
Antimony	ND (<2)	NA	NA
Arsenic	1.39–2.7	NA	NA
Barium	680–1,520	NA	NA
Cadmium	1.3–340	NA	NA
Calcium	12,700–34,000	0.2–0.77	0.58–0.81
Cesium	17–22	NA	NA
Chloride	80,200–184,000	NA	NA
Chromium	ND (<0.1)	NA	NA
Cobalt	0.2–0.3	NA	NA
Copper	1.7–5.9	0.1–1.6	ND
Iron	490–4,540	7.7–53.8	29.0–37.3
Lead	17–300	15.7–40.6	NA
Lithium	210–320	NA	NA
Magnesium	100–200	0.1–0.75	0.1–0.17
Manganese	470–1,860	0.2–0.5	0.47–0.68
Mercury	ND (<0.004)–0.001	NA	NA
Molybdenum	ND (<0.2)	NA	NA
Nickel	ND (<0.2)	NA	NA
Potassium	6,900–15,400	NA	NA
Rubidium	34–83	NA	NA
Selenium	270–320	NA	NA
Silicon	ND (0.6)–0.1	0.55–8.0	19.5–41.70
Silver	<1–1	2.4–6.0	2.12
Sodium	34,800–68,000	NA	NA
Strontium	640–1,510	NA	NA
Sulfur	NA	5.70–18.1	ND (0.05)
Total dissolved solids	128,000–262,000	NA	NA
Tin	ND (<10)	NA	NA
Zinc	190–1,170	0.6–7.2	NA

Table 2-10   General Chemical Characteristics for Brine and Scale
(continued)

| Parameter | Brine (mg/L) | Scale (in wt%)[a] | |
		Well Head Separator Scale	Injection Line Scale
Anions			
Ammonia	69–150		
Bromide	107,000–190,000		
Chloride	ND (<1)		
Fluoride	ND 24–ND (<30)		
Iodide	ND (<5)		
Nitrate	ND (<6)		
Sulfate	53–60		
TDS	559–874		
Thiosulfate	1.1–1.199		

Note: ND = Not detected at analytical detection limit shown in parentheses; NA = Not available.

[a]   Concentration in mg/kg.

BIBLIOGRAPHY

Anderson, P.W. and Dewling, R.T., 1981, Industrial Ocean Dumping in EPA Region II-Regulatory Aspects: In *Ocean Dumping of Industrial Wastes* (Edited by Ketchum, B.H., Kester, D.R. and Park, P.K.); Plenum Press, New York, pp. 25-37.

Arbuckle, J.G. et al., 1991, *Environmental Law Handbook:* Government Institute, Inc., Rockville, MD, 666 p.

Arscott, R.L., 1989, New Directions in Environmental Protection in Oil and Gas Operations: In *Environmental Concerns in the Petroleum Industry* (Edited by Testa, S.M); Pacific Section American Association of Petroleum Geologists, Palm Springs, CA, pp. 217-227.

Berg, M.G. and Callaway, R.J.D., 1989, Consideration of Mining Issues at Superfund Sites: In Proceedings of the Hazardous Material Control Research Institute Sixth National ARCRA/Superfund Hazardous Waste and Hazardous Materials Conference and Exhibition, New Orleans, p. 624-629.

Bowie, C.P., 1918, Oil-Storage Tanks and Reservoirs with a Brief Discussion of Losses of Oil in Storage and Methods of Prevention: U.S. Bureau of Mines Bulletin, No. 155, Petroleum Technology Report, No. 41, 76 p.

Caldwell, J.A. and Truitt, D., 1987, Technical Approaches on the UMTRA Project

Relevant to Superfund Projects: In Proceedings of the Hazardous Materials Control Research Institute 8th National Superfund '87 Conference, Washington, D.C., pp. 449-452.

Chartrand, A.B., et al., 1985, Ocean Dumping Under Los Angeles Regional Water Quality Control Board Permit: A Review of Past Practices, Potential Adverse Impacts, and Recommendations for Future Action: Report of the Los Angeles Regional Water Quality Control Board, 44 p.

Dezfulian, H., 1988, Site of an Oil-Producing Property: In Proceedings of the Second International Conference on Case Histories in Geotechnical Engineering, Vol. 1, Rolla, MO, pp. 43-49.

Dorries, A.M., Conrad, R.C. and Nonno, L.M., 1992, RCRA Facility Investigation for the Townsite of Los Alamos, New Mexico: In Proceedings of the Hazardous Materials Control Research Institute HMC-South '92, New Orleans, pp. 371-376.

Dyroff, G.V., 1989, *Manual on Significance of Test for Petroleum Products,* 5th Edition: American Society for Testing and Materials, Philadelphia, 169 p.

Eger, C.K. and Vargo, J.S., 1989, Prevention, Ground Water Contamination at the Martha Oil Field, Lawrence and Johnson Counties, Kentucky: In *Environmental Concerns in the Petroleum Industry* (Edited by Testa, S.M.); Pacific Section of the American Association of Petroleum Geologists, Palm Springs, CA, pp. 83-105.

Fortuna, R.C. and Lennett, D.J., 1987, Hazardous Waste Regulation—The New Era, An Analysis and Guide to RCRA and the 1984 Amendments: McGraw-Hill, New York, 393 p.

Garcia, D.H. and Henry, E.C., 1989, Environmental Considerations for Real Estate Development of Oil Well Drilling in California: In *Environmental Concerns in the Petroleum Industry* (Edited by Testa, S.M.); Pacific Section of the American Association of Petroleum Geologists, Palm Springs, CA, Symposium Volume, pp. 117-127.

Government Institutes, Inc., 1989, *Environmental Statutes:* Government Institutes, Inc., Rockville, MD, 1169 p.

Haugseth, L.A. and Roline, R.A., 1989, Removal of Toxic Heavy Metals from the Leadville Mine Drainage Tunnel Effluent: In Proceedings of the Hazardous Material Control Research Institute Sixth National ARCRA/Superfund Hazardous Waste and Hazardous Materials Conference and Exhibition, 1989, New Orleans, pp. 664-669.

Hotchkiss, W.O., 1945, *Minerals of Might:* The Jacques Cattell Press, Lancaster, PA, 206 p.

Howar, M., Hayes, D. and Gaskill, B., 1987, Clearing and Abandonment of Deep Water Wells at the Tar Creek Superfund Site: In Proceedings of the Hazardous Materials Control Research Institute 8th National Superfund '87 Conference, Washington, D.C., pp. 439-443.

Hoye, R.L., Hearn, R.L. and Hubbard, S.J., 1987, Heap and Dump Leaching and Management Practices to Minimize Environmental Impacts: In Proceedings of the Hazardous Materials Control Research Institute National Conference on Hazardous Wastes and Hazardous Materials, Washington, D.C., pp. 368-373.

Jones, S.C. and O'Toole, P., 1989, Increasing Environmental Regulation of Oil and Gas

Operations: In *Environmental Concerns in the Petroleum Industry* (Edited by Testa, S.M.), Pacific Section American Association of Petroleum Geologists, Palm Springs, CA, pp. 209-215.

Kaplan, I.R., 1989, Forensic Geochemistry in Characterizational Petroleum Contaminants in Soils and Groundwater: In *Environmental Concerns in the Petroleum Industry* (Edited by Testa, S.M.); Pacific Section, American Association of Petroleum Geologists, Palm Springs, CA, pp. 159-181.

Kiernan, B. and Neukirchner, R.J., 1989, Rational Approach to CERCLA Mine Site Remediation: In Proceedings of the Hazardous Material Control Research Institute Sixth National ARCRA/Superfund Hazardous Waste and Hazardous Materials Conference and Exhibition, New Orleans, p. 636-639.

Lewis, A., 1983, Leaching and Precipitation Technology for Gold and Silver Ores: *Engineering and Mining Journal.*

Lichtenberg, J.J., Winter, J.A., Weber, C.I. and Fradkin, L., 1988, Chemical and Biological Characterization of Municipal Sludges, Sediments, Dredge Spoils, and Drilling Mud: American Society for Testing and Materials (ASTM) Special Publication, No. 976, 512 p.

Lovegreen, J.R., 1989, Environmental Concerns in Oil-Field Areas During Property Transfers: In *Environmental Concerns in the Petroleum Industry* (Edited by Testa, S.M.); Pacific Section, American Association of Petroleum Geologists, Palm Springs, CA, pp. 129-158.

Mernitz, S., Derkics, D.L. and Brown, L.J., 1989, Mining Waste as Hazardous Waste - the Technical and Policy Issues: In Proceedings of the Hazardous Material Control Research Institute Sixth National ARCRA/Superfund Hazardous Waste and Hazardous Materials Conference and Exhibition, New Orleans, pp. 630-635.

Mills, W.B., Gherini, S.A. and Bigham, G., 1987, Mobilization and Transport of Metals from Mine Tailings to an Alluvial Aquifer: In Proceedings of the Hazardous Materials Control Research Institute 8th National Superfund '87 Conference, Washington, D.C., pp. 444-448.

Moody, D.W., Carr, J., Chase, E.B. and Paulson, R.W., 1986, National Water Summary 1986 - Hydrologic Events and Ground-Water Quality: U.S. Geological Survey Water Supply Paper 2325, 560 p.

National Research Council, 1990, Monitoring Southern California's Coastal Waters: National Research Council, 154 p.

Office of Technology Assessment, 1984, Protecting the Nation's Groundwater from Contamination - Vol. I and II: OTA, Washington, D.C., No. OTA-0233, October, 503 p.

Office of Technology Assessment, 1987, Wastes in Marine Environments: Congress of the U.S., Washington, D.C.

Porter, D.C. and Cobb, W.E., 1987, Degree of Cleanup - SARA 121(d), ARARs and Mining Sites: In Proceedings of the Hazardous Materials Control Research Institute 8th National Superfund '87 Conference, Washington, D.C., pp. 436-438.

Rickard, T.A., 1932, A History of American Mining: American Institute of Mining Engineers Series, McGraw-Hill Book Company, New York, 419 p.

Rintoul, W., 1990, *Drilling Through Time:* California Division of Oil and Gas, Sacramento, CA, 178 p.

Robbins, E.A., Robertson, M. and Cesark, D.R., 1990, Chemical Nature of Crude Oil,

Condensate Crude, and Natural Gas: In Proceedings of the Sixth Annual Hazardous Materials Management Conference West, Long Beach, CA, pp. 392-416.

Rowlands, R.D. and Botherus, M.C., 1992, Denver Radium Superfund Site Changes in "Selected Remedy" after Record of Decision: In Proceedings of the Hazardous Materials Control Research Institute HMC-South '92, New Orleans, pp. 367-370.

Rowley, K., 1986, The Rules of the Games in Ground-Water Monitoring: In Proceedings of the Second Annual Hazardous Materials Management Conference West, pp. 365-374.

Sackett, R. L., and Bowman, I., 1905, Disposal of Strawboard and Oil-Well Wastes: U.S. Geological Survey Water-Supply and Irrigation Paper, No. 113, 52 p.

Schoolcraft, M.R., 1819, *A View of the Lead Mines of Missouri:* Charles Wiley and Company, New York, 299 p.

Sheffer, H.W. and Evans, L.G., 1968, Copper Leaching Practices in the Western U.S.: U.S. Bureau of Mines Information Circular 8342.

Singh, P.N., Breeden, K.H. and Hana, S.L.A., 1988, Site Closure and Perpetual Care of a Low-Level Radioactive Waste Disposal Facility in Semi-arid Climate: In Proceedings of the Hazardous Materials Control Research Institute 5th National Conference on Hazardous Wastes and Hazardous Materials, Las Vegas, NV, pp. 351-358.

Smith, O.T., 1991, *Environmental Lender Liability:* John Wiley and Sons, New York, 427 p.

Society of Petroleum Engineers (SPE), 1962, *Petroleum Production Handbook, Volume II, Reservoir Engineering* (Edited by Frick, T.C. and Williams, R.W.); Society of Petroleum Engineers of American Institute of Mining Engineers, Dallas.

Sydow, W.L., Derkics, D.L. and Deichmann, J.W., 1988, State Regulatory Approaches to Management of Mining Wastes: In Proceedings of the Hazardous Materials Control Research Institute 5th National Conference on Hazardous Wastes and Hazardous Materials, Las Vegas, NV, pp. 359-363.

Testa, S.M., Henry, E.C. and Hayes, D., 1988, Impact of the Newport-Inglewood Structural Zone on Hydrogeologic Mitigation Efforts - Los Angeles Basins, California: In Proceedings of the National Water Well Association of Groundwater Scientists and Engineers FOCUS Conference on Southwestern Groundwater Issues, pp. 181-203.

Testa, S.M. and Townsend, D.S., 1990, Environmental Site Assessments in Conjunction with Redevelopment of Oil-Field Properties Within the California Regulatory Framework: In Proceedings of the National Water Well Association of Groundwater Scientists and Engineers Cluster of Conferences.

Testa, S.M. and Winegardner, D.L., 1991, *Restoration of Petroleum-Contaminated Aquifers:* Lewis Publishers, Boca Raton, FL, 269 p.

Ullom, J.W., 1990, Developing Residual Oilfield Property: In Proceedings of the Sixth Annual Hazardous Materials Management Conference West, Long Beach, CA, pp. 417-422.

Ulrich, H.D., Van Munster, J. and Brink, J., 1987, Institutional Realities of Locating a Repository for Low-Level Radium Waste in Colorado: In Proceedings of the Hazardous Materials Control Research Institute National Conference on Hazardous Wastes and Hazardous Materials, Washington, D.C., pp. 356-367.

U.S. Accounting Office, 1984, Federal and State Efforts to Protect Groundwater: Washington, D.C., February, 80 p.

U.S. Environmental Protection Agency, 1985, Report to Congress - Wastes from the

Extraction and Beneficiation of Metallic Ores, Phosphate Rock, Asbestos, Overburden from Uranium Mining, and Oil Shale: USEPA Office of Solid Waste and Emergency Response, EPA/530-SW-85-033.

U.S. Environmental Protection Agency, 1985, National Water Quality Inventory 1984 National Report to Congress: Office of Water Regulations and Standards, Washington, D.C., EPA-440/4-85-029, 173 p.

U.S. Environmental Protection Agency, 1985, Guidance on Feasibility Studies Under CERCLA: EPA/540/G-85/003.

U.S. Environmental Protection Agency, 1986, Amendment to National Oil and Hazardous Substances Contingency Plan National Priorities List, Final Rule and Proposed Rule: Federal Register, Vol. 51, No. 111, June 10, pp. 21053-21112.

U.S. Environmental Protection Agency, 1987, Report to Congress - Management of Wastes from the Exploration, Development and Protection of Crude Oil, Natural Gas, and Geothermal Energy: EPA/530-SW88-003.

U.S. Geological Survey, 1984, National Water Summary 1984 - Hydrologic Events, Selected Water-Quality Trends, and Ground-Water Resources: U.S. Geological Survey Water Supply Paper 2275, 467 p.

3   GEOLOGIC PRINCIPLES

*". . . all former changes of the organic and inorganic creation
are referrible to one uninterrupted succession of physical
events, governed by the laws now in operation."*

(Lyell, 1830)

INTRODUCTION

The overriding concern regarding the release of hazardous and toxic materials in the subsurface is the potential for adverse environmental impact to overall groundwater resources. Groundwater contamination can result from surface or near-surface activities including unauthorized releases or leaching from landfills, repositories, underground and above-ground storage tanks and pipelines, wells, septic systems, and accidental spills. Contamination can also occur from the application of agricultural chemicals to the land surface, or from nonpoint sources.

Comprehensive understanding of the three-dimensional framework of geologic materials is essential to the assessment of groundwater vulnerability to contamination, the lateral and vertical extent or distribution of hazardous and toxic constituents in the subsurface, and design and monitoring of subsurface remedial systems since geology is the prime controlling agent for the movement of groundwater, thus, contaminants. Geology also influences groundwater quality through several physical, chemical, and/or biological processes as further discussed in Chapter 8.

Several models for the regional assessment of aquifer vulnerability have been proposed and support use of geologic information as a primary component of such models. Contaminant potential maps are one of the products of such models. For example, contamination potential maps can be developed based on several parameters including:

55

- Depth to shallow aquifers (i.e., 50 ft or less)
- Hydrogeologic properties of materials between the aquifers and ground surface
- Relative potential for geologic material to transmit water
- Description of surface materials and sediments
- Soil infiltration data
- Presence of deeper aquifers
- Potential for hydraulic intercommunication between aquifers.

Development of other types of geologic and hydrogeologic maps can be used for the preliminary screening of sites for the storage, treatment, or disposal of hazardous and toxic materials. Such maps focus on those parameters which are evaluated as part of the screening process. Maps exhibit outcrop distribution of rock types that may be suitable as host rocks, distribution of unconsolidated, water-bearing deposits, distribution and hydrologic character of bedrock aquifers, and regional recharge/discharge areas. These maps can thus be used to show areas where special attention needs to be given for overall waste management including permitting of new facilities, screening of potential new disposal sites or waste management practices, and the need for increased monitoring of existing sites and activities in environmentally sensitive areas.

Overall understanding of the regional geologic and hydrogeologic framework, characterization of regional geologic structures, and proper delineation of the relationship between various aquifers is essential to implementing both short- and long-term aquifer restoration and rehabilitation programs, and assessing aquifer vulnerability. All too often maximum contaminant levels (action levels) are used to dictate the level of effort required for remediation with no or little conception of risk, thus, misutilizing limited manpower and financial resources. Understanding of the regional hydrogeologic setting can serve in designation of aquifers for beneficial use, determination of the level of remediation warranted, implementation of regional and local remediation strategies, prioritization of limited manpower and financial resources, and overall future management.

This chapter will focus on both regional and local geologic factors. Included is discussion of the concept of porosity and permeability, sedimentary sequences and facies architecture, and fractured media in relation to preferred fluid migration pathways. The influence of seismicity, drainage, and erosion is also discussed.

POROSITY, PERMEABILITY, AND DIAGENESIS

A recurring theme throughout this book is that of the ability of a fluid to migrate through various subsurface media. Two important aspects of subsurface materials in this regard are porosity and permeability. **Porosity** is a measure of pore space per unit volume and can be divided into two types: absolute porosity and effective porosity. **Absolute porosity (n)** is the total void space per unit volume and is defined as the percentage of the bulk volume that is not solid material.

$$n = \frac{\text{bulk volume } - \text{ solid volume}}{\text{bulk volume}} \times 100 \qquad (3\text{-}1)$$

Porosity can be individual open spaces between sand grains in a sediment or fracture spaces in a dense rock. A fracture in a rock or solid material is an opening or a crack within the material. **Matrix** is an important term referring to dominant constituent of the soil, sediment, or rock, and is usually a finer-sized material surrounding or filling the interstices between larger-sized material or features. As an example, gravel may be composed of large cobbles in a matrix of sand. Likewise, a volcanic rock may have large crystals floating in a matrix of glass. The matrix will usually have different properties than the other features in the material. Often, either the matrix or the other features will dominate the behavior of the material, leading to the terms matrix-controlled transport, or fracture-controlled flow.

Effective porosity (N_e) is defined as the percentage of the interconnected bulk volume (i.e., void space through which flow can occur) that is not solid material.

$$N_e = \frac{\text{interconnected pore volume}}{\text{bulk volume}} \times 100 \qquad (3\text{-}2)$$

Effective porosity (N_e) is of more importance and, along with permeability (the ability of a material to transmit fluids), determines the overall ability of the material to readily store and transmit fluids. Where porosity is a basic feature of sediments, permeability is dependent upon the effective porosity, the shape and size of the pores, pore interconnectiveness (throats), and properties of the fluid. Fluid properties include capillary force, viscosity, and pressure gradient.

Porosity can be primary or secondary. Primary porosity develops as the sediment is deposited and includes inter- and intraparticle porosity. Secondary porosity develops after deposition or rock formation and is referred to as diagenesis. Basic porosity types are presented in Figure 3-1. Secondary porosity types on a petrographic scale are presented in Figure 3-2.

Permeability is a measure of the connectedness of the pores. Thus, a glass with many unconnected air bubbles in it may have a high porosity but no permeability, whereas a sandstone with many connected pores will have both a high porosity and a high permeability. Likewise, a fractured, dense basaltic rock may have low porosity but high permeability because of the fracture flow.

The nature of the porosity and permeability in any material can change with time through diagenesis. Diagenetic processes are important to the subsurface

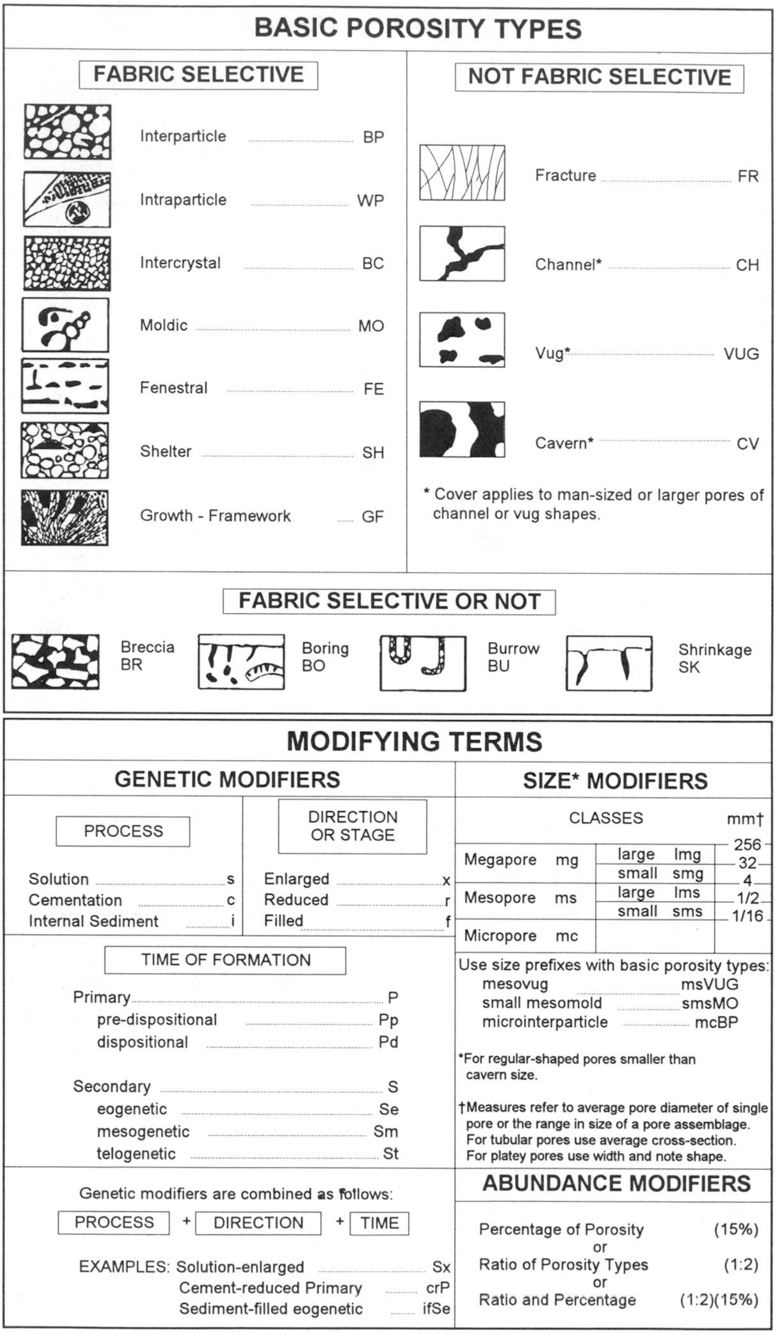

BASIC POROSITY TYPES

FABRIC SELECTIVE

Interparticle .... BP

Intraparticle .... WP

Intercrystal .... BC

Moldic .... MO

Fenestral .... FE

Shelter .... SH

Growth - Framework ... GF

NOT FABRIC SELECTIVE

Fracture .... FR

Channel* .... CH

Vug* .... VUG

Cavern* CV

* Cover applies to man-sized or larger pores of channel or vug shapes.

FABRIC SELECTIVE OR NOT

Breccia BR   Boring BO   Burrow BU   Shrinkage SK

MODIFYING TERMS

GENETIC MODIFIERS

PROCESS

Solution .... s
Cementation c
Internal Sediment i

DIRECTION OR STAGE

Enlarged x
Reduced .... r
Filled .... f

TIME OF FORMATION

Primary .... P
   pre-dispositional .... Pp
   dispositional .... Pd

Secondary .... S
   eogenetic .... Se
   mesogenetic .... Sm
   telogenetic .... St

Genetic modifiers are combined as follows:

PROCESS + DIRECTION + TIME

EXAMPLES: Solution-enlarged .... Sx
          Cement-reduced Primary .... crP
          Sediment-filled eogenetic .... ifSe

SIZE* MODIFIERS

CLASSES			mm†
Megapore	mg	large lmg	256
		small smg	32
Mesopore	ms	large lms	4
		small sms	1/2
Micropore	mc		1/16

Use size prefixes with basic porosity types:
   mesovug .... msVUG
   small mesomold .... smsMO
   microinterparticle .... mcBP

*For regular-shaped pores smaller than cavern size.

†Measures refer to average pore diameter of single pore or the range in size of a pore assemblage. For tubular pores use average cross-section. For platey pores use width and note shape.

ABUNDANCE MODIFIERS

Percentage of Porosity (15%)
   or
Ratio of Porosity Types (1:2)
   or
Ratio and Percentage (1:2)(15%)

Figure 3-1   Classification of porosity types (after Choquette and Pray, 1970).

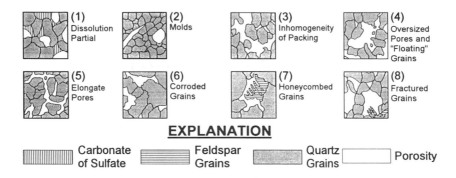

Figure 3-2 Petrographic criteria for secondary porosity (after Schmidt, McDonald, and Platt, 1977).

investigator because such processes can significantly affect the overall porosity and permeability of a sediment (Tables 3-1, 3-2). Once a sediment is deposited, diagenetic processes begin immediately. These processes include compaction, recrystallization, dissolution, replacement, fracturing, authigenesis, and cementation. Compaction occurs by the accumulating mass of overlying sediments. Unstable minerals may recrystallize, changing the crystal fabric but not the mineralogy, or may undergo dissolution and/or replacement by other minerals. Dissolution and replacement processes are common with limestones, sandstones, and evaporites. Authigenesis refers to the precipitation of new mineral within the pore spaces of a sediment. When sufficient in quantity, cementation of the sediment occurs (Figure 3-3). The most important parameters influencing porosity in sandstone are age (time of burial), mineralogy (i.e., detrital quartz content), sorting, and the maximum depth of burial, and to a lesser degree, temperature. Compaction and cementation will reduce porosity, although porosity reduction by cement is usually only a small fraction of the total reduction. The role of temperature probably increases above a geothermal gradient of 4°C/100 m. Uplift and erosional unloading may also be important in the development of fracture porosity and permeability.

SEDIMENTARY SEQUENCES AND FACIES ARCHITECTURE

Analysis of sedimentary depositional environments is important since groundwater resource usage occurs primarily in unconsolidated deposits formed in these environments. Such water-bearing environments are often considered of beneficial use and warrant protection. Most subsurface environmental investigations conducted are also performed in these types of environments. The literature is full of examples where erroneous hydraulic or contaminant distribution information has been relied upon. In actuality, however, (1) the wells were screened across several high-permeability zones or across different zones,

**Table 3-1 Parameters Affecting Permeability of
Sediment Following Deposition**

Depositional Processes	Diagenetic Processes
Texture	
Grain size	Compaction
Sorting	Recrystallization
Grain slope	Dissolution
Grain packing	Replacement
Grain roundness	Fracturing
Mineral composition	Authigenesis
	Cementation

Table 3-2 Summary of Diagenesis and Secondary Porosity

Process	Effects
Leaching	Increase ϕ and K
Dolomitization	Increase K; can also decrease ϕ and K
Fracturing joints, Breccia	Increase K; can also increase channeling
Recrystallization	May increase pore size and K; can also decrease ϕ and K
Cementation by calcite, dolomite, anhydrite, pyrobitumen, silica	Decrease ϕ and K

(2) inadequate understanding of soil-gas surveys and vapor-phase transport in heterogenous environments allowing for an ineffective vapor extraction remedial strategy prevailed, (3) wells screened upward fining sequences with the technically unsound expectation of remediating hydrocarbon-affected soil and groundwater within the upper fine-grained section of the sequence, the focus of this chapter, or (4) the depositional environment was erroneously interpreted. Heterogeneities within sedimentary sequences can range from large-scale features associated with different depositional environments (Figure 3-4) that further yield significant large- and small-scale heterogeneities via development of preferential grain orientation. This results in preferred areas of higher permeability and, thus, preferred migration pathways of certain constituents considered hazardous.

To adequately characterize these heterogeneities, it becomes essential that subsurface hydrogeologic assessment include determination of the following:

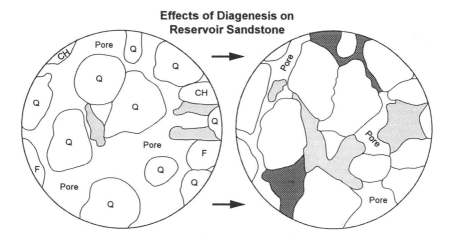

Figure 3-3 **Reduction in porosity in sandstone as a result of cementation and growth of authigenic minerals in the pores affecting the amount, size, and arrangement of pores (modified after Ebanks, 1987).**

- Depositional environment and facies of all major stratigraphic units present
- Propensity for heterogeneity within the entire vertical sequence and within different facies of all major stratigraphic units present
- Potential for preferential permeability (i.e., within sand bodies).

The specific objectives to understanding depositional environments as part of subsurface environmental studies are to (1) identify depositional processes and resultant stratification types that cause heterogeneous permeability patterns, (2) measure the resultant permeabilities of these stratification types, and (3) recognize general permeability patterns that allow simple flow models to be generated. Flow characteristics in turn are a function of the types, distributions, and orientations of the internal stratification. Since depositional processes control the zones of higher permeability within unconsolidated deposits, a predictive three-dimensional depositional model to assess potential connections or intercommunication between major zones of high permeability should also be developed. A schematic depicting the various components of an integrated aquifer description is shown in Figure 3-5.

Understanding of the facies architecture is extremely important to successful characterization and remediation of contaminated soil and groundwater. Defining a hydrogeologic facies can be complex. Within a particular sedimentary sequence, a hydrogeologic facies can range over several orders of magnitude, while other parameters such as storativity and porosity vary over a range of only one order of magnitude. A hydrogeologic facies is defined as a

Figure 3-4 Photograph showing varying depositional environments and sedimentary structures in this outcrop of Dalles conglomerate and underlying tuffaceous sediments and vitric tuff units at the Arlington Facility, North-Central Oregon.

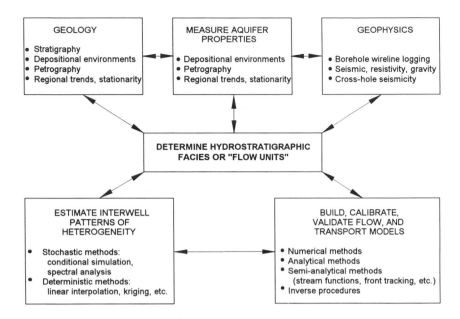

Figure 3-5 **Schematic depicting the various components of an integrated aquifer description.**

homogeneous, but anisotropic, unit that is hydrogeologically meaningful for purposes of conducting field studies or developing conceptual models. Facies can be gradational in relation to other facies, with a horizontal length that is finite but usually greater than its corresponding vertical length. A hydrogeologic facies can also be viewed as a sum of all the primary characteristics of a sedimentary unit. A facies can thus be referred to with reference to one or several factors, such as lithofacies, biofacies, geochemical facies, etc. The importance of facies cannot be understated. For example, three-dimensional sedimentary bodies of similar textural character are termed lithofacies. It is inferred that areas of more rapid plume migration and greater longitudinal dispersion correlate broadly with distribution and trends of coarse-grained lithofacies and are controlled by the coexistence of lithologic and hydraulic continuity. Therefore, lithofacies distribution can be used for preliminary predictions of contaminant migration pathways and selection of a subsurface assessment and remediation strategy. However, caution should be taken in proximal and distal assemblages where certain layered sequences may be absent due to erosion and the recognition of cyclicity is solely dependent on identifying facies based simply on texture. Regardless, the facies reflects deposition in a given environment and possesses certain characteristics of that environment.

Sedimentary structures also play a very important role in deriving permeability-distribution models and developing fluid-flow models. Nearly all depositional environments are heterogeneous, which for all practical purposes restricts the sole use of homogeneous-based models in developing useful hydraulic conductivity distributions data for assessing preferred contaminant migration pathways, and developing containment and remediation strategies. There has been much discussion in the literature regarding the influences of large-scale features such as faults, fractures, significantly contrasting lithologies, diagenesis, and sedimentalogical complexities. Little attention, however, has been given to internal heterogeneity within genetically defined sand bodies caused by sedimentary structures and associated depositional environment and intercalations. In fact, for sand bodies greater variability exists within bedding and lamination pockets than between them. An idealized model of the vertical sequence of sediment types by a meandering stream is shown in Figure 3-6. The various layers illustrated affect the flow of fluids according to their relative characteristics. For example, in a point-bar sequence, the combination of a ripple-bedded, coarser-grained sandstone will result in retardation of flow higher in the bed, and deflection of flow in the direction of dip of the lower trough crossbeds (Figure 3-7).

Hydrogeologic analysis is conducted in part by the use of conceptual models. These models are used to characterize spatial trends in hydraulic conductivity and permit prediction of the geometry of hydrogeologic facies from limited field data. Conceptual models can be either site specific or generic. Site-specific models are descriptions of site-specific facies that contribute to understanding the genesis of a particular suite of sediments or sedimentary rocks. The generic model, however, provides the ideal case of a particular depositional environment or system. Generic models can be used in assessing and predicting the spatial trends of hydraulic conductivity and, thus, dissolved contaminants in groundwater. Conventional generic models include either a vertical profile that illustrates a typical vertical succession of facies, or a block diagram of the interpreted three-dimensional facies relationships in a given depositional system.

Several of the more common depositional environments routinely encountered in subsurface environmental studies are discussed below. Included is discussion of fluvial, alluvial fan, glacial, deltaic, eolian, carbonate, and volcanic-sedimentary sequences. Hydrogeologic parameters per depositional environment obtained from the literature are summarized in Table 3-3.

Fluvial Sequences

Fluvial sequences are difficult to interpret due to their sinuous nature and complexities of their varied sediment architecture reflecting complex depositional environments. Fluvial sequences can be divided into high-sinuosity meandering channels and low-sinuosity braided channel complexes. Meandering stream environments (i.e., Mississippi River) consist of an asymmetric

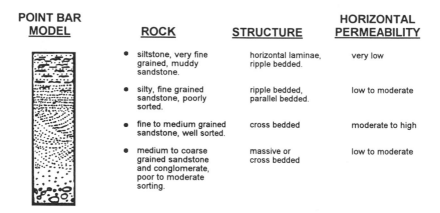

POINT BAR MODEL	ROCK	STRUCTURE	HORIZONTAL PERMEABILITY
	• siltstone, very fine grained, muddy sandstone.	horizontal laminae, ripple bedded.	very low
	• silty, fine grained sandstone, poorly sorted.	ripple bedded, parallel bedded.	low to moderate
	• fine to medium grained sandstone, well sorted.	cross bedded	moderate to high
	• medium to coarse grained sandstone and conglomerate, poor to moderate sorting.	massive or cross bedded	low to moderate

Figure 3-6   Point bar geologic model showing the influence of a sequence of rock textures and structures in a reservoir consisting of a single point bar deposit on horizontal permeability excluding effects of diagenesis (modified after Ebanks, 1987).

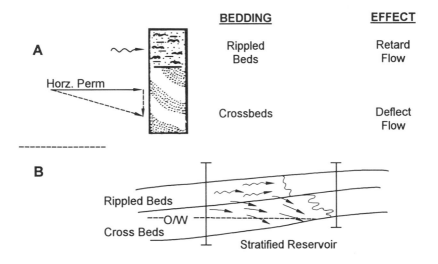

Figure 3-7   Diagram showing the effects of sedimentary structures and textures on the flow of fluids in a point bar sandstone reservoir: The crossbedded unit is coarser grained and is inferred to have better reservoir properties (i.e., permeability) than the overlying rippled unit (A). Uneven advance of injected fluids illustrating permeability variations results from differences in reservoir quality (modified after Ebanks, 1987).

Table 3-3 Summary of Hydraulic Properties for Certain Depositional Environments

Depositional Environment[a]	Hydrogeologic Facies	Hydraulic Conductivity[b,c]		Porosity in percent[c]	Ref.
		Horizontal	Vertical		
Eolian	Dune sand	5–140 (54)		42–55 (49)	Pryor (1973)
	Interdune/extra-erg	0.67–1,800			Chandler et al. (1989)
	Wind-ripple	900–5,200			Chandler et al. (1989)
	Grain flow	3,700–12,000			Chandler et al. (1989)
Fluvial	River point bar	4–500 (93)		17–52 (41)	
	Beach sand	3.6–166 (68)		39–56 (49)	
Glacial	Meltwater streams	10^{-1}–10^{-5}cm/s			Anderson (1989)
	Outwash drift	10^{-3}–10^{-4d}			Sharp (1984)
	Basal till	10^{-4}–10^{-9}cm/s	10^{-11} [e]		Anderson (1989)
	Esker sediment	10^{-1}–10^{-3}cm/s			Caswell (1988a; 1988b); De Gear (1986); Patson (1970)
	Supraglacial sediments	10^{-3}–10^{-7}cm/s			Stephanson et al. (1989)
Deltaic	Distributary channel sandstone	(436)		(28)	Tillman and Jordan (1987)
	Splay channel sandstone	(567)		(27)	Tillman and Jordan (1987)

						References
	Wave-dominated sandstone within prodelta and shelf mudstone	(21)			(21)	Tillman and Jordan (1987)
Volcanic-Sedimentary	Basalt (CRG)[f]	0.002–1,600 (0.65)	10^{-8}–10			Lindholm and Vaccaro (1988)
	Basalt (SRG)[f]	150–3,000				Lindholm and Vaccaro (1988)
	Basalt (CRG)	1×10^{-8}–10^{-5}cm/s	1×10^{-7}–2×10^{-7}cm/s			Testa (1988); Wang and Testa (1989)
	Sedimentary Interbed (SRG)	3×10^{-6}–3×10^{-2} [e]				Lindholm and Vaccaro (1988)
	Tuffaceous siltstone (interbeds; CRG)	1×10^{-6}–2×10^{-4}cm/s	1×10^{-8}–1×10^{-3}cm/s[e]		27–68 (42)	Testa (1988); Wang and Testa (1989)
	Interflow zone (CRG)	2×10^{-4}cm/s				Testa (1988); Wang and Testa (1989)

[a] Carbonates not represented but can have permeabilities ranging over 5 orders of magnitude.

[b] Values are in millidarcy(mD) per day unless otherwise noted; cm/s = centimeters per second; 1 mD = .001 darcy, 1 cm/s = 1.16×10^{-3} darcy.

[c] Values shown in parentheses are averages.

[d] Field.

[e] Laboratory.

[f] CRG = Columbia River Group; SRG = Snake River Group.

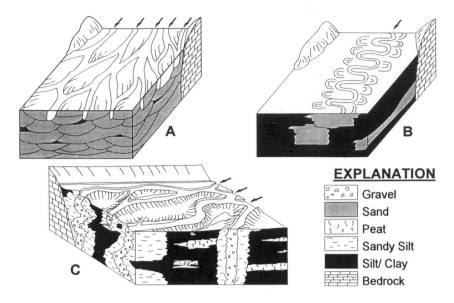

EXPLANATION

- Gravel
- Sand
- Peat
- Sandy Silt
- Silt/ Clay
- Bedrock

Figure 3-8   Fluvial facies model illustrating contrasting patterns of heterogeneity in (A) braided rivers, (B) meandering rivers, and
(C) anastomosing rivers (modified after Allen, 1965 and Smith
and Smith, 1980).

main channel, abandoned channels, point bars, levees and floodplains (Figure
3-8). Usually developed where gradients and discharge are relatively low, five
lithofacies have been recognized:

- Muddy fine-grained streams
- Sand-bed streams with accessory mud
- Sand-bed stream without mud
- Gravelly sand-bed stream
- Gravelly stream without sand.

Meandering streams can also be subdivided into three subenvironments:
floodplain subfacies, channel subfacies, and abandoned channel subfacies.
Floodplain subfacies is comprised of very fine sand, silt, and clay deposited
on the overbank portion of the floodplain, out of suspension during flooding
events. Usually laminated, these deposits are characterized by sand-filled
shrinkage cracks (subaerial exposure), carbonate caliches, laterites, and root
holes. The channel subfacies is formed as a result of the lateral migration of
the meandering channel which erodes the outer concave bank, scours the
riverbed, and deposits sediment on the inner bank referred to as the point bar.
Very characteristic sequences of grain size and sedimentary structures are
developed. The basal portion of this subfacies is lithologically characterized

by an erosional surface overlain by extraformation pebbles and intraformational mud pellets. Sand sequences with upward fining and massive, horizontally stratified and trough cross-bedded sands overlie these basal deposits. Overlying the sand sequences are tabular, planar, cross-bedded sands which grade into microcross-laminated and flat-bedded fine sands, grading into silts of the floodplain subfacies. The abandoned channel subfacies are curved fine-grained deposits of infilled abandoned channels referred to as oxbow lakes. Oxbow lakes form when the river meanders back, short-circuiting the flow. Although lithologically similar to floodplain deposits, geometry and absence of intervening point-bar sequences distinguishes it from the abandoned channel subfacies.

Braided river systems consist of an interlaced network of low-sinuosity channels and are characterized by relatively steeper gradients and higher discharges than meandering rivers. Typical of regions where erosion is rapid, discharge is sporadic and high, and little vegetation hinders runoff, braided rivers are often overloaded with sediment. Because of this sediment overload, bars are formed in the central portion of the channel around which two new channels are diverted. This process of repeated bar formation and channel branching generate a network of braided channels throughout the area of deposition.

Lithologically, alluvium deposits derived from braided streams are typically composed of sand and gravel-channel deposits. Repeated channel development and fluctuating discharge results in the absence of laterally extensive cyclic sequences as produced with meandering channels. Fine-grained silts are usually deposited in abandoned channels formed by both channel choking and branching, or trapping of fines from active downstream channels during eddy reversals.

The degree of interconnectedness is very important in addressing preferred migration pathways in fluvial sequences. Based on theoretical models of sand-body connectedness, the degree of connectedness increases very rapidly as the proportion of sand bodies increase above 50%. Where alluvial soils contain 50% or more of overbank fines, sand bodies are virtually unconnected.

Alluvial Fan Sequences

Alluvial fan sequences accumulate at the base of an upland area or mountainous area as a result of an emerging stream. These resulting accumulations form segments of a cone with a sloping surface ranging from less than 1° up to 25° averaging 5°, and rarely exceeding 10°. Alluvial fans can range in size from less than 100 m to more than 150 km in radius, although typically averaging less than 10 km. As the channels shift laterally through time, the deposit develops a characteristic fan shape in plane view, a convex-upward cross-fan profile, and concave-upward radial profile as shown in Figure 3-9.

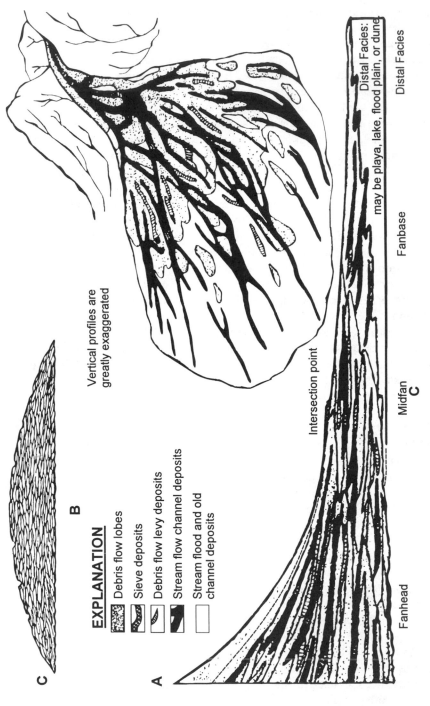

Figure 3-9 Generalized model of alluvial fan sedimentation showing (A) fan surface, (B) cross-fan profile, and (C) radial profile (after Spearing, 1974).

Facies analysis of alluvial fans require data on morphology and sediment distribution, and can be divided into four facies: proximal, distal, and of lesser importance, outer fan and fan fringe facies. Proximal facies are deposited in the upper or inner parts of the fan near the area of stream emergence and are comprised of relatively coarser-grained sediments. The proximal facies comprising the innermost portion of the fan (i.e., apex or fan-head area) contains an entrenched straight valley which extends outward onto the fan from the point of stream emergence. This inner fan region is characterized by two subfacies: a very coarse-grained, broad, deep, deposit of one or several major channels, and a finer-grained channel-margin level and interchannel deposits which may include coarse-grained landslides and debris flows type material. Distal facies are deposited in the lower and outer portion of the fan, and are comprised of relatively finer-grained sediments. The distal facies typically comprises the largest area of most fans, and consists of smaller distributary channels radiating outward and downfan from the inner fan valley. Hundreds of less-developed channels may be present on the fan. Depending on fan gradient, sediment time and supply, and climatic effects among other factors, commonly braided but straight, meandering and anastomosing channel systems may also be present. Outer fan facies are comprised of finer-grained, laterally extensive, sheet-like deposits of nonchannelized or less-channelized deposits. These deposits maintain a very low longitudinal gradient. The fan-fringe facies is comprised of very fine-grained sediments that intertongue with deposits of other environments (i.e., eolian, fluvial, lacustrine, etc.). Most deposits comprising alluvial fan sequences consist of fluvial (streamflow) or debris flow types.

Alluvial fans are typically characterized by high permeability and porosity. Groundwater flow is commonly guided by paleochannels which serve as conduits, and relatively less permeable and porous debris and mud flow deposits. The preponderance of debris-flow and mudflow deposits in the medial portion of fans may result in decreased and less-predictable porosity and permeability in these areas. Aquifer characteristics vary significantly with the type of deposit and relative location within the fan. Pore space also develops as intergranular voids, interlaminar voids, bubble cavities, and desiccation cracks.

Deltaic Sequences

Deltas are abundant throughout the geologic record with 32 large deltas forming at this time and countless others in various stages of growth. A delta deposit is partly subaerial built by a river into or against a body of permanent water. Deltaic sedimentation requires a drainage basin for a source of sediment, a river for transport of the material, and a receiving basin to store and rework it. During formation, the outer and lower parts are constructed below water, and the upper and inner surfaces become land reclaimed from the sea. Deltas form by progradation or a building outward of sediments onto themselves.

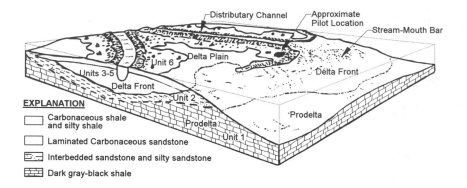

Figure 3-10   Block diagram showing vertical and areal distribution of units in a typical modern delta (modified after Harris, 1975).

Once a particular pathway is no longer available, another delta forms in a different location. Any given delta is thus a composite of conditions reflecting initiation of delta development to its ultimate abandonment of a particular deposition center (Figure 3-10).

Delta sequences reflect condition of source (volume and type of available sediment) and distribution and dispersal processes. Two general classes or end members have been defined: high- and low-energy deltas. High-energy deltas or sand deltas are characterized by few active meandering distributary channels, with shoreline comprised of continuous sand (i.e., Nile, Rhone, or Brazos-Colorado). Low-energy or mud deltas are characterized by numerous bifurcating or branching, straight to sinuous, distributary channels, with shorelines comprised of discontinuous sands and muds.

No two deltas are exactly alike in their distribution and continuity of permeable and nonpermeable sediments. The most important parameters controlling the distribution is size and sorting of grains. All delta systems form two parts during development: a regressive sequence of sediment produced as the shoreline advances seaward and a system of distributary channels. These two parts result in two main zones of relatively high permeability: channel sands and bar sands.

Typical deltaic sequences from top to bottom include marsh, inner bar, outer bar, prodelta, and marine. Depositional features include distributary channels, river-mouth bars, interdistributary bays, tidal flats and ridges, beaches, eolian dunes, swamps, marshes, and evaporite flats. Characteristics per sequence are summarized in Table 3-4. A schematic profile is presented in Figure 3-11.

The geometry of channel deposits reflects delta size, position of the delta in the channel, type of material being cut into, and forces at the mouth of the channel distributing the sediments. For high-energy sand deltas, channels can be filled with up to 90% sand, or clay and silt. For example, those sequences

Table 3-4   Summary of Deltaic Sequences and Characteristics

Sequence	Sediment Type	Permeability
Marsh	Clay, silt, coal with a few reservoir sands of limited extent	Relatively low
Inner bar	Sand with some clay/silt intercalations	Moderate to high
Outer bar	Sand with many clay/silt intercalations	Moderate
Prodelta	Clay and silt	Low
Marine	Clay	Low

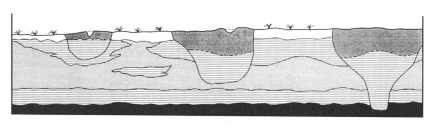

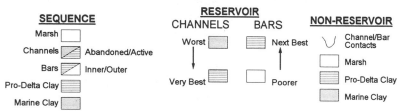

Figure 3-11   Deltaic facies model illustrating heterogeneity and continuity.

of high permeability in bar deposits is at or near the top with decreasing permeability with depth and laterally away from the main sand buildup. Within bar deposits, permeability generally increases upward and is highest at or near the top, and is similar parallel to the trend of the bar (i.e., highest at the top near the shore, decreasing progressively seaward). Porosity is anticipated to be well connected throughout the bar with the exception of the lower part. In channels, however, high-permeability sequences occur in the lower part with decreasing permeability vertically upward. Within channel deposits, porosity and permeability is high at the lower part of the channel, decreasing vertically upward, with an increase in the number, thickness, continuity, and areal extent of clay interbeds.

Low-energy mud deltas differ from sand deltas in that sediments are carried into a basin via numerous channels during flooding events, and the fine-grained silts and clay is not winnowed from the sand before new sediment circulates from the next flooding event. The coarse-grained sediment is thus more discontinuous with more numerous (less than 1 in. thick) and continuous clay and silt interbeds. Along the perimeter or within bar deposits, grain size increases with sorting; improves vertically upward; and clay and silt interbeds decrease in number, thickness, and areal extent. Within individual bars, overall permeability decreases laterally away from the coarse-grained sand depositional pathways. The highest permeability is at or near the top of the bar decreasing vertically downward and in a seaward direction. Sand continuity, thus permeability, is poor due to the numerous shifting distributary channels forming widespread clay and silt interbeds ranging from less than 1 ft to more than 12 ft in thickness. Coarse-grained sands predominate within the lower portion of the distributary channels with clay and silt interbeds typically less than 1 ft in thickness, and range from a few feet to a few tens of feet in maximum areal extent, thus, not providing a barrier to vertical flow. The number, thickness, and areal extent of these fine-grained interbeds generally increases vertically upward depending on how fast and where the channel was abandoned. Overall, permeability and porosity continuity is high only in the upward portion of the bar. Within the channels, however, permeability and porosity continuity is high at the basal portion of the channels, but the amount and quality of coarse-grained sand (high permeability zones) is dependent upon the location and rate of channel abandonment.

Average porosity and permeability based on a broadly lenticular wave-dominated deltaic sandstone (e.g., Upper Cretaceous Big Wells reservoir, which is one of the largest oil fields located in south Texas) increased in prodelta and shelf mudstones, averaging 21% and 6 mD, respectively. Studies on the El Dorado field located in southeastern Kansas, a deltaic sequence containing the 650-ft-thick Admire sandstone, has reported porosity and permeability averaging 28% and 436 mD, respectively, within the distributary channel sandstones. Thinner and discontinuous splay channels sandstones average 27% porosity and 567 mD in permeability. The variation in porosity and permeability reflect diagenetic processes (i.e., deformation, secondary leaching of feldspar, and formation of calcite cement and clay laminae).

Glacial Sequences

Sequences derived from glacial processes include four major types of materials: tills, ice-contact, glacial fluvial or outwash, and delta and glaciolacustrine deposits. Glacial tills make up a major portion of a group of deposits referred to as diamictons which are defined as poorly sorted, unstratified deposits of unspecific origin. Tills and associated glaciomarine drift deposits are both deposited more or less directly from ice without the winnowing effects of water. Till is deposited in direct contact with glacial ice and, although substantial thickness accumulations are not common, tills make up a discontinuous cover totaling up to 30% of the earth's continental surface.

Glaciomarine drift, however, accumulates as glacial debris melts out of ice floating in marine waters. These deposits are similar to other till deposits but also includes facies that do not resemble till or ice-contact deposits. A lesser degree of compaction is evident due to a lack of appreciable glacial loading.

Tills can be divided into two groups based on deposition (basal till or supra glacial till), or three groups based upon physical properties and varying depositional processes: lodgement, ablation, and flow. Lodgement tills are deposited subglacially from basal, debris-laden ice. High shear stress results in a preferred fabric (i.e., elongated stones oriented parallel to the direction of a flow) and high degree of compaction, high bulk densities, and low void ratios of uncemented deposits. Ablation tills are deposited from englacial and superglacial debris dumped on the land surface or the ice melts away. These deposits lack significant shear stresses and thus are loosely consolidated with a random fabric. Flow tills are deposited by water-saturated debris flowing off glacial ice as mudflow. Flow tills exhibit a high degree of compaction, although less than that of lodgement tills, with a preferred orientation of elongated stones due to flowage.

Till is characterized by a heterogeneous mixture of sediment sizes (boulders to clay) and a lack of stratification. Particle size distribution is often bimodal with predominant fractions in the pebble-cobble range and silt-clay range, both types being massive with only minor stratified intercalations. Other physical characteristics of till include glaciofluvial deposits or outwash deposits having strong similarities to sediments formed in fluvial environments due to similar transportation and deposition mechanisms. These types of deposits are characterized by abrupt particle-size changes and sedimentary structures reflecting fluctuating discharge and proximity to glaciers. Characteristics include a down-gradient fining in grain size and down-gradient increase in sorting, therefore a decrease in hydraulic conductivity. Outwash deposits can be divided into three facies: proximal, intermediate or medial, and distal. Outwash deposits are typically deposited by braided rivers, although the distal portions are deposited by meandering and anastomosing rivers. Proximal facies are deposited by gravel-bed rivers while medial and distal facies are deposited by sand-bed rivers. Thus, considerable small-scale variability within each facies assemblage exists. Vertical trends include fining-upward sequences as with meandering fluvial sequences. Within the medial portion, series of upward fining or coarsening cycles are evident depending on whether the ice front was retreating or advancing, respectively. Layered sequences within the gravel-dominant proximal facies and sand-dominant distal facies are either absent or hindered by the relatively large-grain size component of the proximal facies. A hydrogeologic facies model and their respective vertical profile is shown in Figure 3-12.

Delta and glaciolacustrine deposits are formed when meltwater streams discharge into lakes or seas. Ice-contact delta sequences produced in close proximity to the glacier margins typically exhibit various slump-deformation

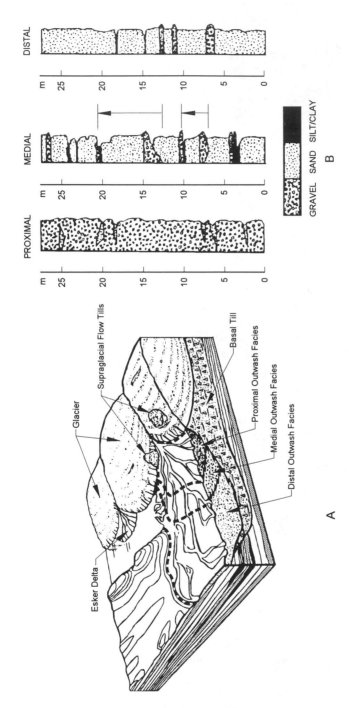

Figure 3-12   Hydrogeologic facies model for glacial depositional environment.

structures. Delta sequences produced a considerable distance from the glacier margins exhibit no ice-collapse structures, variable sediment discharge, and particle-size distribution and structures (i.e., graded bedding, flow rolls, varies, etc.) similar to that of meltwater streams.

Also associated with till deposits are ice-contact deposits which form from meltwater on, under, within, or marginal to the glacier. Detritus deposits formed on, against, or beneath the ice exhibit better sorting and stratification, a lack of bimod particle-size distribution, and deformational features such as collapse features (i.e., tilting, faulting, and folding).

Hydrogeologically, hydraulic conductivity of basal tills facies are on the order of 10^{-4} cm/s with horizontal hydraulic conductivities on the order of 10^{-3} to 10^{-7} cm/s reflecting locations and degree of interconnected sand and gravel channel deposits contained within the till. Drift deposits can vary from about 10^{-11} m/s (laboratory tests) to 10^{-6} to 10^{-7} (field) when permeable sand lenses or joints are intersected.

Eolian Sequences

Eolian or wind-deposited sediments are complex, highly variable accumulations. They are characterized as well-sorted, matrix-free, well-rounded sediments with a dominance of sand-sized fractions, and are perceived as essentially lithologically homogeneous with irregular plan and cross-sectional geometrics with exception to the linear trends of coastal dunes. Unlike many other sedimentary facies, eolian deposits have no predictable geometry and/or cyclic motif of subfacies. Recent studies have provided a better understanding of the stratigraphic complexity and thus flow regime within these deposits.

Small-scale forms of eolian deposits include wind sand ripples and wind granule ripples. Wind sand ripples are wavy surface forms on sandy surfaces whose wavelength depends on wind strength and remains constant with time. Wind granule ripples are similar to wind sand ripples but are usually produced in areas of erosion. Excessive deflation produces a large concentration of grains 1 to 3 mm in diameter which are too big to be transported via saltation under the existing wind conditions.

Larger-scale eolian sand forms include sand drifts, sand shadows, gozes, sand sheets, and sand dunes. Sand drifts develop by some fixed obstruction which lies in the path of a sand-laden wind. When sand accumulates in the lee of the gap between two obstacles, a tongue-shaped sand drift develops. As the wind velocity is reduced by an obstruction, a sand shadow develops. Gozes are gentle large-scale undulatory sand surfaces associated with sparse desert vegetation. Sand sheets are more or less flat, with slight undulations or small dune tile features, and encompassing large areas.

Sand dunes are the most impressive features and develop whenever a sand-laden wind deposits in a random patch. This patch slowly grows in height as a mound, until finally a slip face is formed. The sand mound migrates forward as a result of the advance of the slip face, but maintains its overall shape providing wind conditions do not change. Sand dunes are characterized by wind conditions, sand type, and sand supply as summarized in Figure 3-13.

Eolian deposits are stratigraphically complex because of (1) differing spatial relationships of large-scale forms such as dunes, interdunes, and sand-sheet deposits relative to one another and to ectradune (noneolian) sediments and (2) varying dune types, each with its own cross-bedding patterns and different degrees of mobility; thus, there are different fluid-flow properties when consolidated or lithified. Sedimentary structures within eolian deposits include ripples, contorted sand bedding, cross-bedding with great set heights, normally graded beds, inversely graded beds, evenly laminated beds, discontinuously laminated beds, nongraded beds, and lag deposits along boundary surfaces and sets. Basic eolian bed forms as related to a number of slip facies are shown in Figure 3-14.

Relatively recent eolian deposits are presumed to have high porosity and permeability, and are typically well rounded, well sorted, and generally only slightly cemented. Regional permeability is usually good due to a lack of fine-grained soils, shales, interbeds, etc., and thus constitutes important aquifers. Studies conducted on several large eolian deposits (i.e., Page sandstone of Northern Arizona) have shed some light on preferred fluid migrations in such deposits. For example, fluid flow is directional dependent because of inverse grading within laminae. Permeability measured parallel to wind-ripple laminae has been shown to be from two to five times greater than that measured perpendicular to the laminae. Four common cross-set styles based on bulk permeability and directional controls of each stratum type for the Page sandstone in Northern Arizona are shown in Figure 3-15. Page sandstone is poorly cemented, has high porosity, and has a permeability outcrop which has never been buried. In Case A, the cross set is composed exclusively of grain-flow strata; thus, the permeability of each grain-flow set is high in all directions with significant permeability contrasts, flow barriers, or severely directional flow. Due to inverse grading, fluid flow is greater parallel to the grain-flow strata than across it. In Case B, laminae occuring in wind-ripple cross sets have low permeability throughout, thus inhibiting flow and imparting preferred flow paths. Case C illustrates bulk permeability based on the ratio of grain-flow strata to wind-ripple strata in the cross set. Higher ratios are indicative of a greater capacity to transmit fluids, while low permeability logs created by wind-ripple sets act to orient fluid migration parallel to the stratification. Case D shows that a cross set which exhibits grain-flow deposits grading into wind-ripple laminae is more permeable at the top of the set due to the dominance of grain-flow strata. Fluid flow is thus reduced downward throughout the set due to the transition from high-permeability grain-flow cross strata to low permeability wind-ripple laminae, with greater ease of flow occurring in the cross set from the wind-ripple laminae into the grain-flow strata.

Dune Type	Definition and Occurence

Barchan Dune   A crescent-shaped dune with horns pointing downward. Occurs on hard, flat floors of desserts. Constant wind and limited sand supply. Height 1m to more than 30m.

Transverse Dune   A dune forming a asymmetrical ridge transverse to wind direction. Occurs in area with abundant sand and little vegetation. In places grades into barchans.

Parabolic Dune   A dune of U-shape with the open end of the U facing upwind. Some form by piling of sand along leeward and lateral margins of areas of deflation in older dunes.

Linear Dune   A long, straight, ridge-shaped dune parallel with wind direction. As much as 100m high and 100km long. Occurs in deserts with scanty sand supply and strong winds varying within one general direction. Slip faces vary as wind shifts direction.

Star Dune   An isolated hill of sand having a base that resembles a star in plan view. Ridges converge from basal points to a central peak as high as 100m. Tends to remain fixed in place in an area where wind blows from all directions.

Figure 3-13   Dune types based on form (modified after McKee, 1979).

Figure 3-14   Basic eolian bed forms as related to the number of slip facies.

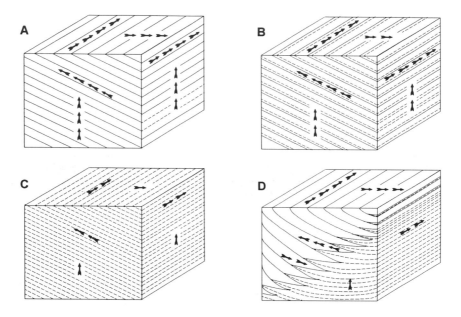

Figure 3-15   Common styles of cross-strata in Page sandstone: (A) grain-flow set, (B) wind-ripple set, (C) interlaminated grain-flow and wind-ripple set, and (D) grain-flow foresets toeing into wind-ripple bottom sets. Directional permeability indicated by arrows; more arrows denote higher potential flow.

Overall, interdune or extra-erg deposits are least permeable (0.67 to 1800 mD), wind-ripple strata moderately permeable (900 to 5200 mD), and grain-flow strata the most permeable (3700 to 12,000 mD). A diagrammatic eolian sequence exhibiting some typical pattern of eolian stratification with potential fluid flow, assuming a vertical pressure field exists, is shown in Figure 3-16. Compartmentalization develops due to bounding surfaces between the cross sets, with flow largely channeled along the sets. Fluid flow is especially great

EXPLANATION

- Wind-Ripple Strata
- Grain-Flow Strata
- Interdune Deposits
- Fluid Movement Focus

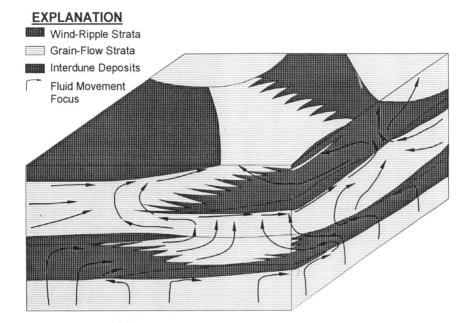

Figure 3-16   **Fluid flow through idealized eolian sequence based on relative permeability values of stratification types and bounding surfaces, assuming a vertical pressure field (after Chandler et al., 1989).**

where low permeability interdune or extra-erg deposits overlie bounding surfaces (i.e., along sets). Flow windows, however, occur where low permeability strata were eroded or pinched out. Because of high-permeability grain-flow deposits relative to wind-ripple strata, fluid migration between adjacent grain-flow sets would be more rapid than across bounding surfaces separating sets of wind-ripple deposits. Flow through the sets themselves would have been dictated by internal stratification types.

Carbonate Sequences

Carbonate sequences are important in that aquifers within such sequences are often heavily depended upon for drinking water, irrigation, and other uses. Carbonate rocks are exposed on over 10% of the earth's land area. About 25% of the world's population depends on fresh water retrieved from Karst aquifers. The Floridan aquifer of Cenozoic age, for example, is the principal aquifer in the southeastern U.S., and is encountered in Florida, Georgia, Alabama, and South Carolina. However, although its major resource is as a potable water supply, the nonpotable part of the aquifer in southern Florida serves as a disposal zone for municipal and industrial wastewater via injection wells (refer to Chapter 11).

Karst terrains are the foremost examples of groundwater erosion. Karst terrains can be divided into two types: well developed and incipient. Well-developed karst terrains are marked by surface features such as dolines or sinkholes which can range from 1 to 1000 m in maximum dimension. Other features are closed depressions, dry valleys, gorges, and sinking streams and caves, with local groundwater recharge via both infiltration and point sinks. The subsurface systems of connected conduits and fissure openings, enlarged by solution (i.e., cave systems), serve a role similar to that of stream channels in fluvial systems, and reflects the initial phases of subsequent surface karst landscape features (i.e., sinkholes and depressions). Cave systems actually integrate drainage from many points for discharge at single, clustered, or aligned springs. Incipient karst terrains differ from well-developed terrains in that few obvious surficial features exist and recharge is limited primarily to infiltration. Several other approaches to classification of karst terrains exist: holokarst, merokarst or fluviokarst, and parakarst. In holokarst terrains, all waters are drained underground, including allogenic streams (i.e., those derived from adjacent nonkarst rocks), with little or no surface channel flow. In fluviokarst terrains, major rivers remain at the surface, reflecting either large flow volumes that exceed the aquifer's ability to adsorb the water or immature subsurface development of underground channels. Parakarst terrains are a mixture of the two reflecting mixtures of karst and nonkarst rocks. Covered karst reflects the active removal of carbonate rocks beneath a cover of other unconsolidated rocks (i.e., sandstone, shales, etc.); whereas, mantled karst refers to deep covers of unconsolidated rocks or materials. Paleokarst terrains are karst terrains or cave systems that are buried beneath later strata and can be exhumed or rejuvenated. Pseudokarst refers to karst-like landforms that are created by processes other than rock dissolution (i.e., thermokarst, vulcanokarst, and mechanical piping). An idealized authigenic karst profile is shown in Figure 3-17.

Carbonate sediments can accumulate in both marine and nonmarine environments. The bulk of carbonate sediments are deposited in marine environments, in tropical and subtropical seas, with minimal or no influx of terrigenous or land-derived detritus. Marine depositional environments include tidal flat, beach and coastal dune, continental shelf, bank, reef, basin margin and slope, and deeper ocean or basin. Lakes provide the most extensive carbonate deposits on land, regardless of climate, although carbonate can occur or caliche (i.e., soil-zone deposits) and travertine (i.e., caves, karst, and hot-spring deposits).

Carbonate rocks are defined as containing more than 50% carbonate minerals. The most common and predominant carbonate minerals are calcite $(CaCO_3)$ and dolomite $[CaMg(CO_3)_2]$. Other carbonate minerals include aragonite $(CaCO_3)$, siderite $(Fe^{12}CO_3)$, and magnesite $(MgCO_3)$. The term limestone is used for those rocks in which the carbonate fraction is composed primarily of calcite, whereas the term dolomite is used for those rocks composed primarily of dolomite.

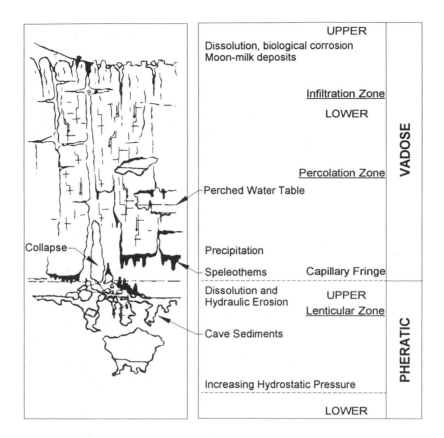

Figure 3-17   Idealized authigenic karst profile.

Overall, carbonate rocks serve as significant aquifers worldwide and are not limited by location or age of the formation. Carbonate rocks show a total range of hydraulic conductivities over a range of ten orders of magnitude. The broad diversity in hydrogeologic aspects of carbonate rocks reflects the variable combination of more than 60 processes and controls. Hydrogeologic response is related to rock permeability, which is affected most by interrelated processes associated with dynamic freshwater circulation and solution of the rock. Dynamic freshwater circulation is controlled and maintained primarily by the hydraulic circuit: maintenance of the recharge, flowthrough, and discharge regime. Without these regimes, the overall system is essentially stagnant and does not act as a conduit. The primary controls on solutions include rock solubility and chemical character of the groundwater; secondary controls include digenetic, geochemical, and chronologic aspects.

Carbonate reservoirs or aquifers are characterized by extremely heterogeneous porosity and permeability, reflecting the wide spectrum of depositional environments for carbonate rocks and subsequent diagenetic alteration of the original rock fabric (Table 3-2). Pore systems can range from thick, vuggy reservoirs in the coarse-grained skeletal-rich facies of reef or platform margin, to highly stratified, often discontinuous reservoirs in reef and platform interiors, and nearshore facies.

Karst terrains are highly susceptible to groundwater contamination. When used for waste disposal, these areas are susceptible to potential failure due to subsidence and collapse, which in turn can result in reservoir compartmentalization. To assess secondary porosity and potential contamination susceptibility, characterization of carbonate or karst aquifers include generation of data regarding percent rock core recovery, mechanical response during drilling, drilling fluid loss, and drilling resistance.

Volcanic-Sedimentary Sequences

Volcanic-sedimentary sequences are prevalent in the northwestern U.S. The Columbia Lava Plateau, for example, encompasses an area of 366,000 km² and extends into northern California, eastern Oregon and Washington, southern Idaho, and northern Nevada; the Snake River Group encompasses 40,400 km² in southern Idaho. The geology of this region consists of a thick, accordantly layered sequence of basalt flows and sedimentary interbeds.

The basalt flows can range from a few tens of centimeters to more than 100 m in thickness, averaging 30 to 40 m. From bottom to top, individual flows generally consist of a flow base, colonnade, and entablature (Figure 3-18). The flow base makes up about 5% of the total flow thickness and is typically characterized by a vesicular base and pillow-palagonite complex of varying thickness if the flow entered water. The colonnade makes up about 30% of the flow thickness and is characterized by nearly vertical 3- to 8-sided columns of basalt, with individual columns about 1 m in diameter and 7.5 m in length. The colonnade is usually less vesicular than the base. The entablature makes up about 70% of the flow thickness. This upper zone is characterized by small-diameter (averaging less than 0.5 m) basalt columns which may develop into a fan-shape arrangement; hackly joints, with cross joints less consistently oriented and interconnected, may be rubbly and clinkery, and the upper part is vesicular. Following extrusion, flows cool rapidly, expelling gases and forming vesicles and cooling joints. These upper surfaces are typically broken by subsequent internal lava movement resulting in brecciated flow tops. The combination of the superposed flow base and vesicular upper part of the entablature is referred to as the interflow zone (Figure 3-19). Interflow zones generally make up 5 to 10 % of the total flow thickness.

Groundwater occurrence and flow within layered sequences of basalt flows and intercalated sedimentary interbeds is complex. Such sequences typically consist of multiple zones of saturation with varying degrees of interconnection. Principle aquifers or water-bearing zones are associated with interflow zones

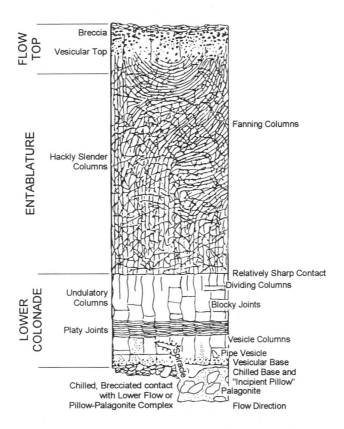

Figure 3-18   **Generalized schematic diagram showing intraflow structure of a basalt flow (modified after Swanson and Wright, 1978).**

between basalt flows. These interflow zones commonly have high to very high permeability and low storativity because of the open nature, but limited volume, of joints and fractures.

Furthermore, because of the generally impervious nature of the intervening tuffaceous sediments and dense basalt, stratigraphically adjacent interflow zones may be hydraulically isolated over large geographic areas. This physical and hydraulic separation is commonly reflected by differences in both piezometric levels and water quality between adjacent interflow aquifers, as illustrated in the conceptual model shown in Figure 3-20. Recharge occurs mainly along outcrops and through fractures which provide hydraulic communication to the surface.

Interflow zones generally have the highest hydraulic conductivities and can form a series of superposed water-bearing zones. The colonnade and entablature

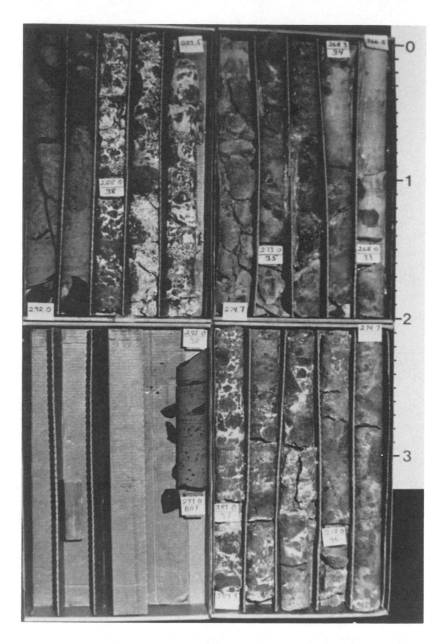

Figure 3-19 Rock core of brecciated interflow zone between the upper and lower Priest Rapids basalt beneath the Arlington Facility, North-Central Oregon.

are connected better vertically rather than horizontally, which allows for the movement of groundwater between interflow zones, although overall flow is three-dimensional. Multiple interflow zones can result in high total horizontal transmissivity. Position of the basalt flow within the regional flow system and varying hydraulic conductivities create further head differences with depth that can be very large in comparison to other sedimentary sequences. Horizontal hydraulic conductivities range from 0.65 up to 1600 or even 3000 m/day, whereas vertical hydraulic conductivities range from 10^{-8} to 10 m/day depending on the structural elements present (i.e., degree of fracturing joints, presence of sedimentary interbeds, etc.).

Sedimentary interbeds are typically comprised of tuffaceous sediments of varying thickness, lateral extent, lithology, and degree of weathering. These interbeds usually impede groundwater movement in many areas. Groundwater flow within the more prominent interbeds are affected by the thickness and anisotropy of each hydrostratigraphic unit, and the position and continuity of each layer within the units. As with any layered media, the hydrostratigraphic unit with the lowest vertical hydraulic continuity is the controlling factor for groundwater flow in the vertical direction (normal to bedding). In the horizontal direction (parallel to bedding), groundwater flow is controlled by the hydrostratigraphic unit with the highest horizontal hydraulic continuity. Horizontal hydraulic conductivities based on pumping tests range from about 1×10^{-6} to 1×10^{-4} cm/s. Vertical hydraulic conductivities based on laboratory tests range from about 1×10^{-6} to 1×10^{-4} cm/s. Both methods showed a variance of two orders of magnitude. A more detailed discussion of aquifer characteristics of a sedimentary interbed is provided in Chapter 9.

STRUCTURAL STYLE AND FRAMEWORK

Structural geologic elements which can play a significant role in dealing with subsurface environment-related issues include faults, fractures, joints, and shear zones. Faults can be important from a regional perspective in understanding their impact on the regional groundwater flow regime, and delineation and designation of major water-bearing strata. Faults are usually less important in most site-specific situations. Fractures, joints, and shear zones, however, can have a significant role both regionally and locally in fluid flow and assessment of preferred migration pathways of dissolved contaminants in groundwater in consolidated and unconsolidated materials. Regional geologic processes that produce certain structural elements, notably fracture porosity, include faulting (seismicity), folding, uplift, erosional unloading of strata, and overpressing of strata.

Fracture classification is outlined in Table 3-5. Tectonic and possibly regional fractures result from surface forces (i.e., external to the body as in

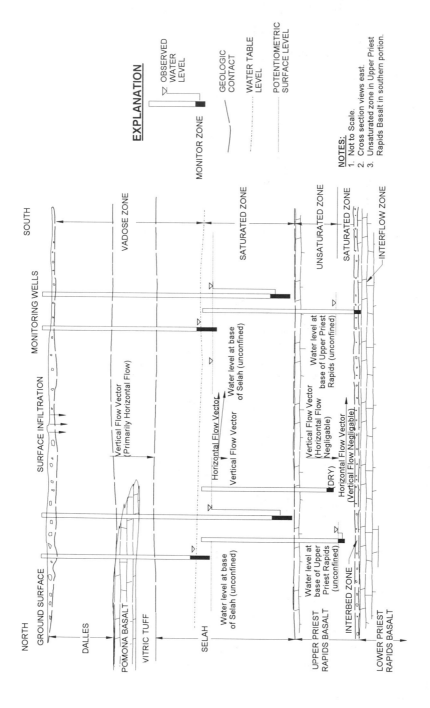

Figure 3-20   Conceptual flow model showing zones of saturation flow vectors in relationship to observed water levels (from Testa, 1988).

Table 3-5 Classification of Fractures

Fracture Type	Classification	Remarks
Experimental	Shear	
	Extension	
	Tensile	
Natural	Tectonic	Due to surface forces
	Regional	Due to surface forces (?)
	Contractional	Due to body force
	Surface-related	Due to body force

tectonic fractures); contractional and surface-related fractures result from body forces (i.e., internal to the forces). Contractional fractures are of varied origin resulting from desiccation, syneresis, thermal gradients, and mineral phase changes. Desiccation fractures develop in clay and silt-rich sediments upon a loss of water during subaerial drying. Such fractures are typically steeply dipping, wedge-shaped openings that form cuspate polygons of several nested sizes. Syneresis fractures result from a chemical process involving dewatering and volume reduction of clay, gel, or suspended colloidal material via tension or extension fractures. Associated fracture permeability tends to be isotropically distributed since developed fractures tend to be closely and regularly spaced. Thermal contractional fractures are caused by the cooling of hot rock, as with thermally induced columnar jointing in fine-grained igneous rocks (i.e., basalts). Mineral phase-change fracture systems are composed of extension and tension fractures related to a volume reduction due to a mineral phase change. Mineral phase changes are characterized by irregular geometry. Phase changes, such as calcite to dolomite or montmorillonite to illite, can result in about a 13% reduction in molar volume. Surface-related fractures develop during unloading, release of stored stress and strain, creation of free surfaces or unsupported boundaries, and general weathering. Unloading fractures or relief joints occur commonly during quarrying or excavation operations. Upon a one-directional release of load, the rock relaxes and spalls or fractures. Such fractures are irregular in shape and may follow topography. Free or unsupported surfaces (i.e., cliff faces, banks, etc.) can develop both extension and tensional fractures. These types of fractures are similar in morphology and orientation to unloading fractures. Weathering fractures are related to mechanical and chemical weathering processes such as freeze-thaw cycles, mineral alluation, diagenesis, small-scale collapse and subsidence, and mass-wasting processes.

Faults

Faults are regional structures that can serve as barriers, partial barriers, or conduits to groundwater flow. The influence and effect of faults on fluid flow entrapment depend on the rock properties of strata that are juxtaposed and the attitude or orientation of the strata within their respective fault blocks. The influence of regional structural elements, notably faults, can have a profound effect on groundwater occurrence, regime, quality and usage, and delineation and designation of water-bearing zones of beneficial use.

The Newport-Inglewood Structural Zone in southern California exemplifies this important role. The structural zone is characterized by a northwesterly trending line of gentle topographic prominences extending about 40 mi (Figure 3-21). This belt of domal hills and mesas, formed by the folding and faulting of a thick sequence of sedimentary rocks, is the surface expression of an active zone of deformation. An important aspect of this zone is the presence of certain fault planes that serve as effective barriers to the infiltration of seawater into the severely downdrawn groundwater reservoirs of the coastal plain. These barriers also act as localized hydrogeologic barriers for freshwater on the inland side of the zone, reflected in the relatively higher water level elevations and enlarged effective groundwater reservoirs.

The structural zone separates the Central groundwater basin to the northeast from the West Coast groundwater basin to the southwest. In the West Coast Basin area, at least four distinct water-bearing zones exist. In descending stratigraphic position, these zones are the shallow, unconfined Gaspur aquifer, the unconfined Gage aquifer of the upper Pleistocene Lakewood Formation, the semiconfined Lynwood Aquifer, and the confined Silverado aquifer of the lower Pleistocene San Pedro Formation.

Groundwater conditions are strikingly different on opposing sides of the structural zone and are characterized by significant stratigraphic displacements and offsets, disparate flow directions, as much as 30 ft of differential head across the zone, and differences in overall water quality and usage. Shallow water-bearing zones situated in the area south of the structural zone have historically (since 1905) been recognized as being degraded beyond the point of being considered of beneficial use due to elevated sodium chlorides. Ground-water contamination is also evident by the localized but extensive presence of light nonaqueous phase liquid (LNAPL) hydrocarbon pools and dissolved hydrocarbons due to the presence of 70 years of industrial development including numerous refineries, terminals, bulk liquid-storage tank farms, pipelines, and other industrial facilities on opposing sides of the structural zone.

The underlying Silverado aquifer has a long history of use, but has not been significantly impacted thus far by the poor groundwater quality conditions which have existed for decades in the shallower water-bearing zones where the Lynwood aquifer serves as a "guardian" aquifer. This suggests a minimal potential for future adverse impact of the prolific domestic-supply groundwater encountered at depths of 800 to 2600 ft below the crest of the structural zone. South of the structural zone no direct communication exists between the historically degraded shallow and deeper water-bearing zones. The exception is in areas where intercommunication or leakage between water-bearing zones

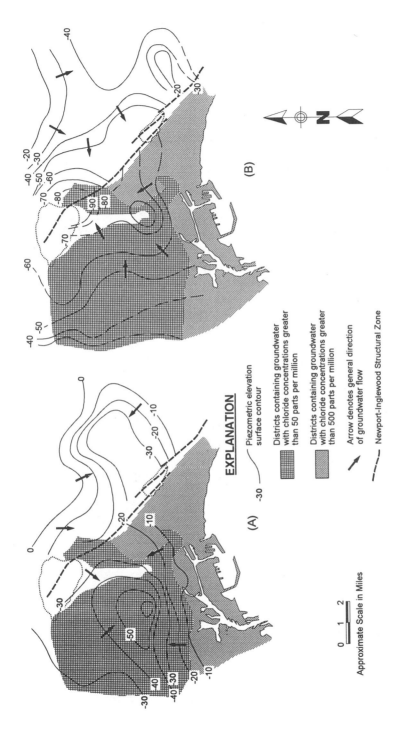

Figure 3-21 **Regional groundwater contour maps showing the Newport-Inglewood Structural Zone in relation to major water-bearing units.**

or heavy utilization of groundwater resources may exist (i.e., further to the northwest within the West Coast Basin). In contrast, north of the structural zone, shallow groundwater would be considered beneficial as a guardian aquifer due to the inferred potential for leakage into the deeper water-supply aquifers.

The beneficial use and clean-up standards thus are different north and south of the structural zone, with lower standards to the south. The overall environmental impact on groundwater resources, regardless of the ubiquitous presence of LNAPL pools and dissolved hydrocarbon plumes in certain areas relative to the structural zone, is minimal to nil. Within the structural zone, structures such as folds and faults are critical with respect to the effectiveness of the zone to act as a barrier to the inland movement of saltwater. An early continuous set of faults is aligned along the general crest of the structural zone, notably within the central reach from the Dominguez Gap to the Santa Ana Gap. The position, character, and continuity of these faults are fundamental to the discussion of groundwater occurrence, regime, quality, and usage. In addition, delineation and definition of aquifer interrelationships with a high degree of confidence is essential. The multifaceted impact of the structural zone is just one aspect of the level of understanding required prior to addressing certain regional groundwater issues. Another important issue is the assessment of which aquifers are potentially capable of beneficial use vs those that have undergone historic degradation. Those faults that do act as barriers with respect to groundwater flow may, in fact, be one of several factors used in assigning a part of one aquifer to beneficial-use status as opposed to another. A second issue, based on the beneficial-use status, is the level of aquifer rehabilitation and restoration deemed necessary as part of the numerous aquifer remediation programs being conducted in the Los Angeles Coastal Plain. This example illustrates that, relative to aquifer remediation and rehabilitation efforts, clean-up strategies should not be stringent, nor should they be applied uniformly on a regional basis. Clean-up strategies should, however, take into account the complex nature of the hydrogeologic setting, and clean-up standards should be applied appropriately.

Fractured Media

Fractured media in general can incorporate several structured elements including faults, joints, fractures, and shear zones. These structural elements, as with faults, can serve as a barrier, partial barrier, or conduit to the migration of subsurface fluids. Most fractured systems consist of rock or sediment blocks bounded by discrete discontinuities. The aperture can be open, deformed, closed, or a combination thereof. The primary factors to consider in the migration of subsurface fluids within fractured media are fracture density, orientation, effective aperture width, and nature of the rock matrix.

Fracture networks are complex three-dimensional systems. The analysis of fluid flow through a fractured media is difficult since the only means of evaluating hydraulic parameters is by means of hydraulic tests. The conduct of such tests requires that the geometric pattern or degree of fracturing formed by the structural elements (i.e., fractures) be known. Fracture density (or the number of fractures per unit volume of rock) and orientation are most important in assessing the degree of interconnection of fracture sets. Fracture spacing is influenced by mechanical behavior (i.e., interactions of intrinsic properties). Intrinsic and environmental properties include load-bearing framework, grain size, porosity, permeability, thickness, and previously existing mechanical discontinuities. Environmental properties of importance include net overburden, temperature, time (strain rate), differential stress, and pore fluid composition. Fracturing can also develop under conditions of excessive fluid pressures. Clay-rich soils and rocks, for example, are commonly used as an effective hydraulic seal. However, the integrity of this seal can be jeopardized if excessive fluid pressures are induced, resulting in hydraulic fracturing. Hydraulic fracturing in clays is a common feature in nature at hydrostatic pressures ranging from 10 KPa up to several MPa. Although hydraulic fracturing can significantly decrease the overall permeability of the clay, the fractures are likely to heal in later phases due to the swelling pressure of the clay.

Several techniques have been used to attempt to characterize fracture networks. These techniques have included field mapping (i.e., outcrop mapping, lineation analysis, etc.), coring, aquifer testing, tracer tests, borehole flowmeters, statistical methods, geophysical approaches, and geochemical techniques to evaluate potential mixing. Vertical parallel fractures are by far the most difficult to characterize for fluid flow analysis due to the likelihood of their being missed during any drilling program. This becomes increasingly important because certain constituents such as solvents and chlorinated hydrocarbons, which are denser than water, are likely to migrate vertically downward through the preferred pathways, and may even increase the permeability within these zones.

Within a single set of measured units of the same lithologic characteristics, a linear relationship is assumed between bed thickness and fracture spacing. Thus, as shown in Figure 3-22, a typical core will intercept only some of the fractures. In viewing the schematic block diagram of a well bore through fractured strata of varying thicknesses, the core drilled in the upper and lower beds would intersect fractures, but the cores drilled within the two central beds do not encounter any fractures. Closer fracture spacing is, however, evident in the two upper thinner beds.

The probability of intercepting a vertical fracture in a given bed is given by

$$P = \frac{D}{S} = \frac{D\,I}{T(ave)}$$

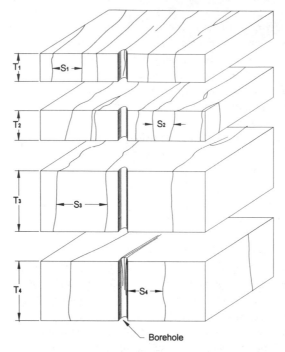

Figure 3-22   **Schematic block diagram of well bore through fractured strata of two different thicknesses. Cores cut in upper and lower beds would intersect fractures, but core from two central beds is unfractured. Note closer fracture spacing in two upper, thinner beds. (After Narr and Lerche, 1984).**

where P = probability, D = core diameter, S = distance between fractures, T(ave) is average thickness, and I is a fracture index given by

$$I = \frac{T_i}{S_i}$$

where the subscript i refers to the properties of the bed. In other words, T_i is thickness of the ith bed and S_i is fracture spacing in the ith bed. A fracture index must also be determined independently for each set of fractures in the core, and must be normal to the bedding.

Based on probability, a core has a finite chance of intersecting a vertical fracture in a bed of a given thickness depending on core diameter, bed thickness, and the value of the fracture index (I). Thus, a sparsely fractured region has a small value of I (i.e., large spacing between parallel fractures), and thicker beds have a larger spacing between fractures for a given I.

SEISMICITY

Earthquakes can cause significant changes in water quality and water levels, ultimately enhancing permeability. During the Loma Prieta, CA earthquake of October 17, 1989 (magnitude 7.1), ionic concentrations and the calcite salination index of streamwater increased, streamflow and solute concentrations decreased significantly from within 15 min to several months following the earthquake, and groundwater levels in the highland parts of the basins were locally lowered by as much as 21 m within weeks to months after the earthquake. The spatial and temporal character of the hydrologic response sequence increased rock permeability and temporarily enhanced groundwater flow rates in the region as a result of the earthquake.

BIBLIOGRAPHY

Allan, U.S., 1989, Model for Hydrocarbon Migration and Entrapment Within Faulted Structures: *American Association of Petroleum Geologists Bulletin,* Vol. 73, No. 7, pp. 803-811.

Allen, J.R.L., 1977, *Physical Processes of Sedimentation:* Allen & Unwin, London, 248 p.

Allen, J.R.L., 1978, Studies in Fluviatile Sedimentation: An Exploratory Quantitative Model for the Architecture of Avulsion-Controlled Alluvial Suites: *Sedimentary Geology,* Vol. 21, pp. 129-147.

Anderson, M.P., 1989, Hydrogeologic Facies Models to Delineate Large-Scale Spatial Trends in Glacial and Glaciofluvial Sediments: *Geological Society of America Bulletin,* Vol. 101, pp. 501-511.

Aquilera, R., 1988, Determination of Subsurface Distance between Vertical Parallel Natural Fractures based on Core Data: *American Association of Petroleum Geologists Bulletin,* Vol. 72, No. 7, pp. 845-851.

Back, W., Rosenshein, J.S. and Seaber, P.R., 1988, *Hydrogeology, The Geology of North America:* Vol. 0-2, Geological Society of America, Boulder, CO, 524 p.

Barker, J.F., Barbash, J.E. and Labonte, M., 1988, Groundwater Contamination at a Landfill Sited on Fractured Carbonate and Shale: *Journal of Contaminant Hydrology,* Vol. 3, pp. 1-25.

Beck, B.F., 1989, *Engineering and Environmental Impacts of Sinkholes and Karst:* A.A. Balkema, Rotterdam, Netherlands, 384 p.

Berg, R.C. and Kempton, J.P., 1984, Potential for Contamination of Shallow Aquifers from Land Burial of Municipal Wastes: Illinois State Geological Survey Map, Scale 1:500,000.

Berg, R.C., Kempton, J.P. and Cartwright, K., 1984, Potential for Contamination of Shallow Aquifers in Illinois: Illinois State Geological Survey Circular 532, 30 p.

Caswell, B., 1988a, Time-of-Travel in Glacial Aquifers: *Water Well Journal,* March, pp. 48-51.

Caswell, B., 1988b, Esker Aquifers: *Water Well Journal,* July, pp. 36-37.

Cayeux, L., 1970, *Carbonate Rocks:* (translated by A.T. Carozzi), Hafner Publishing, Darien, CT, 506 p.

Cehrs, D., 1979, Depositional Control of Aquifer Characteristics in Alluvial Fans, Fresno County, California: *Geological Society of America Bulletin,* Vol. 90, No. 8, Part I, pp. 709-711, Part II, pp. 1282-1309.

Chandler, M.A., Kocurek, G., Goggin, D.J. and Lake, L.W., 1989, Effects of Stratigraphic Heterogeneity on Permeability in Eolian Sandstone Sequence, Page Sandstone, Northern Arizona: *American Association of Petroleum Geologists Bulletin,* Vol. 73, No. 5, pp. 658-668.

Coleman, J.M. and Wright, L.D., 1975, Modern River Deltas: Variability of Processes and Sand Bodies: In *Deltas: Models for Exploration* (Edited by Broussard, M.), Houston Geological Society.

Currie, J.B., 1977, Significant Geologic Processes in Development of Fracture Porosity: *American Association of Petroleum Geologists Bulletin,* Vol. 61, No. 7, pp. 1086-1089.

Davidson, K.S., 1988, Geologic Controls on Oil Migration Within a Coastal Limestone Formation: In Proceedings of the Association of Ground Water Scientists and Engineers and the American Petroleum Institute Conference on Petroleum Hydrocarbons and Organic Chemicals in Ground Water: Prevention, Detection and Restoration, Houston, pp. 233-251.

DeGear, J., 1968, Some Hydrogeological Aspects on Aquifers, Especially Eskers: In *Ground Water Problems* (Edited by Erickson, E., Gustafsson, Y. and Nilson, K.), Pergamon Press, New York, pp. 358-364.

Ebanks, W.J., Jr., 1987, Geology in Enhanced Oil Recovery: In *Reservoir Sedimentology* (Edited by Tillman, R.W. and Weber, K.J.); Society of Economic Paleontologists and Mineralogists, Special Publication No. 40, pp. 1-14.

Fogg, G.E., 1989, Emergence of Geologic and Stochastic Approaches for Characterization of Heterogeneous Aquifers: In Proceedings of the U.S. Environmental Protection Agency, Robert S. Kerr Environmental Research Laboratory Conference on New Field Techniques for Quantifying the Physical and Chemical Properties of Heterogeneous Aquifers, March 20-23, 1989, Dallas, pp. 1-17.

Folk, R.L., 1974, *Petrology of Sedimentary Rocks:* Hemphill Publishing, Austin, TX, 159 p.

Fryberger, S.G., 1986, Stratigraphic Traps for Petroleum in Wind-Laid Rocks: *American Association of Petroleum Geologists Bulletin,* Vol. 70, No. 12, pp. 1765-1776.

Gregory, K.J. (Editor), 1983, *Background to Paleohydrology - A Perspective:* John Wiley & Sons, New York, 486 p.

Grim, R.E., 1968, *Clay Mineralogy:* McGraw-Hill, New York, 596 p.

Haberfeld, J.L., 1991, Hydrogeology of Effluent Disposal Zones, Floridan Aquifer, South Florida: *Ground Water,* Vol. 29, No, 2, pp. 186-190.

Halbouty, M.T., 1982, The Deliberate Search for the Subtle Trap: American Association of Petroleum Geologists Memoir No. 32, Tulsa, OK, 351 p.

Harman, Jr., H.D., 1986, Detailed Stratigraphic and Structural Control: The Keys to Complete and Successful Geophysical Surveys of Hazardous Waste Sites: In Proceedings of the Hazardous Materials Control Research Institute Conference on Hazardous Wastes and Hazardous Materials, Atlanta, pp. 19-21.

Harris, D.G., 1975, The Role of Geology in Reservoir Simulation Studies: *Journal of Petroleum Technology,* May, pp. 625-632.

Hitchon, B., Bachu, S. and Sauveplane, C.M. (Editors), 1986, Hydrogeology of Sedimentary Basins - Application to Exploration and Exploitation: In Proceedings of the National Water Well Association Third Canadian/American Conference on Hydrogeology, June 22-26, 1986, Banff, Alberta, Canada, 275 p.

Hitchon, B. and Bachu, S. (Editors), 1988, Fluid Flow, Heat Transfer and Mass Transport in Fractured Rocks: In Proceedings of the National Water Well Association Fourth Canadian/American Conference on Hydrogeology, Banff, Alberta, 283 p.

Jardine, D. and Wilshart, J.W., 1987, Carbonate Reservoir Description: In *Reservoir Sedimentology* (Edited by R.W. Tillman and K. J. Weber), Society of Economic Paleontologists and Mineralogists, Special Publication No. 40, p. 129.

Keller, B., Hoylmass, E. and Chadbourne, J., 1987, Fault Controlled Hydrology at a Waste Pile: *Ground Water Monitoring Review,* Spring, pp. 60-63.

Kerans, C., 1988, Karst-controlled Reservoir Heterogeneity in Ellenburger Group Carbonates of West Texas: *American Association of Petroleum Geologists Bulletin,* Vol. 72, No. 10, pp. 1160-1183.

Larkin, R.G. and Sharp, J.M., Jr., 1992, On the Relationship Between River-Basin Geomorphology, Aquifer Hydraulics, and Ground-Water Flow Direction in Alluvial Aquifers: *Geological Society of America Bulletin,* Vol. 104, pp. 1608-1620.

Lindholm, G.F. and Vaccaro, J.J., 1988, Region 2, Columbia Lava Plateau: In *The Geology of North America, Vol. 0-2, Hydrogeology,* Geological Society of America, Boulder, CO, pp. 37-50.

Magner, J.A., Book, P.R. and Alexander, Jr., E.C., 1986, A Waste Treatment/Disposal Site Evaluation Process for Areas Underlain by Carbonate Aquifers: *Ground Water Monitoring Review,* Spring, pp. 117-121.

Mancini, E.A., Mink, R.M., Bearden, B.L. and Wilkerson, R.P., 1985, Norphlet Formation (Upper Jurassic) of Southwestern and Offshore Alabama: Environments of Deposition and Petroleum Geology: *American Association of Petroleum Geologists Bulletin,* Vol. 69, No. 6, pp. 881-898.

Miall, A.D., 1988, Reservoir Heterogeneities in Fluvial Sandstones: Lessons from Outcrop Studies; *American Association of Petroleum Geologists Bulletin,* Vol. 72, No. 6, pp. 682-697.

Millot, G., 1970, *The Geology of Clays:* Springer-Verlag, New York, 429 p.

Moran, M.S., 1980, Aquifer Occurrence in the Fort Payne Formation: *Ground Water,* Vol. 18, No. 2, pp. 152-158.

Narr, W. and Lerche, I., 1984, A Method for Estimating Subsurface Fracture Density in Core: *American Association of Petroleum Geologists Bulletin,* Vol. 68, No. 5, pp. 637-648.

Nelson, R.A., 1979, *American Association of Petroleum Geologists Bulletin,* Vol. 63, No. 12, pp. 2214-2232.

Nilsen, T.H., 1982, Alluvial Fan Deposits: In *Sandstone Depositional Environments* (Edited by P.A. Scholle and D. Spearing): American Association of Petroleum Geologists, Memoir No. 31, Tulsa, OK, pp. 49-86.

Osciensky, J.L., Winter, G.V. and Williams, R.E., 1983, Monitoring and Mathematical Modeling of Contaminated Ground-Water Plumes in Paleofluvial Environments for Regulatory Purposes: In Proceedings of the National Water Well Association Third National Symposium on Aquifer Restoration and Ground-Water Monitoring, pp. 355-364.

Pettijohn, F.J., Potter, P.E. and Siever, R., 1973, *Sand and Sandstone:* Springer-Verlag, New York, 618 p.

Poeter, E. and Gaylord, D.R., 1990, Influence of Aquifer Heterogeneity on Contaminant Transport at the Hanford Site: *Ground Water,* Vol. 28, No. 6, pp. 900-909.

Potter, P.E. and Pettijohn, F.J., 1977, *Paleocurrents and Basin Analysis:* Springer-Verlag, New York, 296 p.

Pryor, W.A., 1973, Permeability-Porosity Patterns and Variations in Some Holocene Sand Bodies: *American Association of Petroleum Geologists Bulletin,* Vol. 57, No. 1, pp. 162-189.

Reading, H.G. (Editor), 1978, *Sedimentary Environments and Facies:* Blackwell Scientific Publications, Oxford, 557 p.

Reineck, H.E. and Singh, I.B., 1975, *Depositional Sedimentary Environments With Reference to Terrigenous Clastics:* Springer-Verlag, New York, 439 p.

Rojstaczer, S. and Wolf, S., 1992, Permeability Changes Associated with Large Earthquakes: An Example from Loma Prieta, California: *Geology,* Vol. 20, pp. 211-214.

Scherer, M., 1987, Parameters Influencing Porosity in Sandstones: A Model for Sandstone Porosity Prediction: *American Association of Petroleum Geologists Bulletin,* Vol. 71, No. 5, pp. 485-491.

Scholle, P.A., 1979, *A Color Illustrated Guide to Carbonate Rock Constituents, Textures, Cements and Porosities of Sandstones and Associated Rocks:* American Association of Petroleum Geologists Memoir No. 27, Tulsa, OK, 241 p.

Scholle, P.A., 1979, *A Color Illustrated Guide to Constituents, Textures, Cements and Porosities of Sandstones and Associated Rocks:* American Association of Petroleum Geologists Memoir No. 28, Tulsa, OK, 201 p.

Scholle, P.A. and Spearing, D., 1982, *Sandstone Depositional Environments:* American Association of Petroleum Geologists Memoir No. 31, Tulsa, OK, 410 p.

Scholle, P.A., Bebout, D.G. and Moore, C.H., 1983, *Carbonate Depositional Environments:* American Association of Petroleum Geologists Memoir No. 33, Tulsa, OK, 708 p.

Sciacca, J., 1991, Essential Applications of Depositional Analysis and Interpretations in Hydrogeologic Assessments of Contaminated Sites: In Proceedings of the Hazardous Materials Control Research Institute HWHM/HMC-South '91 Conference, Houston, pp. 294-298.

Sciacca, J., Carlton, C. and Gios, F., 1991, Effective Application of Stratigraphic Borings and Analysis in Hydrogeologic Investigations: In Proceedings of the Hazardous Materials Control Research Institute HWHM/HMC-South '91 Conference, Houston, pp. 69-74.

Schmelling, S.G. and Ross, R.R., 1989, Contaminant Transport in Fractured Media: Models for Decision Makers: U.S. Environmental Protection Agency EPA/54014-89/004, 8 p.

Selley, R.C., 1976, *An Introduction to Sedimentology:* Academic Press, London, 408 p.

Selley, R.C., 1978, *Ancient Sedimentary Environments:* Chapman and Hall, London, 287 p.

Sharp, J.M., 1984, Hydrogeologic Characteristics of Shallow Glacial Drift Aquifers in Dissected Till Plains (North-Central Missouri): *Ground Water,* Vol. 22, No. 6, pp. 683-689.

Sneider, R.M., Tinker, C.N. and Meckel, L.D., 1978, Deltaic Environment Reservoir Types and Their Characteristics: *Journal of Petroleum Technology,* pp. 1538-1546.

Soller, D.R. and Berg, R.C., 1992, A Model for the Assessment of Aquifer Contamination Potential Based on Regional Geologic Framework: *Environmental Geology and Water Science,* Vol. 19, No. 3, pp. 205-213.

Spearing, D.R., 1974, Alluvial Fan Deposits - Summary Sheet of Sedimentary Deposits: Sheet 1, Geological Society of America, Boulder, CO.

Stephenson, D.A., Fleming, A.H. and Mickelson, D.M., 1989, The Hydrogeology of Glacial Deposits: In *Hydrogeology* (Edited by Back, W., Rosenshein, J.S. and Seaber, P.R.), Geological Society of America Decade of North American Geology, Boulder, CO, Vol. 0-2, pp. 301-304

Testa, S.M., Henry, E.C. and Hayes, D., 1988, Impact of the Newport-Inglewood Structural Zone on Hydrogeologic Mitigation Efforts - Los Angeles Basin, California: In Proceedings of the Association of Groundwater Scientists and Engineers FOCUS Conference on Southwestern Groundwater Issues; Albuquerque, NM, pp. 181-203.

Testa, S.M., 1989, Regional Hydrogeologic Setting and its Role in Developing Aquifer Remediation Strategies: In Proceedings of the Geological Society of America Annual Meeting, Abstracts with Programs, November 6-9, 1989, St. Louis, pp. A96.

Testa, S.M., 1991, Site Characterization and Monitoring Well Network Design, Columbia Plateau Physiographic Province, Arlington, North-Central Oregon: Geological Society of America Abstract, 325 p.

Tyler, N., Gholston, J.C. and Ambrose, W.A., 1987, Oil Recovery in a Low Permeability, Wave-Dominated, Cretaceous, Deltaic Reservoir, Big Wells (San Miguel) Field, South Texas: *American Association of Petroleum Geologists Bulletin,* Vol. 71, No. 10, pp. 1171-1195.

Wang, C.P. and Testa, S.M., 1989, Groundwater Flow Regime Characterization, Columbia Plateau Physiographic Province, Arlington, North-Central Oregon: In Proceedings of the Association of Groundwater Scientists and Engineers Conference on New Field Techniques for Quantifying the Physical and Chemical Properties of Heterogeneous Aquifers, pp. 265-291.

Weber, K.J., 1980, Influence on Fluid Flow on Common Sedimentary Structures in Sand Bodies: Society of Petroleum Engineers, SPE Paper 9247, 55th Annual Meeting, pp. 1-7.

Weber, K.J., 1982, Influence of Common Sedimentary Structures on Fluid Flow in Reservoir Models: *Journal of Petroleum Technology,* March, pp. 665-672.

Yaniga, P.M. and Demko, D.J., 1983, Hydrocarbon Contamination of Carbonate Aquifers: Assessment and Abatement: In Proceedings of the National Water Well Association Third National Symposium on Aquifer Restoration and Ground-Water Monitoring, pp. 60-65.

4 HYDROGEOLOGIC PRINCIPLES

*". . . the science of hydrology would be relatively simple if
water were unable to penetrate below the earth's surface."*
(H.E. Thomas, 1952)

INTRODUCTION

When rainwater soaks below the surface of the earth, it enters a complex
three-dimensional system which is controlled by a wide variety of physical and
chemical processes. In a general sense, the water will continue to migrate in the
direction of lowest pressure or be retained by the soil until conditions of energy
equilibrium are satisfied. Forces controlling water movement through soil
include gravity, adhesion to soil particles, viscosity, and surface tension of the
water itself. Each of these forces will vary in importance depending on several
factors, including dissolved content of the water, temperature, minerals present,
and soil pore size.

Groundwater can be found in the traditional sense (at the water table) below
which the soil pore spaces are essentially saturated and free to move, and also
in the unsaturated zone above the water table. It is possible for water to migrate
through both of these zones, transporting dissolved components (or contaminants). The interaction of the various forces involved will determine the
direction and rate of migration.

Presented in this chapter is discussion of relationships that govern migration
in the subsurface, including the flux equation, Darcy's Law, and behavioral
aspects of gases and vapors. The occurrence and migration of water within the
saturated system, including types of aquifers and flow under steady-state and
nonsteady-state flow; unsaturated systems and capillary barriers, and groundwater chemistry and quality are subsequently discussed.

THE FLUX EQUATION

The average macroscopic flow velocity of a fluid is the average of all the microscopic flow velocities, and is usually observed by tracking some chemical tracer or fluid flow front through the material and is a bulk property of the media. There are several bulk properties of the media that are important and that are more readily determined than their microscopic components. These bulk properties describe the system itself and can be used to compare different systems under similar conditions and the same system under different conditions. Certain bulk properties describe the migration of fluids, dissolved molecules, and gases through the porous media and are called the transport parameters. The transport equations are the mathematical relationships used to relate the transport parameters, among other things, to the overall migration of a species of interest.

The flux equation is one of the most powerful relationships for describing migration through a conducting medium, e.g., water though a soil, electricity through a wire, molecules through a membrane, heat through a wall, gas through a pore space, etc. The flux equation works for those situations where the flow rate is proportional to the driving force. The flux equation has many manifestations depending upon what is moving, and is consequently given different names, like Fick's Law for diffusing molecules or Darcy's Law for flowing water. But the three essential components of the equation are always the same:

$$\text{Flux Density} = \text{Conductivity} \times \text{Driving Force}$$

The flux density is the flow rate per unit of cross-sectional area, or the amount of whatever is flowing per unit time divided by the area through which it is flowing. For a garden hose it can be described as the number of gallons per minute per square inch of hose cross-sectional area. For a thermos, it might be the number of joules or calories transferred per second per square centimeter of surface area. The conductivity is the parameter that describes the bulk property of the system with respect to whatever is flowing and describes how well or how poorly that material conducts that water, or that heat, or that electricity. For flowing water, it is called the hydraulic conductivity, for molecular diffusion it is the diffusion coefficient, for heat it is the thermal conductivity. The driving force is usually a potential gradient. It might be a pressure gradient, a gravitational acceleration, a temperature gradient, or a voltage.

The two flux equations of importance to subsurface transport are Darcy's Law for the advective flow of water and other liquids and Fick's Law for the diffusive flow of molecules and gases. These laws are independently discussed below.

$$Darcy's\ Law: \quad q = -K\nabla h \qquad\qquad (4\text{--}1)$$

$$Fick's\ Law: \quad J = -D\ \nabla C \qquad\qquad (4\text{--}2)$$

where q is the flux density of water, K is the hydraulic conductivity, ∇h is the potential gradient (hydraulic gradient when the fluid is water, J is the flux density of molecules, D is the diffusion coefficient, and ∇C is the concentration gradient of the molecular species. The negative sign means that flow will occur in the direction of decreasing potential (i.e., flow will occur downhill or downgradient).

Darcy's Law

The hydraulic conductivity, K, is a parameter that includes the behavior of both the porous media and the fluid. It is often desirable to know the behavior of just the porous media or its intrinsic permeability, k, which is theoretically independent of the fluid. The relationship between the two parameters is given by

$$K = k\,\frac{\rho g}{\eta} \qquad\qquad (4\text{-}3)$$

where ρ is the fluid density, g is the gravitational acceleration, and n is the viscosity of the fluid. The proportionality between K and k is the term $\rho g/\eta$, called the fluidity. Equation 4-3 can be used to determine the Darcy behavior of one fluid in a porous media, such as oil, from the measured behavior of another fluid, such as water, in the same porous media.

It is often easy to measure the flux density, e.g., using a flow meter, and then determine the hydraulic conductivity or diffusion coefficient by dividing the flux by the driving force. One of the most difficult problems is determining how to represent the driving force. The symbol ∇ is called an operator, which signifies that some mathematical operation is to be performed upon whatever function follows. ∇ means to take the gradient with respect to distance. For Darcy's Law under saturated conditions in the subsurface, this can often be easy because the potential gradient can be described by the hydraulic head gradient. The hydraulic head, h, is the height, z, of a free water surface above an arbitrary reference point plus any additional pressures on the system, Ψ, such that h = Ψ + z. The hydraulic head gradient, designated $\nabla h/\nabla l$, is the difference between two hydraulic heads with respect to the distance between them. Therefore, in Equation 4-1, ∇h can be given by $\Delta h/\Delta l$. For mathematical convenience in working with functions and infinitesimal changes, the differences are symbolized as a differential, dh/dl.

Fick's Law

Fick's Law relates mainly to diffusion of dissolved chemicals with respect to differences in concentration gradients in the material or fluid, or migration of gases or vapors in response to differences in gas pressure. However, before discussing Fick's Law, we need to define the term ion. An ionic species, or an ion, is defined as a species that carries a charge (i.e., Na^+, Cl^-, Fe^{2+}, Sr^{2+}, NO_3^-, CO_3^{2-}, etc.), as opposed to dissolved SiO_2 and H_2CO_3, which are not ionic species, but rather are neutral species, sometimes designated as SiO_2^0. A monovalent ion has a single charge or a charge number of one (i.e., Na^+ or Cl^-). A divalent ion has two charges or a charge number of two (i.e., Ca^{2+} or SO_4^{2-}). A trivalent ion has three charges or a charge number of three (i.e., Ce^{3+}, PO_4^{3-}, etc.). Often, shorthand notations are used to depict complicated ions and use an abbreviation along with an important aspect or functional group (i.e., HOAc means acetic acid and OAc^- is the acetate ion formed by loss of H^+). Ionization is the process by which a neutral species becomes an ion, or a preexisting ion becomes further ionized. The energy required to ionize a molecule or species is called the ionization potential, and must be exceeded in order to remove or add electrons to produce an ion. There are as many ionization mechanisms as there are energy sources such as chemical reactions, energy deposited by light or high-energy particles in the case of ionizing radiation, thermal ionization using heat, etc.

Keeping in mind Fick's Law and Equation 4-2, consider diffusion of chloride ion (Cl^-) through a 0.0001-cm membrane separating freshwater with a Cl concentration of 100 mg/L (0.1 g/1000 cm^3 or 0.0001 g/cm^3), from saltwater with a Cl concentration of 10,000 mg/L (10 g/1000 cm^3 or 0.01 g/cm^3). The driving force is the concentration gradient, $\nabla C / \nabla 1$, or $[(0.01 - 0.0001) \div 0.0001]$ $g/cm^3/cm$ or 99 g/cm^4. If the diffusion coefficient of the membrane is 0.00001 cm^2/s, then the flux is 0.00001(99) = 0.00099 g/cm^2 s, meaning that 0.00099 g of Cl will pass through each square centimeter of membrane surface per second.

Gases and Vapors

For vapors and gases, the driving force is a pressure gradient in the gas, often resulting from changes in temperature in different regions of the subsurface. Mixtures of gases can behave more or less ideally, and each gas can be described by its partial pressure (i.e., the pressure that the gas would exert if it alone occupied the total space of the mixture). Therefore, all the individual partial pressures of each gas in a gas mixture add up to the actual pressure of the whole mixture. Partial pressures are very sensitive to temperature. A common reference pressure for gases is the atmospheric pressure at sea level, called an atmosphere (atm), which is equal to approximately 14.5 psi. This is the basal pressure generated by a column of air the height of the atmosphere at sea level, a column of mercury 760 mm high, or a column of water 1020 cm high. Ordinary air is 78% nitrogen, 21% oxygen, and 0.03% carbon dioxide, with the rest argon and some trace gases. The partial pressures for each gas are, therefore, PN_2 = 0.78 atm, P_{O_2} = 0.21 atm, and P_{CO_2} = 0.0003 atm. Alternatively, pressure can be given in the mass of gas per volume of air (g/cm^3).

Consider a shallow waste disposal vault made of cement buried in a sandy loam zone. The vault contains highly alkaline, liquid low-level defense waste with a pore water composition equivalent to a *3 M* NaNO$_3$ solution (these units are further discussed under chemical processes in Chapter 8). The vault is surrounded by 30 cm of a coarse gravel. The porewater of the soil is normal dilute groundwater with a composition of about 0.005 *M* NaHCO$_3$. The partial pressure of water vapor decreases as the solute concentration in the water increases. The partial pressure of water vapor in the soil water at 21°C is 1.88 × 10^{-5} g/cm^3. If the pore water eventually saturates the cement, then the partial pressure of water vapor at the vault wall at 21°C is 1.56 × 10^{-5} g/cm^3. Therefore, a pressure gradient, ΔP = 0.19 × 10^{-5} g/cm^3, exists that will drive water from the soil through the gravel toward the vault, where it can condense out on the wall of the vault and potentially leach out contaminants. If the vapor diffusion coefficient of the gravel is 0.081 cm^2/s, then the flux of water vapor according to Fick's Law is (0.081)(0.19 × 10^{-5}) = 5.1 × 10^{-9} g/cm^2 s or 0.16 g of water diffusing through each cross-sectional square centimeter of gravel per year.

Of course, more complicated situations and conditions will require more sophisticated mathematical treatment, especially for the driving force, but the basic flux relationships are similar for any liquid and gas migration through the subsurface. If the hydraulic conductivities and diffusion coefficients are known for the materials and each migrating fluid of interest, then predictive computer models can often handle the difficult calculations associated with multiple fluids, multiple pressures, and multiple types of materials.

SATURATED SYSTEMS

Water table is a traditional term that describes the level at which water surface will stabilize when it freely enters a boring (or well) at atmospheric pressure. Below this level the water occupies all the pore spaces between mineral grains, cracks, and other openings (i.e., fractures, joints, cavities, etc.) and the formation is considered saturated.

All voids in the subsurface medium are classified as porosity. When pore spaces are interconnected so that water can flow between them, the medium is said to be permeable. The actual openings which permit water flow are referred to as effective porosity. Effective porosity is calculated as the ratio of the void spaces through which water flow can occur to the bulk volume of the medium (expressed as a percentage) as follows:

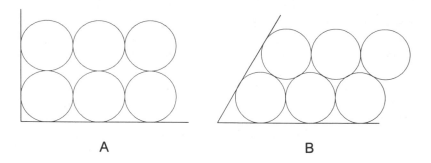

A   B

Figure 4-1   Cubic packing of spheres with a porosity of 47.65% (A) vs rhombohedral packing of spheres with a porosity of 25.95%.

$$n = \frac{Vt - Vs}{Vt} = \frac{Vv}{Vt} \qquad (4\text{-}4)$$

where Vt = total volume of soils, Vs = volume of solid, and Vv = volume of voids.

Porosity which includes the voids between mineral (or soil) grains is referred to as primary porosity. When the porosity is the result of cracks, fractures, or solution channels, it is known as secondary porosity. The porosity of soft clay is often over 50%, but clay typically has low permeability because the pores are either not interconnected or are too small to permit easy passage of water. On the other extreme, nonfracture igneous rock often has a porosity of less than 0.1% but again, low permeability.

Within the two extremes, porosity is dependent upon the organization of the mineral or soil grains. If perfect spheres (i.e., marbles) of uniform size are packed into a cube, the open volume (porosity) can vary from a maximum of 48% to a minimum of 26% depending on how the spheres are organized (Figure 4-1). If several different sizes of spheres are placed in the cube, the porosity will be dependent on the sphere-size distribution and the packing arrangement. In nature, porosity values are also dependent upon cementing of the grains by minerals such as carbonates or silicates. Typical total porosity values for various soils and rocks are listed in Table 4-1; density values are provided in Table 4-2.

The quantity of water that can be retrieved from a medium is related to size and shape of the connected pore spaces within that medium. The quantity of water that can be freely drained from a unit volume of porous medium is referred to as the specific yield. The volume of water retained in the medium by capillary and surface active forces is called the specific retention. The sum of specific retention and specific yield is equal to the effective porosity (Table 4-1). Neither term has a time value attached. Drainage can occur over long periods (i.e., weeks or months).

Table 4-1 Selected Values (% by Volume) of Porosity, Specific Yield, and Specific Retention

Material	Porosity	Specific Yield	Specific Retention
Soil	55	40	15
Clay	50	2	48
Sand	25	22	3
Gravel	20	19	1
Limestone	20	18	2
Sandstone (semiconsolidated)	11	6	5
Granite	0.1	0.09	0.01
Basalt (young)	11	8	3

Table 4-2 Densities (g/cm³) of Sediments and Sedimentary Rocks

Rock Type	Range	Average (wet)	Range	Average (dry)
Alluvium	1.96–2.0	1.98	1.5–1.6	1.54
Clays	1.63–2.6	2.21	1.3–2.4	1.70
Glacial drift	—	1.80	—	—
Gravels	1.7–2.4	2.0	1.4–2.2	1.95
Loess	1.4–1.93	1.64	0.75–1.6	1.20
Sand	1.7–2.3	2.0	1.4–1.8	1.60
Sands and clays	1.7–2.5	2.1	—	—
Silt	1.8–2.2	1.93	1.2–1.8	1.43
Soils	1.2–2.4	1.92	1.0–2.0	1.46
Sandstones	1.61–2.76	2.35	1.6–2.68	2.24
Shales	1.77–3.2	2.40	1.56–3.2	2.10
Limestones	1.93–2.90	2.55	1.74–2.76	2.11
Dolomite	2.28–2.90	2.70	2.04–2.54	2.30

The rate at which a porous medium will allow water to flow through it is referred to as permeability. Henry Darcy was the engineer who performed the first time-rate studies of water flowing through a sand filter. Darcy determined that, for a given material, the rate of flow is directly proportional to the driving forces (head) applied (hence, Darcy's Law).

Hydraulic conductivity is defined as volume units per square unit of medium face per unit of time under a unit hydraulic gradient (often expressed as units³/units²/time). However, many convenient variations of this definition are used for convenience. For example, in the U.S. hydraulic conductivity is referred to in terms of gallons per day per square foot — or, by the U.S. Geological Survey, as square feet per day.

Darcy's work was confined to the quantity of water discharged from a sand filter. Four examples of the application of Darcy's Law as applied through a sand filter (actually a permeameter) is shown in Figure 4-2. Notice that the orientation of the cylinder has no effect on permeability. An example calculation of hydraulic conductivity (K) is presented in Figure 4-2 using the equation below:

$$K = \frac{Q}{IA} \qquad (4\text{-}5)$$

where Q = discharge (units³/units²/time), I = hydraulic gradient (units/units), and A = cross-sectional area (units²). Later studies found that other factors were also involved, such as viscosity and density of the liquid. However, for our purposes, we will assume that the density and viscosity of underground fresh-water are relatively uniform the world over. In order to reduce term confusion, the ability of a porous medium to convey water flow will be referred to as hydraulic conductivity for the remainder of this chapter. Typical ranges of hydraulic conductivity for various soil and rock types is presented in Figure 4-3. Conversion factors for hydraulic conductivity and permeability is contained in Appendix A.

In practical field work, the idea of expressing hydraulic conductivity in units of square area is cumbersome. A more common convention is to recognize that the entire thickness of the water-bearing formation will allow water flow. If the aquifer is divided into vertical unit width segments (i.e., 1 ft wide), each segment then has a water-conducting capacity of the hydraulic conductivity times the thickness, which is called transmissivity (formerly transmissibility) which may be expressed in units such as gallons per day per foot or, in USGS terms, square feet per day (actually, $\dfrac{ft^3/ft^2}{day} \times ft = ft^2/day$).

Although hydraulic conductivity is often expressed in length units per unit time (cm/s or ft/day), these units do not express the actual linear flow velocity through the porous medium. Because hydraulic conductivity is measured as a discharge rate (i.e., cm³/cm², or ft³/ft²), it would only represent the linear velocity of flow through a unit area of the medium if it consisted of 100% porosity. In reality, a cross section of a porous medium usually exhibits less than 30% open area; therefore, the actual linear velocity of water molecules must be greater in order to supply the discharge from the entire surface area. The "tortuous path" that the water must travel to reach the discharge surface is shown in Figure 4-4. Determination of the actual linear velocity can be calculated by dividing the hydraulic conductivity value by the effective porosity.

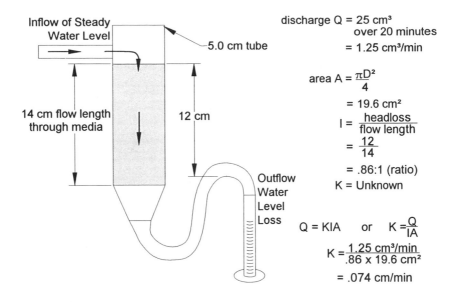

Inflow of Steady Water Level

5.0 cm tube

14 cm flow length through media

12 cm

Outflow Water Level Loss

discharge Q = 25 cm³ over 20 minutes = 1.25 cm³/min

$$\text{area A} = \frac{\pi D^2}{4}$$

$$= 19.6 \text{ cm}^2$$

$$I = \frac{\text{headloss}}{\text{flow length}}$$

$$= \frac{12}{14}$$

$$= .86:1 \text{ (ratio)}$$

K = Unknown

$$Q = KIA \quad \text{or} \quad K = \frac{Q}{IA}$$

$$K = \frac{1.25 \text{ cm}^3/\text{min}}{.86 \times 19.6 \text{ cm}^2}$$

$$= .074 \text{ cm/min}$$

Figure 4-2   Examples for determination of seepage velocity using a permeameter.

Determination of the actual velocity of the water through the permeameter used in Figure 4-5 is presented below:

$$v = \frac{KI}{n} \tag{4-6}$$

If K = .074 cm/min, I = 0.86, v = velocity (actual) and effective porosity (n) = 0.30, then

$$v = \frac{(0.074 \text{ cm} / \text{min}) \, (0.84)}{0.30} = 0.21 \text{ cm} / \text{min}$$

Thus, the actual forward movement of water is 0.21 cm/min.

The horizontal migration rate of water at contaminated sites is a primary concern for evaluation. Example 2 describes the procedure for determining the actual migration rate of water at a site. The hydraulic conductivity and effective porosity are the same as presented in Figure 4-5. The gradient (difference in height/length of flow) is determined from the water table contour map as shown in Figure 4-6.

Two hydraulic head gradients determined from wells or piezometers are shown in Figure 4-7. Piezometers are basically pipes or wells put in the ground

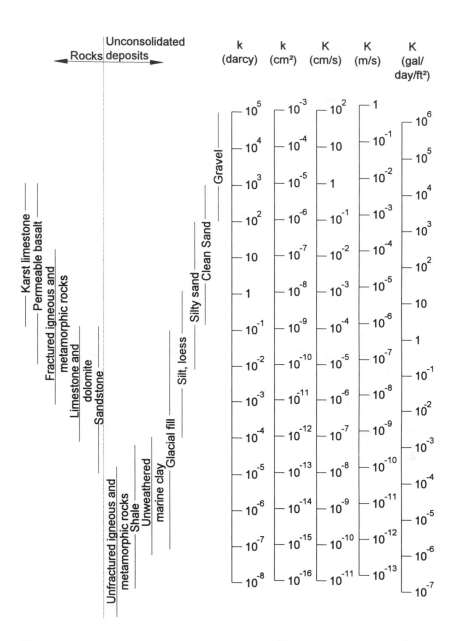

Figure 4-3   Range of values of hydraulic conductivity and permeability (after Freeze and Cherry, 1979).

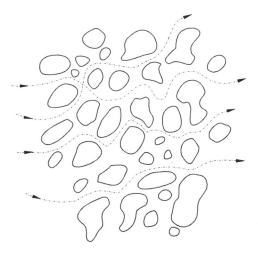

Figure 4-4   Torturous flow paths.

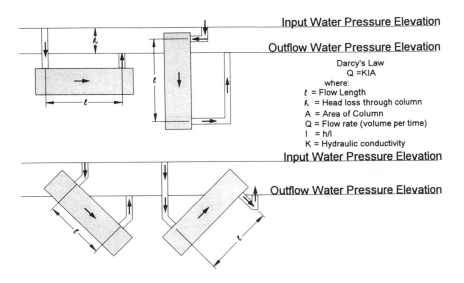

Figure 4-5   Examples of the application of Darcy's Law for the movement of water through a sand filter (or permeameter).

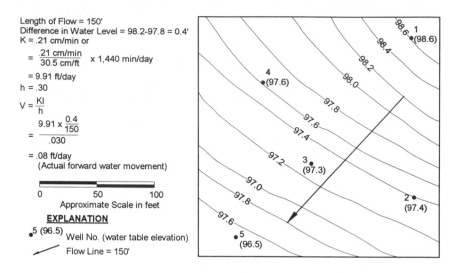

Length of Flow = 150'
Difference in Water Level = 98.2-97.8 = 0.4'
K = .21 cm/min or

$$= \frac{.21 \text{ cm/min}}{30.5 \text{ cm/ft}} \times 1{,}440 \text{ min/day}$$

= 9.91 ft/day

h = .30

$$V = \frac{KI}{h}$$

$$= \frac{9.91 \times \frac{0.4}{150}}{.030}$$

= .08 ft/day
(Actual forward water movement)

0   50   100
Approximate Scale in feet

EXPLANATION

•5 (96.5) Well No. (water table elevation)

⟋   Flow Line = 150'

Figure 4-6   Example determination of gradient from a water table map.

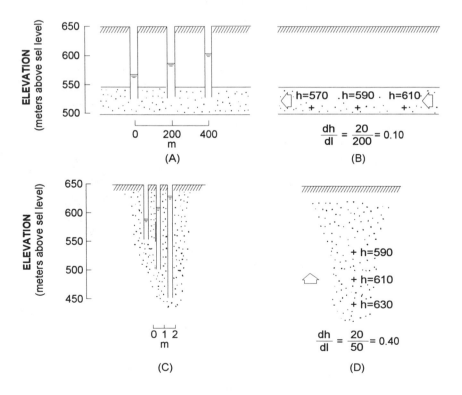

$$\frac{dh}{dl} = \frac{20}{200} = 0.10$$

(B)

h=570 .h=590 . h=610

$$\frac{dh}{dl} = \frac{20}{50} = 0.40$$

(D)

+ h=590

+ h=610

+ h=630

(A)

(C)

Figure 4-7   Determination of hydraulic gradients from piezometer installations (after Freeze and Cherry, 1979).

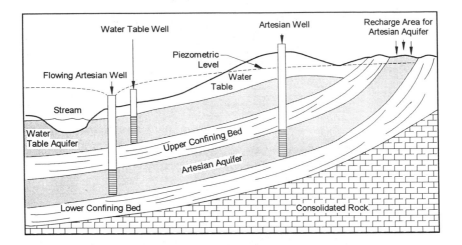

Figure 4-8　Schematic showing various types of aquifers.

which are sealed along their lengths and open to water flow at the bottom and open to atmospheric pressure at the top. In Figures 4-7A and B, the gradient determined from three widely spaced piezometers is seen to be 0.1 and flow will occur to the left, downgradient. In Figures 4-7C and D, the gradient is 0.4, determined from three closely spaced, or *nested,* piezometers that are tapping three different depths, probably three different geologic formations. These piezometers show that the flow is actually directed upward, which, even so, is still downgradient. Unusual subsurface conditions can cause unusual flow patterns, but flow will always occur downgradient.

Types of Aquifers

Water existing below the water table has the ability to migrate in the subsurface reflecting the hydraulic conductivity of the formation it is found in and the hydraulic gradient (driving force) of its position. A formation which can supply a "reasonable" quantity of water to a borehole or well within a fairly short period of time for sampling purposes is called an aquifer. An aquitard is a geological unit that is permeable enough to transmit water in significant quantities over large areas or over long periods of time, but of sufficiently low permeability to restrict its use for commercial or public use. An aquiclude is an impermeable geologic unit that does not transmit water in significant quantities, and differs in permeability relative to adjacent units by several orders of magnitude.

Aquifers exist in a wide variety of forms and can be classified based on lithic characteristics or hydrogeologic behavior (Figure 4-8). Aquifers classified on the basis of lithic character are referred to as either unconsolidated or rock (or consolidated) aquifers.

Unconsolidated aquifers are generally composed of sand and gravel deposit with subordinant or minor amounts of silt and clay. Although compacted to some degree, lithification due to cementation is minor or absent, with much of the original intergranular porosity retained. Groundwater flow is controlled and channeled through zones of relatively higher permeability. Consolidated or rock aquifers are comprised of grain that are cemented or crystallized into a firm and cohesive mass. Groundwater movement in consolidated aquifers is primarily via secondary porosity. In igneous and some metamorphic rocks, secondary porosity occurs as highly vesicular zones, joints, and fractures. In carbonate rocks, secondary porosity is prevalent by chemical dissolution. Clastic sedimentary rocks may maintain some primary porosity, but tends to be much lower than that of unconsolidated aquifers.

Unconfined or water table aquifers maintain a saturated surface that is exposed directly to the atmosphere. These are often similar to a bathtub full of sand or gravel to which water has been added. A well drilled through the water table would fill with water to the common water elevation in the tub. Thus, the potentiometric head in the aquifer is at the elevation of the water table. Unconfined aquifers are also characterized by a fluctuating water table which responds seasonally. With unconfined aquifers, the water table is at atmospheric pressure, and only the lower portion of the aquifer is saturated. Recharge to a water table aquifer comes from rainfall which seeps downward to the water table. The water table level in this type of aquifer rises in direct proportion to the effective porosity. If the equivalent of 2 in. of rainfall seeps into the water table (actually reaches the water table) in an aquifer with an effective porosity of 0.3, the water table would rise 6.7 in. Alternatively, if the same water is pumped and removed from a well, the water table aquifer is then derived from the "storage" in the formation in the immediate vicinity of the well. Natural migration through a water table aquifer is toward a lower elevation discharge. Most streams derive their base flow from water table aquifers (Figure 4-9).

A confined aquifer is a saturated, hydrogeologic unit that is bounded top and bottom by relatively low permeability or impermeable units through which groundwater flow is nonexistent or negligible. All voids within the aquifer are filled with water at a pressure greater than atmospheric, thus, the potentiometric head in the confined aquifer is at a level higher than the top of the aquifer. The classic example is the sand layer which is sandwiched between two clay layers. When a well penetrates through the upper confining layer, the water rises in the bore hole above the top of the sand layer to the elevation which is equal to the pressure of water in the sand formation. This type of aquifer is also called an artesian aquifer. Recharge to a pure artesian aquifer occurs at some location where the confining layer is nonexistent and the aquifer formation is exposed to infiltration at a higher elevation. In many cases, the recharge area is a water table aquifer at a higher elevation. The driving force that causes water flow through a confined aquifer is the pressure caused by the standing water

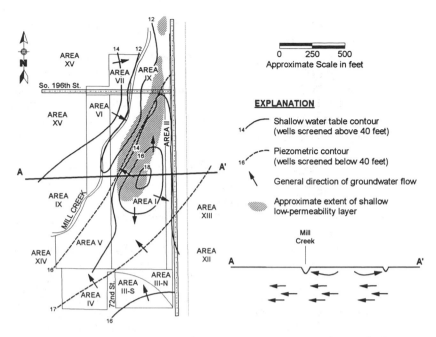

Figure 4-9   Example of water table contour map and hydrogeologic cross-section showing two groundwater flow regimes beneath Western Processing, Kent, WA (after Testa, 1989).

at the higher elevation. Any disturbance to water pressure in an artesian aquifer (such as starting a pump) is felt very quickly as a shock wave over a wide area of aquifer. This rapid pressure response is due to the pressurized nature of the system. If water is pumped from an artesian aquifer, only a very small quantity of the water retrieved from the well is derived from "storage" near the well. Water flows in toward the well from a wide area to replace that removed by the well.

Semiconfined or leaky aquifers are similar to confined aquifers with the exception that the overlying and/or underlying units are not impermeable and some leakage occurs. Groundwater is free to migrate upward or downward. If the water level in a well tapping a semiconfined aquifer equals that of the water table, then the aquifer is considered to be in hydrological equilibrium.

More complex arrangements of aquifers, aquiclude and aquitards, notably in deep sedimentary basins, are systems of interbedded geologic units of variable permeability. These systems are referred to as a multilayered aquifer system. Such systems are considered more of a succession of semiconfined aquifers separated by aquitards.

Above the water table, groundwater can also occur in perched aquifer conditions. In these instances, groundwater occurs in relatively permeable soil which is suspended over a relatively low permeability layer of limited lateral extent and thickness at some elevation above the water table. Perched ground-water occurrences are common within the vadose zone, high-permeability zones overlie low permeability zones of limited lateral extent in unconsolidated deposits. However, perched conditions can also occur within low permeability units overlying zones of higher permeability in both unconsolidated and consolidated deposits. In the latter case, for example, a siltstone or claystone overlies jointed and fractured bedrock such that groundwater presence reflects the inability of the water to drain at a rate which exceeds replenishment from above.

In subsurface characterization studies and detection-monitoring design, defining the horizontal (K_h) vs vertical (K_v) hydraulic conductivities are of prime importance in detecting and monitoring dissolved constituents in groundwater. Aquifers can thus be characterized as homogeneous, heterogeneous, isotropic, and/or anisotropic.

Aquifers are rarely homogeneous and isotropic; that is, hydraulic conductivity is the same throughout the geologic unit and is the same in all directions, respectively (i.e., $K_h = K_v$). Since most geologic units vary both horizontally and vertically in their physical and structural characteristics, they are seldom considered homogeneous but rather heterogeneous. For example, geologic units rarely maintain uniform thickness, and individual layers may pinch out, grain size will vary, and stratification and layering is common — such that the horizontal hydraulic conductivity is greater than the vertical by several orders of magnitude (i.e., $K_h > K_v$). Such heterogeneity is commonplace in nature. When the hydraulic conductivity is significantly greater in one direction relative to another, as within fractured rock where K is greater parallel to the fracture than normal to it, this aquifer is considered anisotropic (i.e., in lieu of isotropic). When a well is drilled into a water-bearing formation, its purpose is to remove water from that formation. Wells are constructed to be holes below the water table which allow easy entrance of water, but prevent inflow of formation solids. When a well is pumped, the standing water level (static level) in the well is lowered, the well becomes a low-pressure area (Figure 4-10). Water in the surrounding area naturally flows in the direction of lowest pressure, thereby recharging the well. This same basic phenomenon occurs whether the producing formation is under artesian or water table conditions. If the pumping rate is increased, the lowered pumping level will increase the inflow.

Darcy's law (Q = KIA) is the operating factor in well flow. The only significant difference in the application of this law to wells rather than to a laboratory permeameter is that the water flows to the well from a radial pattern rather than via linear flow as shown in the permeameter. The flow paths found in both artesian and water table wells are shown in Figure 4-11. The circular lines ("potential") are really contour lines drawn to represent the elevation of the surface of the water table. The surface of the water table is the elevation that water would rise in a well drilled at that location and thus represent the "driving force" which causes water flow to the well.

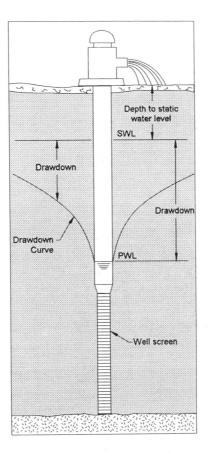

Figure 4-10   Terms relating to well performance (after Driscoll, 1986).

At a given pumping rate, a more permeable formation will have less drawdown as less effort is required to move the water. Low permeability formations often offer so much resistance to flow that it is not possible for the well to produce water at a rate equal to the pump capacity, and the well goes "dry". After a period of recovery, the static level of water in the well will rise to the original level.

Steady State Flow

If a well is pumped continuously at a constant, reasonable rate for a sufficiently long period of time, the water level in the well will stabilize with the inflow rate equal to the pumping rate. Alternatively, the radius of influence (as measured by observation wells at some distance from the pumped well

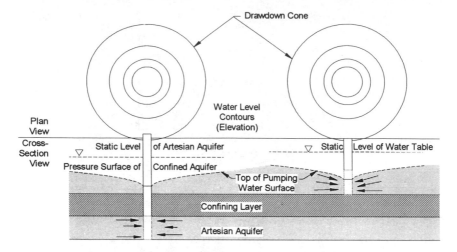

Figure 4-11   **Schematic showing flow paths encountered in both artesian and water table wells.**

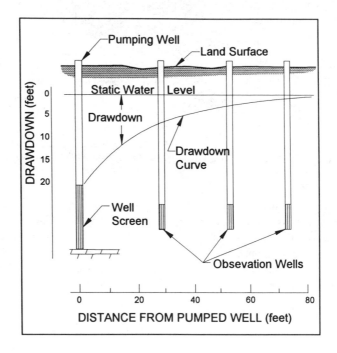

Figure 4-12   **Trace of half a cone of depression showing variations in drawdown with distance from a pumped well.**

shown in Figure 4-12) will expand until adequate recharge is captured to balance the pumping rate of the well. When the water level in a well reaches a stable level while the pumping rate remains constant, the well is said to be at "steady state". The mathematical relationships that describe a classic steady-state relationship between the aquifer characteristics, the pumping rate, and the well construction are presented in Figure 4-13.

Aquifer test analyses can be conducted by one or several methods depending upon whether the solution is applied to a specific aquifer type and with certain assumptions being made on the hydrogeologic nature of the aquifer and nature of the test. The initial analytical methods employed assume that the aquifer is:

- Confined
- Homogeneous
- Isotropic
- Areally infinite
- Uniform in thickness; and
- Monitoring and observation wells fully penetrate the aquifer and are fully screened
- Pumping rate is constant
- Water storage is negligible
- Potentiometric surface prior to pumping is horizontal
- Water removed from storage is immediately discharged once the head declines.

Use of these assumptions is necessary to limit the number of variable parameters which must be considered in the equations. Calculation of the response of an aquifer which is not homogeneous, isotropic, or infinite in extent becomes very complex. Many complex situations are better suited to sophisticated computer simulations.

Certain factors must, however, be considered in choosing the appropriate analytical solution: unconsolidated vs consolidated conditions, fully vs partially penetrating wells, variable discharge rules, delayed yield, and aquifer boundaries. Most methods are best suited for unconsolidated aquifers with well-defined overlying and underlying boundaries; whereas, with consolidated aquifers, the effective aquifer thickness is uncertain. A pumping well that fully penetrates a confined aquifer (i.e., screened through 80% or more of the aquifer's saturated thickness) would be analyzed differently from that which is partially penetrating. In the latter case, vertical flow in the aquifer may affect water levels in nearby observation wells; whereas, with fully penetrating wells, groundwater flow toward the pumping well is assumed to be horizontal. To avoid problems associated with vertical flow near the well screen, observation wells should be at a distance of generally 1.5 times the saturated thickness of the aquifer from a partially penetrating well. It is difficult to achieve a constant

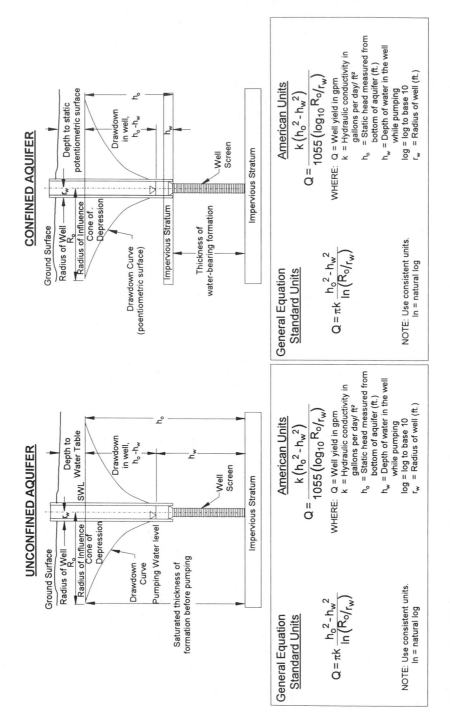

Figure 4-13   Mathematical relationships for steady-state confined and unconfined conditions.

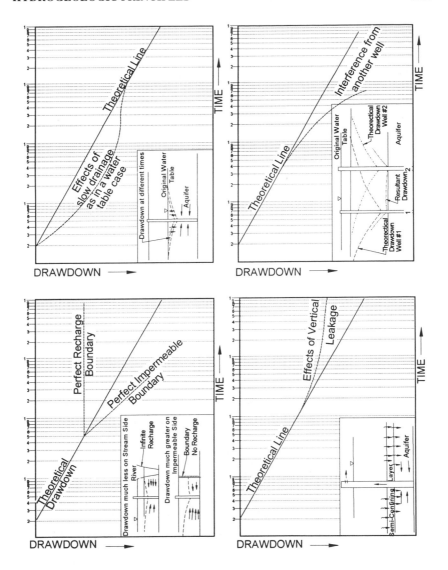

Figure 4-14   Typical curves showing effects of delayed yield, vertical leakage, no-flow boundary, and constant head boundary.

pumping rate during aquifer testing; however, if the pumping rate does not vary more than 10% during conduct of the test, then the rate can be considered constant. Delayed yield is the process wherein water in an unconfined aquifer is yielded to the pumping well after the water level has declined and before steady state has been reached. Delayed yield of water is reflected in a flattening of the drawdown curve simulating a steady-state condition; however, once over, water levels may continue to drop (Figure 4-14). Boundary conditions (i.e., boundary of an aquifer) can be characterized as no-flow or constant-head boundaries. A no-flow boundary (i.e., impermeable fault zone, pinching-out of aquifer, etc.) contributes no groundwater flow and is reflected by a sudden water-level decline in one or more observation wells; whereas, a constant-head boundary (i.e., perennial stream or lake) may contribute significant recharge to the aquifer (thus lessening drawdown) or sudden stabilization of water levels (Figure 4-14). A listing of several of the most common analytical solutions for aquifer analysis in unconsolidated, fractured rock, and karst environments is presented in Tables 4-3 and 4-4, respectively.

In consolidated or fractured-rock aquifers, analytical techniques focus on a dominant characteristic such as a well-developed fracture or void. Boundaries and other aquifer properties are assigned constant values. Some approaches as presented in Table 4-4 assume that the aquifer (i.e., fracture or void) is important for groundwater movement, but is relatively insignificant as reservoirs of groundwater storage. Other approaches consider groundwater storage in both the fractures and in the aquifer matrix. Some of the methodologies listed address-specific features associated with fractured-rock aquifers such as anisotropy, storage-release effects, and transmissivity contrasts of the bulk aquifer matrix and fractures. Such an approach is applicable where test conditions do not significantly compromise boundary conditions as specified via conventional analysis of transmissivity and storativity. Other methods noted address phenomena typically observed in fractured-rock aquifers. For example, double-porosity models emphasize the role of fractures and the aquifer matrix as sources of groundwater storage. Time-drawdown response of groundwater released from these sources are similar to the delayed-yield response of unconfined aquifers. The single-fractured models emphasize the interaction between the aquifer matrix and a fracture penetrated by a production well. Time-drawdown responses for wells located on a fracture or fracture system often exhibit a diagnostic half-slope (0.5) on a log-log plot.

Fractured-rock aquifers may be analyzed in a similar manner as unconfined aquifers, providing the responses exhibit the same general characteristics. During the initial testing, the fracture contributes water to the well, followed by primarily pores and smaller fractures being dewatered, exhibiting the effects of delayed yield. Groundwater migrates to the well from fractures situated farther away during the latter part of the test.

Carbonate aquifers (i.e., limestone and dolomite) are typically characterized by cavernous zones via chemical dissolution (Chapter 3). Some analytical methods that emphasize long, well-developed fractures may thus be applicable to solution-channeled aquifers. Notably, the block-and-fissure model applicable to fractured-rock aquifers may also be useful in carbonate aquifers where significant solution channel development has occurred. Carbonate rocks can be highly anisotropic and nonhomogeneous on a localized scale, but behave more homogeneously on a regional scale. As such, analytical methods that recognize water table and/or leaky artesian conditions generally may also be useful in

Table 4-3 Types of Aquifer-Test Analyses for Unconsolidated Environments

Aquifer	Type of Solution	Name of Solution	Method of Solution	Assumption
Confined	Steady state	Thiem	Calculation	a
	Unsteady state	Theis	Curve fitting	
		Chow	Nomogram	
		Jacob	Straight line	
		Theis recovery	Straight line	
Semiconfined	Steady state	De Glee	Curve fitting	a
		Hantush Jacob	Straight line	
		Ernst modification of Thiem method	Calculation	
	Unsteady state	Walton	Curve fitting	a
		Hantush I	Inflection point	
		Hantush II	Inflection point	
		Hantush III	Curve fitting	
Unconfined, with delayed yield	Unsteady state	Boulton	Curve fitting	a
Semiunconfined, with delayed yield	Unsteady state	Boulton	Curve fitting	a

Table 4-3   Types of Aquifer-Test Analyses for Unconsolidated Environments (continued)

Aquifer	Type of Solution	Name of Solution	Method of Solution	Assumption
Unconfined	Steady state	Thiem-Dupuit	Calculation	Solutions for the confined, unsteady-state case can be applied to the unconfined, unsteady-state case only if the drawdown is modified by an appropriate factor
	Unsteady state	Thiem Theis Chow Jacob		
Confined or unconfined	Steady state	Dietz	Calculation	Aquifer crossed by one or more fully penetrating recharge or barrier boundaries
	Unsteady state	Stallman Hantush image	Curve fitting Straight line	
Confined or unconfined	Unsteady state	Hantush	Calculation	Aquifer homogeneous, anisotropic, and of uniform thickness
Semiconfined	Unsteady state	Hantush-Thomas Hantush	Calculation Calculation	

Aquifer	State	Method	Analysis	Remarks
Confined	Unsteady state	Hantush	Curve fitting	Aquifer homogeneous and isotropic but thickness varies exponentially
Unconfined	Steady state	Culmination point	Calculation	Prior to pumping the potentiometric surface slopes
Confined or unconfined	Unsteady state	Hantush	Curve fitting	Discharge rate variable
	Unsteady state	Cooper-Jacob	Straight line	
		Aron-Scott	Straight line	
		Sternberg	Straight line	
		Sternberg recovery	Straight line	
Confined	Steady state	Huisman correction I and II	Calculation	Partially penetrating pumping well
		Jacob correction	Calculation	
Semiconfined	Steady state	Huisman correction I and II	Calculation	
Unconfined	Steady state	Hantush correction	Calculation	
Confined	Unsteady state	Hantush modification of Theis	Curve fitting	
		Hantush modification of Jacob	Straight line	
Semiconfined	Steady state	Huisman-Kemperman	Nomograph and curve fitting	Two-layered aquifer with semi-pervious dividing layer
		Bruggeman	Straight line	

[a] Aquifer is homogeneous, isotropic, areally infinite, and of uniform thickness; pumping and observation wells fully penetrate and screen the aquifer; prior to pumping the piezometric surface is horizontal; discharge rate is constant and storage in the well can be neglected; water removed from storage is discharged instantaneously with decline of head (modified after Kruseman and DeRidder, 1979.)

Table 4-4 Summary of Analytical Solutions for Anisotropy in Fractured Rock and Karst Environments

Aquifer Type(s)	Phenomenon Modeled	Method of Solution	Wells Required for Calculations	Remarks	Ref.
Confined, homogeneous	Two-dimensional anisotropy	Curve fitting, or straight line with calculation	4	Horizontal aquifer[a]	Papadopulos (1964)
Leaky and nonleaky, homogeneous	Two-dimensional anisotropy	Curve fitting, or straight line with calculation	4	Horizontal aquifer[a]	Nantush (1966)
Leaky and nonleaky, homogeneous	Two-dimensional anisotropy	Curve fitting, or straight line with calculation	3	Horizontal aquifer[a]	Neuman, et al. (1984)
Homogeneous and heterogeneous, horizontal and vertical anisotropy, partial penetration	Three-dimensional anisotropy	Curve fitting with calculation	4	Horizontal aquifer[a]	Way and McKee (1982)
Confined, homogeneous, isotropic	Double porosity block and fissure storage	Curve fitting	1	Fractured rock or karst aquifers	Boulton and Streitsova (1977)
Unconfined, homogeneous, isotropic	Double porosity block and fissure storage	Curve fitting	1	Fractured rock or karst aquifers	Boulton and Streitsova (1978)
Confined, matrix is homogeneous and isotropic fracture and aquifer system strongly anisotropic	Pumping well penetrates vertical fracture or horizontal fracture	Curve fitting	1	Analysis for pumping well data only	Gringarten and Witherspoon (1972); Gringarten (1962)

| Confined, matrix is homogeneous and isotropic fracture and aquifer system strongly anisotropic | Pumping well penetrates vertical fracture | Straight line | 1 | Analysis for hydraulic diffusivity (T/S), estimate of S from other methods needed to solve for T | Jenkins and Prentice (1982) |

[a] For application in fractured rock aquifers, it is assumed the aquifer's behavior approximates that of a porous medium. Standard methodologies and their applicable assumptions are used to obtain values of transmissivity and storage, from which anisotropy is calculated.

evaluating aquifer characteristics in fractured and solution-channeled carbonate rock aquifers.

Important to any aquifer restoration program is the radius of influence or capture zone to be anticipated during pumping. The radius of influence (r_o) for a vertical pumping well can be calculated using the following equation.

$$r_o = 300 \, (h_o - h_w)\sqrt{K} \qquad (4\text{-}7)$$

where h_o = thickness of aquifer at radius of influence, h_w = height of water in well (or water equivalent), and K = hydraulic conductivity.

The zone of influence of a horizontal well can be calculated either by the rectangular or elliptical approach. The rectangular approach assumes the radius of influence is constant along the length of the well and is equal to the extent of the zone of influence from the end points of the well measured in the plane containing the well. Roughly rectangular, the area of drainage (A_r) is

$$A_r = 2LR_{ev} + \pi R_{ev} \qquad (4\text{-}8)$$

where L = length of well screen, and R_{ev} = effective drainage radius of the well calculated from the pumping tests conducted in a vertical well.

The elliptical approach assumes that the zone of influence is elliptical (A_e) with the well end points constituting the face of the ellipse, and the minor semiaxis equal to the radius of influence. The area of drainage (A_e) is

$$A_e + \pi R_{ev} c / 2 \qquad (4\text{-}9)$$

where c = minor axis of the ellipse (screen length or L) and the radius of influence of the well is

$$c = 2\sqrt{[(L/2)^2 + R_{ev}^2]} \qquad (4\text{-}10)$$

Nonsteady Flow

In many situations, only very limited geologic information is available for a site which is under investigation. The published literature may include only limited or approximate hydrologic information. At these sites, the site investigation will necessarily include a drilling program which is designed to sample subsurface materials. An important part of this field program is an evaluation of the hydraulic properties of the subsurface formation.

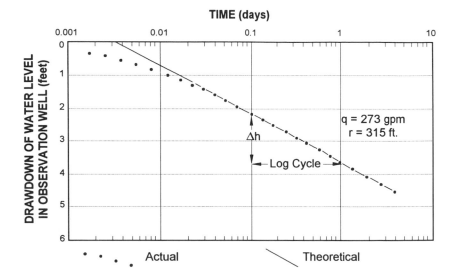

Figure 4-15   Time drawdown graph for a pumping test.

After the borings are drilled and the wells installed, pumping tests or slug tests can be conducted to evaluate the aquifer parameters. These tests disrupt steady state flow equilibrium; thus, nonsteady state. Pumping tests are conducted by pumping a single well at a uniform rate and measuring the time-rate of water level decline in the pumped well and/or adjacent wells. A diagram of time drawdown for a pumping test is shown in Figure 4-15. Comparison of field data with theoretical data results in the definition of aquifer characteristics. Pumping tests are best suited for permeable formations which are capable of producing more than 1 gal of water per minute. Slug tests are made by "instantaneously" adding (or withdrawing) a measured quantity of water from a well and monitoring the time-rate of water level recovery in the same well. Because of the mechanics of measuring the water level response in a timely manner, slug tests are best suited for less-permeable formations. Interpretation of the results of either a slug or pumping test is theoretically a rather straightforward process. In practice, however, the task often becomes more of an art than science.

UNSATURATED SYSTEMS

There are some important differences between the behavior and flow of water in the unsaturated zone (vadose zone) above the water table and the saturated zone below the water table. The surface tension of water or other fluids becomes important when there is a gas phase in contact with the fluid phase and the solid phase. If the total volume of a porous medium (V_T) is divided up into the volume of the solid portion (V_s), the volume of the water (V_w), and the volume of the air or gas (V_a), then the volumetric water content, θ, is equal to V_w/V_T. Like the porosity (n), the volumetric water content may be reported as a fraction or as a percent. For saturated flow, $\theta = n$ and all pores are filled with liquid; for unsaturated flow, $\theta < n$ and $n = V_w + V_a$.

As previously defined, the pressure head, Ψ, can be either a positive hydraulic pressure as would be below a standing column of water, or a negative pressure that would be generated by the suction of a dry soil. Therefore, $\Psi > 0$ below the water, $\Psi = 0$ at the water table, and $\Psi < 0$ above the water table in the vadose zone. This negative pressure in the vadose zone results from the surface tension of water at the air/water interface, often called the capillary pressure or the matric suction. When $\Psi < 0$, Ψ is called the tension head or suction head. Therefore, because $h = \Psi + z$, the gradient can become complicated in the vadose zone with various negative pressures or suctions operating in addition to gravity. Some important aspects of the vadose zone/saturated zone characteristics are illustrated in Figure 4-16. Note in Figure 4-16D that the pressure head, Ψ, varies from some negative value (–ve) in the vadose zone, through zero at the water table, to some positive value (+ve) below the water table. The hydraulic head, h, is the pressure head plus the height above the arbitrarily chosen reference point, called the datum. Because flow occurs downgradient, Figure 4-16E shows that flow is downward for this normal situation. Remember that the negative capillary pressure or matric suction operates in three dimensions, whereas gravity operates vertically in only one direction.

Darcy's Law under saturated conditions uses the saturated hydraulic conductivity (K_s) for K in the above equation. However, K is a strong, nonlinear function of the volumetric water content, $K(\theta)$, meaning that the hydraulic conductivity decreases drastically as the water content decreases. The $K(\theta)$ relationship is often called the characteristic behavior of the sample with respect to hydraulic conductivity. Several types of sediments having different characteristic behaviors, along with a saturated caliche (a calcite-cement soil that in this case has become a rock) that shows different behavior along each of its three orthogonal direction is shown in Figure 4-17. For example, the sample from a depth of 154.2 ft in the Ringold Formation has a saturated K of 10^{-4} cm/s at a 31% volumetric water content, but the same material has a K of only 10^{-8} cm/s at a 12% volumetric water content, a four order of magnitude difference! The exact shape and magnitude of the $K(\theta)$ relationship depends upon the pore-size distribution which results mainly from the distribution of grain sizes in aggregate materials (soils and sediments) and upon fracture aperture sizes in fractured media (rock).

Hydraulic steady state is achieved in a porous media when the flux density and driving forces are unchanging with time. For unsaturated systems, the volumetric water content will also be constant and unchanging in a homogeneous

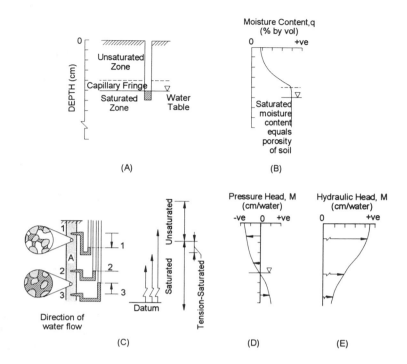

Figure 4-16 Groundwater conditions near the ground surface. (A) Saturated and unsaturated zones; (B) profile of moisture content vs depth; (C) pressure-head and hydraulic-head relationships: insets = water retention under pressure heads less than (top) and greater than (bottom) atmospheric; (D) profile of pressure head vs depth; (E) profile of hydraulic head vs depth. (after Freeze and Cherry, 1979).

material. However, under the same hydraulic steady state, each different material will attain its own steady-state water content depending upon the texture and pore-size distribution in the material. Because there is no suction gradient ($\Delta\Psi = 0$) at unsaturated hydraulic steady state, the only driving force is gravity, a situation referred to as unigradient conditions.

Water movement in the unsaturated zone is dependent on the factors discussed previously: driving force, hydraulic conductivity, viscosity, and density, but is also strongly influenced by additional forces including surface tension and adhesion. The basic atomic organization of the water molecule and electrical charge of the oxygen and hydrogen are responsible for these additional factors.

A water molecule is composed of one oxygen and two hydrogen atoms. These three atoms are organized so that the two hydrogens are at an angle of 105° from each other (Figure 4-18). The weak attraction between positive

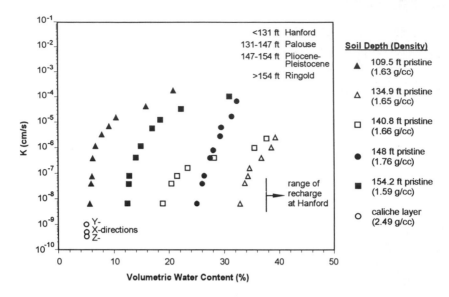

Figure 4-17   Graphs of hydraulic conductivity vs volumetric water content showing characteristic curves for different sediments from the VOC-arid site integrated demonstration at Hanford site.

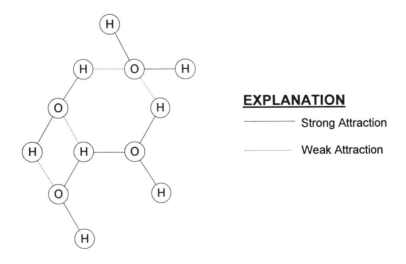

Figure 4-18   Attraction of water molecules.

Table 4-5 Surface Tension of Water

Temperature (C°)	Surface Tension	
	lb/ft	g/cm
0	0.0052	0.076
10	0.0051	0.074
20	0.0050	0.073
30	0.0049	0.071
40	0.0048	0.070

(oxygen) and negative (hydrogen) atoms of adjacent molecules are the forces that hold water together in a liquid state. Without this attraction, water would only exist as a gas. The impact of this angular arrangement is most evident when water freezes. As water is cooled to the temperature where the molecules begin to assemble into crystalline form, the only structure that they can readily assume is one which has a void in the center. This void causes the bulk ice to have a density of less than water, and thus it floats. In its liquid state, water has a definite volume but no definite form. At the boundary of the water mass, the three-dimensional attraction between molecules is no longer possible. Water molecules at the uncontained surface are forced to contort themselves into a two-dimensional form. The result is a molecular organization which is in tension — "surface tension". Pushed to its extreme, the molecules try to associate into the form with least surface area per volume, which is a sphere. A raindrop falling in a vacuum would hold a spherical form. A smaller radius results in a more stressed molecular organization. Surface tension of pure water is summarized in Table 4-5.

Addition of a small quantity of electrolyte (such as minerals dissolved from adjacent soil particles) increases surface tension. A small quantity of soluble organic compound (alcohol, soap, or acid) decreases the surface tension. The addition of glycerine to water reduces surface tension and thus makes it possible to "stretch" water film into bubbles, as with a child's bubble-blowing game.

A second important concept is that of adhesion. Attractive forces between hydrogen and oxygen in the water molecules are also effective between similar molecules in other substances. For example: the beading of raindrops onto a windshield is the result of the attraction of atoms of oxygen and hydrogen to the silicon and oxygen atoms in the glass. Solid surfaces that exhibit adhesion to water molecules are said to be hydrophilic (water attracting). Other molecules, such as paraffin, which do not attract water are hydrophobic (water repelling). The relationship between water and hydrophobic and hydrophilic surfaces is shown in Figure 4-19. The size of the alpha angle indicates the

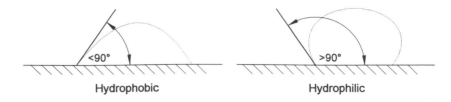

Hydrophobic   Hydrophilic

**Figure 4-19   Relationship between water and hydrophobic and hydro-
philic surfaces.**

preference. If the angle is <90°, the water is attempting to maintain its spherical form, although somewhat squashed by the force of gravity. Quartz and most mineral particles in soil have a contact angle of >90°. Crystalline quartz and water containing some impurities are about 25°, which indicates a rather tight bond. In the subsurface, cohesive forces between soil particles and water are often so large that great energy (such as evaporation) is required to separate water from the soil.

A combination of adhesion and surface tension gives rise (pardon the pun) to capillary action. By its adhesion to the solid surface of the soil particles, the water wants to cover as much solid surface as possible. However, by the effect of surface tension, the water molecules adhering to the solid surface are connected with a surface film in which the stresses cannot exceed the surface tension. As water is attracted to the soil particles by adhesion, it will rise upward until attractive forces balance the pull of gravity (Figure 4-20). Smaller-diameter tubes force the air-water surface into a smaller radius, with a lower solid surface to volume ratio, which results in a greater capillary force. Typical height of capillary rise for several soil types is presented in Table 4-6. The practical relationship between "normal" subsurface water and capillary rise is presented below.

$$h = \frac{2T\cos\theta}{rpg} \qquad (4\text{-}11)$$

where h = rise in centimeters; T = surface tension in dynes; θ = well-liquid interface angle; p = density of fluid; g = acceleration of gravity; and r = radius of tube in centimeters.

In the subsurface, capillary forces can be active in all three dimensions depending on the location of the water source. Regardless of the direction of movement, the water is moving in the direction of least pressure. This phenomena is often called "soil suction". Dry soil particles always have a tendency to retain water molecules on their surface.

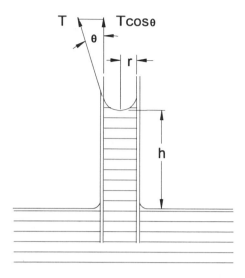

Figure 4-20   Rise of liquid in a capillary tube (after Testa and Winegardner, 1991).

Table 4-6 Capillary Rise in Various Soils

Soil Type	Average Grain Size in Millimeters	Capillary Rise in Meters
Sand	2–0.5	0.03–0.1
	0.5–0.2	0.1–0.3
	0.2–0.1	0.3–1.0
Silt	0.1–0.05	1.0–3.0
	0.05–0.02	
Silt clay	0.02–0.006	3–10
	0.006–0.002	10–30
Clay	<0.002	30–300

Even after a large supply of water has migrated downward through a soil zone, under conditions where gravity is the dominant force, some water will be retained on and between the soil particles as "residual saturation". The relationship between several soil horizons in the unsaturated zone is shown in Figure 4-21. Water in each of these zones is held according to the local conditions of soil suction. It is important to note that water in the liquid form cannot be held by soil if the soil suction is >0.7 atmospheres (10 psi). The zones of unsaturated soil are listed below:

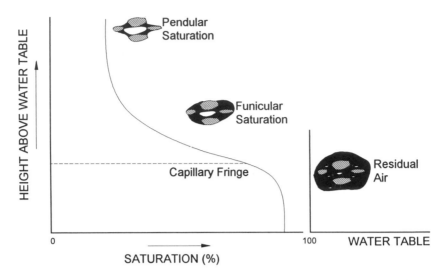

Figure 4-21   Graph of height above water table vs percent saturation.

- *Residual saturation* (hygroscopic water or pendular saturation): water is held under tension of 31 to 10,000 atm force
- *Funicular saturation*: similar to residual saturation, except that each grain is surrounded by a water film and has a large air content
- *Capillary fringe*: the zone immediately above the water table which is essentially saturated; this is the height that can support saturated conditions at negative pressure (due to soil suction).

Water flow through unsaturated soil is controlled by the same forces as capillary action and water retention (i.e., adhesion, gravity, and surface tension). Flow can occur only when the water phase is continuous from pore to pore. If gravity is the controlling force, downward flow will occur according to Darcy's Law in direct proportion to the percentage of water-filled, connected pores. For example, if only half the pores in a cross section are water filled, the flow through that section will be half of that predicted by $Q = KIA$.

When water enters the subsurface from the surface, a great percentage initially flows through macropores (cracks, rootholes, and worm bores). After these larger openings are filled (or plugged by washed-in debris, or soil swelling), the primary flow paths become the soil pores.

If the source of water is relatively unlimited (long-term flooding), the downward migrating water will displace air from the pores and migrate according to Darcy's Law. As the percentage of saturated pores increases, the downward flow will increase. The limit of this process is saturated downward flow.

During "normal" rainfall events, the quantity of water available for infiltration is finite. As water moves from pore to pore, some is retained along the pore walls by adhesion and surface tension in the throat between the pores. After

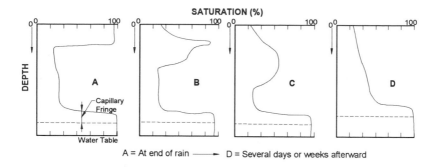

Figure 4-22   Downward moving wave of infiltrating water through the vadose zone.

these forces are satisfied, and most of the air displaced, the flow can continue to the next pore. If the unsaturated zone is thick, it is possible that all the water will be retained as residual saturation and none will reach the water table. The retention capability is often observed at the surface when a garden sprinkler sprays only a small quantity of water on sandy soil. The water will "bead up" on the surface until sufficient water has been applied to fully "wet" the surface soil, then infiltration will start. Following a major rainstorm, when the quantity of percolating water is greater than can be absorbed by the soil, a downward-moving "wetting front" occurs. This phenomenon is the result of the water moving downward, pore by pore. When the wave has passed, some water is retained as residual saturation. A graphic representation of the moving wetting front is shown in Figure 4-22.

Actual migration of a wetting front is highly dependent on the nature and size of the soil pore spaces. If all of the pores are relatively similar, the migration rate of the wetted front is uniform. If the pore spaces become smaller, the capillary forces are greater and the rate of movement faster. When significantly larger pores are encountered (such as a sand lens) the capillary forces may not be sufficient to cause flow into the lens. In effect, it is possible for a highly permeable sand to act as a confining layer which restricts downward water flow.

Capillary Barriers

Two important, nonintuitive behaviors of unsaturated systems should be pointed out. First, under unsaturated conditions, water will not move from a smaller pore into a larger pore unless the smaller pore is saturated. In the previously discussed gravel/vault example, water will not flow from the sand across the boundary with the gravel unless the sand is saturated. If the boundary (called a capillary barrier) is horizontal, saturation is likely to occur as recharge pools in the sand until it finally saturates and flows into the underlying gravel. However, if the boundary is at an angle, even as little as 5°, then the water will not flow into the underlying gravel, but will flow along the sand/gravel boundary remaining within the sand, unless the recharge is high enough to saturate the sand. The greater the difference in relative grain size, the better the barrier will perform. Of course, as previously indicated, the capillary barrier is definitely not a vapor diffusion barrier.

As a corollary to this, the larger pores will empty first during desaturation of a sample. This is especially important for unsaturated or desaturating fractured rock; the larger fractures will desaturate first while the smaller fractures will still be completely saturated. The situation can be complicated when a contaminated fluid encounters many layers of different fracture/pore-size materials having different water contents. Knowledge of these kinds of unsaturated behavior is being used to engineer innovative and effective barriers to contaminant migration in the vadose zone.

GROUNDWATER CHEMISTRY

The analytical determination of water quality is conducted to evaluate potential sources of surface and groundwater, groundwater flow directions, and relative age. As part of subsurface environmental studies, water analyses are also performed to assess the presence of dissolved constituents considered potentially hazardous or toxic, determine the lateral and vertical extent of dissolved contaminant plumes, and evaluate processes and chemical reactions which may significantly influence the fate and transport of certain dissolved constituents (i.e., organic chemicals), and potential influence on the proposed remediation strategy. For example, a nonreactive ion such as chloride (Cl) is not retarded and its direction may reflect water quality changes. The presence of certain dissolved solutes may indicate a man-induced origin since only 4 of the 21 major dissolved inorganic ions regulated under the Safe Drinking Water Act occur in natural waters. A cation/anion balance can also be used to detect the presence of unsuspected inorganic constituents. Many of the chemical reactions further discussed in Chapter 8 are governed by the presence and concentration of major inorganic ions. In remediation systems for the treatment of contaminated groundwater, many common problems such as scaling due to excessive hardness, fouling by iron and magnesium, corrosion due to low pH or sulfides, and biofouling from growth of either iron, manganese or sulfur bacteria solutes usually reflects concentrations of less than 1.0 mg/L. Major ions account for more than 95% of the dissolved inorganic constituents in natural water. As shown in Table 4-7, major ion analysis for the most part represents naturally occurring dissolved inorganic constituents with a concentration of 1.0 mg/L or greater. Other dissolved constituents present (i.e., organics) are more difficult to assess. This reflects their limited presence or

Table 4-7 Summary of Major Dissolved Inorganic Constituents in Natural Water

Ion State	Dissolved Inorganic Constituents[a]	Chemical Symbol
Anions	Bicarbonate (alkalinity)	HCO_3
	Carbonate (alkalinity)	CO_3
	Chloride	Cl
	Fluoride	F
	Nitrate	NO_3
	Nitrate	NO_2
	Phosphorous	PO_4
	Silica	SiO_2 or $SI(OH)_4$
Cations	Ammonia	NH_4
	Calcium	Ca
	Iron	Fe
	Magnesium	Mg
	Manganese	Mn
	Potassium	K
	Sodium	Na

[a] During major ion analysis, not shown is the residue on evaporation (ROE) which provides a good indicator of overall water quality and can be used as a check on the accuracy of the analysis.

naturally occurring solutes, and man-induced site-specific factors such as the nature and quantities of chemicals used, the manner of release (i.e., episodic vs continuous), and the geochemical matrix beneath the site.

In addition to major ion analysis, certain physical properties are measured. These properties include pH, temperature, and specific conductance or conductivity. The oxidation-reduction potential (E_h) is also occasionally measured on a site-specific basis. Other dissolved chemical constituents that may be measured on a site-specific basis include ferrous iron (Fe^{+2}) and hydrogen sulfide (H_2S).

Although all analyses are routinely conducted at a certified laboratory, certain field-measured parameters may be required due to the rapid changes these parameters may undergo once collected. These field-measured parameters include pH, temperature, specific conductance, Fe^{+2}, S^{2-}, E_h, NO_2, and NH_4. In analyzing water quality data, the accuracy of the analysis should be checked by calculating the cation/anion balance.

The analytical data from a typical major ion analysis, along with the anion/cation balance, are shown below in Table 4-8. The solute concentrations are given as both milligrams per liter (mg/L) and milliequivalents per liter (meq/L).

**Table 4-8 Analytical Statement for a
Major Ion Analysis[a,b,c]**

Ion	Constituent	mg/L	meq/L
Anions	Bicarbonate (HCO_3)	440	7.212
	Chloride (C1)	8.0	0.226
	Sulfate (SO_4)	139	2.894
	Fluoride (F)	0.7	0.037
	Nitrate (NO_3)	0.2	0.003
	Silica (SiO_2)	20	<u>0.526</u>
			10.898
Cations	Potassium (K)	2.1	0.054
	Sodium (Na)	13	0.566
	Calcium (Ca)	126	6.287
	Magnesium (Mg)	43	3.538
	Iron (Fe)	2.3	<u>0.124</u>
			10.569

[a] From Francks (1991).
[b] Calculated dissolved solids = 594 mg/L; residue on evaporation (ROE) – 571 mg/L.
[c] Anion/cation balance = $\dfrac{(\text{anions} - \text{cations})}{(\text{anions} + \text{cations})}$ = 1.5%.

If the major ion analysis is complete and the analyses are accurate, the cations/anions should balance between ±2 and 5%. This rule holds true for relatively dilute solutions. For more brackish waters, the allowable cation/anion balance goes up to ±10%. If the balance is greater than ±5% for a dilute solution, a comparison of the calculated dissolved solids content with the ROE will indicate the possibility of ions being present that were not analyzed. Tables showing the conversion factors for mg/L to meq/L are given in many readily available references. Other less common methods include comparisons between the sum of the conductivities from the analysis and field-measured results, comparison of ion ratios, or other statistical methods.

Graphical methods of interpretation include the use of bar graphs, pie charts, stiff diagrams and patterns, and trilinear diagrams. Of the many graphical methods of data interpretation, the Piper trilinear diagram is the most useful (Figure 4-23). Cations (in meq/L) are plotted in the bottom left triangle (plotting field), grouping together the sodium and potassium. The anions (in meq/L) are plotted in the bottom right plotting field, grouping carbonate and bicarbonate together.

Next, the plotted values are projected onto the diamond-shaped plotting field in the center. One variation of the Piper method is to represent the points plotted in the diamond-shaped field as circles, with the size of the circle being

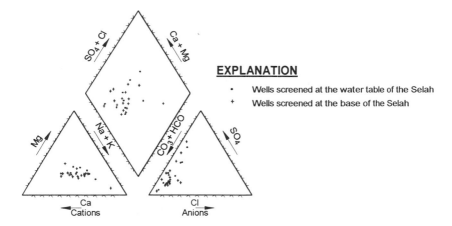

Figure 4-23   Piper trilinear diagram showing calcium carbonate (bicarbonate) type groundwater with minor amounts of chloride and potassium from beneath the Arlington Facility (refer to Chapter 10, Case Histories).

proportional to the ionic strength. The Piper trilinear diagram can thus be used to determine if a water is a simple mixture of two or three sources. Data points aligning along a straight line in all three plotting fields may represent a two-way mixing system. Any point that falls between the end members of the straight lines represents a mixture of the two end members. For a three-way mixing system, the water that is a result of mixing will be represented in each of the three fields by a point located inside a triangle defined by the three end members. Major ion chemical data for groundwater within the Selah interbed, a volcanic-sedimentary sequence localized in north-central Oregon (see Chapters 3 and 9), is shown in Figure 4-23. The groundwater is characterized as a calcium carbonate (bicarbonate) type with minor chloride and potassium. In some locations of depth, sulfate is dominant over carbonate, sodium becomes dominant over calcium, and chloride increases, reflecting greater maturity from increased reaction between groundwater and the tuffaceous siltstone. Magnesium is also a significant component of the groundwater and likely derived from weathering of basalt or volcanic clasts.

Other analyses also have merit when conducting subsurface environmental studies. The presence of tritium on the order of 10 tritium units or less would indicate that post-1953 "modern or bomb tritium" water is absent; that recharge to the water-bearing unit is greater than 40 years. Tritium units greater than 10 would indicate a more rapid rate of recharge is occurring, and that surface water from underflow, recharge, or a release from a waste management unit has reached the uppermost water-bearing zone within the past 40 years. Tritium analysis can also be important in evaluating anticipated recharge rates at potential new waste management sites.

Isotope analysis such as C^{14}, O^{18}, and deuterium can also be useful in evaluating groundwater origin (i.e., meteoric, juvenile, formation, etc.), chemistry (i.e., bicarbonate, sulfate, or chloride), and total salinity (i.e., freshwater or brine). The isotopic composition of groundwater may also be used to show whether there is a correlation to that of precipitation, thus indicating source area(s) of recharge.

BIBLIOGRAPHY

Bear, J., 1979, *Hydraulics of Groundwater:* New York, McGraw-Hill International, 567 p.

Bentall, R., (Compiler), 1963a, Methods of Determining Permeability, Transmissibility and Drawdown: U.S. Geological Survey Water-Supply Paper 1536-I, pp. I243-I341.

Bentall, R., (Compiler), 1963b, Shortcuts and Special Problems in Aquifer Tests: U.S. Geological Survey Water-Supply Paper 1545-C, pp. C1-C117.

Boulton, N.S. and Streltsova, T.D., 1977, Unsteady Flow to a Pumped Well in a Fissured Water-bearing Formation: *Journal of Hydrology,* Vol. 35, pp. 257-269.

Boulton, N.S. and Streltsova, T.D., 1978, Unsteady Flow to a Pumped Well in an Unconfined Fissured Aquifer: *Journal of Hydrology,* Vol. 37, pp. 349-363.

Bouwer, H., 1978, *Groundwater Hydrology:* New York, McGraw-Hill, 480 p.

Clark, L., 1979, The Analysis and Planning of Step Drawdown Tests: *Quarterly Journal of Engineering Geology,* Vol. 10, No. 2, pp. 125-143.

Darcy, H., 1856, *Les Fontaines Publiques de la Ville de Dijon:* V. Dalmont, Paris, 674 p.

Driscoll, F.G., 1986, *Groundwater and Wells,* Johnson Division, St. Paul, MN, 1088 p.

Ferris, J.G., Knowles, D.B., Brown, R.H., and Stallman, R.W., 1962, Theory of Aquifer Tests: U.S. Geological Survey Water-Supply Paper 1536-E, pp. E69-E174.

Fetter, C.W., Jr., 1980, *Applied Hydrogeology:* Charles E. Merrill, Columbus, OH, 488 p.

Francks, P.L., 1991, Using Major Ion Analyses at Hazardous Waste Site Investigations: In Proceedings of the National Research and Development Conference on the Control of Hazardous Materials, Anaheim, CA, pp. 506-510.

Freeze, R.A. and Back, W. (editors), 1983, *Physical Hydrogeology, Benchmark Papers in Geology;* Vol. 72: Hutchinson Ross, Stroudsburg, PA, 431 p.

Freeze, R.A. and Cherry, J.A., 1979, *Groundwater:* Prentice-Hall, Englewood Cliffs, NJ, 604 p.

Greenbury, A.E., Connors, J.J. and Jenkins, D., 1981, Standard Methods for the Examination of Water and Wastewater: American Public Health Association, 15th Edition, Washington, D.C.

Gringarten, A.C., 1982, Flow-test Evaluation of Fractured Reservoirs: Recent Trends in Hydrogeology, Geological Society of America Special Paper 189, pp. 237-263.

Gringarten, A.C. and Witherspoon, P.A., 1972, A Method of Analyzing Pumping Test Data from Fractured Aquifers: Proceedings of the Symposium on Percolation through Fissured Rock, International Society for Rock Mechanics and the International Society of Engineering Geology, Stuttgart, T3, pp. B1-B9.

Grubb, S., 1993, Analytical Model for Estimation of Steady-State Capture Zones of Pumping Wells in Confined and Unconfined Aquifers: *Ground Water,* Vol. 31, No. 1, pp. 27-32.

Hantush, M.S., 1966, Analysis of Data from Pumping Tests in Anisotropic Aquifers: *Journal of Geophysical Research,* pp. 421-426.

Heath, R.C., 1983, Basic Ground-Water Hydrology: U.S. Geological Survey Water-Supply Paper 2220, 85 p.

Hem, J.D., 1985, Study and Interpretation of the Chemical Characteristics of Natural Water: U.S. Geological Survey Water Supply Paper 2254, 3rd Edition.

Houlden, L.J., 1984, Analysis of Aquifer Test Data with Special Reference to Fractured Rock Aquifers in South Africa: Proceedings of the International Conference on Ground Water Technology, Johannesburg, South Africa, Vol. 2, pp. 656-702.

Jenkins, J.D. and Prentice, J.K., 1982, Theory for Aquifer Test Analysis in Fractured Rock Under Linear (non-radial) Flow Conditions: *Ground Water,* Vol. 20, pp. 12-21.

Kruseman, G.P. and DeRidder, N.A., 1970, Analysis and Evaluation of Pumping Test Data: International Institute for Land Reclamation and Improvement, Bull. 11, Wageningen, The Netherlands, 200 p.

Labadie, J.W. and Helweg, O.J., 1975, Step Drawdown Test Analysis by Computer, *Ground Water,* Vol. 13, No. 5, pp. 438-444.

Lohman, S.W., 1972, Ground-Water Hydraulics: U.S. Geological Survey Professional Paper 708, 70 p.

Marshack, J.B., 1989, A Compilation of Water Quality Goals: Staff Report of the California Regional Water Quality Control Board - Central Valley Region, November 1989, 15 p.

Morris, M.D., Berk, J.A., Krulik, J.W. and Ecstein, Y., 1983, A Computer Program for a Trilinear Diagram Plot and Analysis of Water Mixing Systems: *Ground Water,* Vol. 21.

Nahm, G.Y., 1980, Estimating Transmissivity and Well Loss Constant Using Multi-Rate Test Data from a Pumped Well: *Ground Water,* Vol. 18, No. 3, pp. 281-285.

Neuman, S.P., Walter, G.R., Bentley, H.W., Ward, J.J., and Gonzalez, D.D., 1984, Determination of Horizontal Aquifer Anisotropy with Three Wells: *Ground Water,* Vol. 22, No. 1, pp. 66-72.

Papadopulos, I.S., 1965, Nonsteady Flow to a Well in an Infinite Anisotropic Aquifer: Proceedings of the Symposium on Hydrology of Fractured Rocks, Dubrovnik, IAHS, pp. 21-31.

Sheahan, N.T., 1971, Type Curve Solution of Step Drawdown Tests: *Ground Water,* Vol. 9, No. 1, pp. 25-29.

Stanley, D., 1988, Where Are the Rest of the Analyses: *Ground Water,* Vol. 26, No. 1, pp. 2-5.

Suaveplane, C., 1984, Pumping Test Analysis in Fracture Aquifer Formations; State of the Art and Some Perspectives: Ground Water Hydraulics, *Water Resources Monograph Series 9,* pp. 171-206.

U.S. Department of Interior, 1985, Ground Water Manual: Bureau of Reclamation, Engineering and Research Center, Denver, CO, 480 p.

U.S. Geological Survey, 1977, *National Handbook of Recommended Methods for Water-Data Acquisition:* Office of Water Data Coordination, Reston, VA.

Way, S.C. and McKee, C.R., 1982, In-situ Determination of Three-Dimensional Aquifer Permeabilities: *Ground Water,* Vol. 20, No. 5, pp. 594-603.

5 TECHNIQUES FOR SUBSURFACE CHARACTERIZATION

"When in doubt, measure it; when not in doubt, measure it anyway."

(Testa and Winegardner, 1991)

INTRODUCTION

Subsurface geologic, hydrogeologic, and environmental conditions can be explored by several techniques, usually in combination. These techniques can essentially be divided into two general types: noninvasive and invasive. Noninvasive techniques include review of historic aerial photography, published and unpublished maps and reports, as well as remote sensing, surficial vapor surveys, and surface geophysics. Invasive techniques pertain to direct subsurface penetration and include drilling of soil borings and rock cores with subsequent sampling, installation of groundwater monitoring wells, piezometers or vapor probes, downhole geophysics, and cone penetration testing. Samples retrieved are typically classified and chemically field-screened for certain constituents of concern (i.e., hydrocarbons), with selected samples submitted to a certified laboratory for chemical testing in accordance with a required regulatory protocol as outlined in Appendix C. As with all environmental studies, a high probability for future litigation with subsequent identification of responsible parties to incur the financial responsibilities of assessment and mitigation related activities always persist. Thus, sample integrity is of utmost importance and stringent protocols must be followed.

This chapter will focus on some of the more important and conventional techniques used for the purpose of subsurface characterization, design of subsurface remediation strategy, and ultimate monitoring. Geophysical applications and techniques will not be discussed herein, but rather under Geophysical Applications in Chapter 6.

SUBSURFACE EXPLORATION

Subsurface geologic, hydrogeologic, and environmental conditions can be directly evaluated by the drilling of soil or rock borings, and subsequent installation and construction of monitoring wells, piezometers, or soil vapor probes. Several techniques are available for the drilling of soil or rock borings regardless of their eventual use.

Soil and Rock Borings

Different drilling techniques are utilized depending on whether unconsolidated or consolidated deposits are anticipated, and consideration for sample type (undisturbed vs disturbed), sample integrity, and economics. The two most common approaches for unconsolidated deposits are hollow-stem augering and cable-tool drilling techniques. For consolidated or semiconsolidated deposits, continuous wire-line or conventional rock coring techniques are commonly used. The major drilling techniques are summarized in Table 5-1, with the more conventional methods used for environmental-related studies further discussed below.

Hollow-Stem Auger Drilling

Hollow-stem auger drilling techniques are commonly used for subsurface environmental purposes. During drilling, a series of interconnected, hollow, auger flights, usually 5 to 10 ft in length with a lower cutting head, are hydraulically pressed downward (Figure 5-1). During this downward motion, cuttings are rotated up the outside of the continuous flighting. Soil sampling is achieved by passing a smaller diameter drill rod with a sampler attached at the bottom. The sampler is typically either a thin-wall or modified split-spoon sampler with brass inserts. Samples can be continuously retrieved, although in practice samples are routinely retrieved at 5-ft intervals or significant changes in lithology. A standard California modified split-spoon sampler is lined by three brass inserts each 6 in. in length, although up to 18 1-in. brass rings have been used for conventional geotechnical studies (Figure 5-2). Conventionally, the upper 6-in. insert is used for descriptive purposes and physical testing (i.e., permeability, sieve analysis, etc.). The middle insert is used for field-screening for hydrocarbon vapors or other constituents of concern, and the bottom insert is used for chemical testing. The sampler is driven into the soil at the desired sampling interval ahead of the auger bit. Using a 140-lb hammer with a 30-in. drop, a standard penetration test is performed, providing quantitative soil density information as discussed later in this chapter.

Other augering techniques include the use of labor-extensive hand augers for shallow depths, and machine-operated solid-flight augers. Solid-flight augers include solid-stem, solid-core, or continuous-flight augers. As the augers are rotated, cuttings, as with hollow-stem augers, are rotated upward to the surface by outside, upward movement along the continuous flighting. Sample retrieval may be difficult where loosely consolidated, unstable soil conditions persist. Since undisturbed depth discrete samples can only be retrieved if the entire string of augers is removed, this drilling method is seldom used for environmental purposes.

Cable-Tool Drilling

Cable-tool drilling is the oldest drilling technique available and, although commonly used in glacial environments in the Pacific Northwest portion of the U.S., the availability of cable-tool drilling rigs is restrictive in most parts of the country. With use of the cable tool, the hole is continuously cased with an unperforated, 8-in.-diameter steel casing with a drive shoe (Figure 5-3). The casing is attached on top by means of a rope socket to a cable which is suspended through a pulley from the mast of the drill rig. The process of driving the casing downward about 3 to 5 ft, followed by bailing, is referred to as "drive and bail". Cuttings are removed by periodically bailing the borehole; thus, water must be added to the borehole to create a slurry when drilling in nonwater-bearing material.

When drilling under saturated conditions, the drive and bail process also allows for depth-discrete groundwater samples which can be field screened using a portable organic vapor analyzer in the gas chromatograph mode to assess the potential presence of dissolved constituents in groundwater.

Rotary Drilling

Rotary drilling techniques include direct mud rotary, air rotary, air rotary with a casing driver, and dual-wall reverse circulation. With direct mud rotary, drilling fluid is pumped down through the bit at the end of the drill rods, then is circulated up the annular space back to the surface. The fluid at the surface is routed via a pipe or ditch to a sedimentation tank or pit, then to a suction pit where the fluid is recirculated back through the drill rods. Air rotary drilling is similar to that of direct mud rotary, except that air is used as a circulation medium instead of water. Although the air helps cool the bit, small quantities

Table 5-1 Drilling Techniques for the Construction and Installation of Wells

Drilling Technique	Material	Limitations (in feet)	Advantages	Disadvantages
Augering Hand auger	Unconsolidated	15	Shallow soils investigations Soil samples Shallow water-bearing zone identification Piezometer, lysimeter, and small-diameter monitoring well installation Labor intensive, but inexpensive No casing material restriction	Limited to very shallow depths Unable to penetrate extremely dense or rocky soil Borehole stability difficult to maintain Labor intensive
Bucket auger	Unconsolidated	100	Shallow soils investigations Soil samples Vadose zone monitoring wells (lysimeters) Monitoring wells in saturated, stable soils Identification of depth to bedrock Fast and mobile	Unacceptable soil samples unless split-spoon or thin-wall samples are taken Soil sample data limited to areas and depths where stable soils are predominant Unable to install monitoring wells in most unconsolidated aquifers because of borehole caving upon auger removal Depth capability decreases as diameter of auger increases

Method	Formation		Applications	Limitations
Solid-flight auger	Unconsolidated	180	Shallow soils investigations Soil samples Vadose zone monitoring wells (lysimeters) Monitoring wells in saturated, stable soils Identification of depth to bedrock Fast and mobile	Monitoring well diameter limited by auger diameter Unacceptable soil samples unless split-spoon or thin-wall samples are taken Soil sample data limited to areas and depths where stable soils are predominant Unable to install monitoring wells in most unconsolidated aquifers because of borehole caving upon auger removal Depth capability decreases as diameter of auger increases
Hollow-stem auger	Unconsolidated	180	All types of soil investigations Permits good soil sampling with split-spoon or thin-wall samplers Water-quality sampling Monitoring well installation in all unconsolidated formations Can serve as temporary casing for coring rock Can be used in stable formations to set surface casing	Monitoring well diameter limited by auger diameter Difficulty in preserving sample integrity in heaving formations Formation invasion by water or drilling mud if used to control heaving Possible cross contamination of aquifers where annular space not positively controlled by water or drilling mud or surface casing Limited diameter of augers limits casing size Smearing of clays may seal off aquifer to be monitored

Table 5-1 Drilling Techniques for the Construction and Installation of Wells (continued)

Drilling Technique	Material	Limitations (in feet)	Advantages	Disadvantages
Driven	Unconsolidated	50	Water-level monitoring in shallow formations Water samples can be collected Dewatering Water supply Low cost encourages multiple sampling points	Depth limited to approximately 50 ft (except in sandy material) Small diameter casing No soil samples Steel casing interferes with some chemical analysis Lack of stratigraphic detail creates uncertainty regarding screened zones and/or cross contamination Cannot penetrate dense and/or some dry materials No annular space for completion procedures
Jet percussion	Unconsolidated	200	Allows water-level measurement Sample collection in form of cuttings to surface Primary use in unconsolidated formations, but may be used in some softer consolidated rock Best application is 4-in. borehole	Drilling mud may be needed to return cuttings to surface Diameter limited to 4 in. Installation slow in dense, bouldery clay/till or similar formations Disturbance of the formation possible if borehole not cased immediately

Method	Formation	Depth	Advantages	Limitations
Cable tool	Unconsolidated	1000	with 2-in. casing and screen installed, sealed, and grouted Drilling in all types of geologic formations Almost any depth and diameter range Ease of monitoring well installation Ease and practicality of well development Excellent samples of coarse grained materials	Drilling relatively slow Heaving of unconsolidated materials must be controlled Equipment availability more common in central, northcentral, northeast, and northwest sections of the U.S.
Rotary Reverse circulation	Unconsolidated or consolidated	2000+	Very rapid drilling through both unconsolidated and consolidated formations Allows continuous sampling in all types of formations Very good representative samples can be obtained with minimal risk of contamination of sample and/or water-bearing zone In stable formations, wells with diameters as large as 6 in. can be installed in open-hole completions	Limited borehole size that limits diameter of monitoring wells In unstable formations, well diameters are limited to approximately 4 in. Equipment availability more common in the southwest Air may modify chemical or biological conditions; recovery time is uncertain Unable to install filter pack unless completed open hole

Table 5-1 Drilling Techniques for the Construction and Installation of Wells (continued)

Drilling Technique	Material	Limitations (in feet)	Advantages	Disadvantages
Mud rotary	Unconsolidated or consolidated		Rapid drilling of clay, silt, and reasonably compacted sand and gravel Allows split-spoon and thin-wall sampling in unconsolidated materials Allows core sampling in consolidated rock Drilling rigs widely available Abundant and flexible range of tool sizes and depth capabilities Very sophisticated drilling and mud programs available Geophysical bore-hole logs	Difficult to remove drilling mud and well cake from outer perimeter of filter pack during development Bentonite or other drilling fluid additives may influence quality of groundwater samples Circulated (ditch) samples poor for monitoring well-screen selection Split-spoon and thin-wall samplers are expensive and of questionable cost effectiveness at depths greater than 150 ft Wireline coring techniques for sampling both unconsolidated and consolidated formations often not available locally Difficult to identify aquifers Drilling fluid invasion of permeable zones may compromise validity of subsequent monitoring well samples
Air rotary	Unconsolidated or consolidated	2000+	Rapid drilling of semi-consolidated and consolidated rock	Surface casing frequently required to protect top of hole

Method	Formation	Depth	Advantages	Disadvantages
Air rotary (with casing driver)			Good quality/reliable formation samples (particularly if small quantities of water and surfactant are used) Equipment generally available Allows easy and quick identification of lithologic changes Allows identification of most water-bearing zones Allows estimation of yields in strong water-producing zones with short "down time"	Drilling restricted to semiconsolidated and consolidated formations Samples reliable but occur as small particles that are difficult to interpret Drying effect of air may mask lower yield water producing zones Air stream requires contaminant filtration Air may modify chemical or biological conditions Recovery time is uncertain
Air Percussion	Unconsolidated or consolidated	2000+	Rapid drilling of unconsolidated sands, silts and clays Drilling in alluvial material (including boulder formations) Casing supports borehole thereby maintaining borehole integrity and minimizing inter-aquifer cross contamination Eliminates circulation problems Rapid drilling	Thin, low pressure waterbearing zones easily overlooked if drilling not stopped at appropriate places to observe whether or not water levels are recovering Samples pulverized as in all rotary drilling Air may modify chemical or biological conditions; recovery time is uncertain
Wire-line	Consolidated	2000+	Excellent sample retrieval Maintains borehole integrity Allows continuous casing of borehole	Poor sample recovery Air emission concerns Not applicable to unconsolidated formations

Table 5-1 Drilling Techniques for the Construction and Installation of Wells (continued)

Drilling Technique	Material	Limitations (in feet)	Advantages	Disadvantages
Horizontal	Unconsolidated or consolidated	2000+	Allows access beneath structures Larger zone of influence Utilization for variety of remedial purposes Maintain higher specific capacities Likely to intersect vertical fractures	Maintaining directional orientation may present problems Potential caving in loose materials

Figure 5-1   Conventional truck-mounted hollow-stem auger drilling rig.

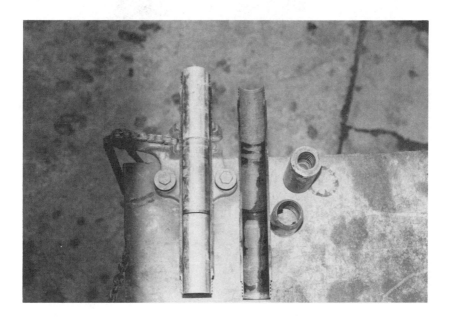

Figure 5-2   Conventional California-modified split-spoon sampler with brass inserts.

Figure 5-3   Conventional truck-mounted cable-tool drilling rig.

of water or foaming surfactants are used to facilitate sampling. In unconsolidated deposits, direct mud or air rotary can be used, providing that a casing is driven as the drill bit is advanced. In dual-wall reverse circulation, the circulating medium (mud or air) is pumped downward between the outer casing and inner drill pipe, out through the drill bit, then up the inside of the drill pipe.

Rotary drilling techniques are commonly limited to consolidated deposits of rocks and typically not used in subsurface environmental studies due to poor sampling capabilities. Sample integrity is questioned since the added water, mud, or surfactant may chemically react with the formation water. In addition, thin water-bearing zones are often missed. With mud rotary, the mud cake that develops along the borehole wall can adversely affect permeability of the adjacent formation materials. With air rotary, dispersion of potentially hazardous and toxic particulates in the air during drilling is also a concern. However, air rotary drilling techniques are fast and, where the subsurface geology is relatively well characterized or a resistant stratum (i.e., overlying basalt flows or conglomerate strata) exists at shallow levels within the vadose zone and above the depth of concerns, utilization of air rotary to a predetermined depth followed by another more suitable drilling technique may have merit.

Wire-Line Coring

Coring is the drilling of consolidated deposits or rock and subsequent retrieval of the subsurface material in the form of a cylindrically shaped core. A rotary rig is used in conjunction with water, drilling mud, or air. During coring, an annular opening is made by the bit. The barrel gradually slides down into the annular opening. The core is then separated from the rest of the formation mass and the barrel containing the core is retrieved. Both single and double-tube core barrels exist. With double-tube core barrels, the inner tube retains the core while the tube rotates. Core diameter varies depending on bit size, ranging from about 27.0 to 151.6 mm. Cutting is accomplished by drill bits located at the end of the rotating barrel or tube.

Horizontal Drilling

Horizontal or lateral radial wells are fast becoming popular in subsurface environmental studies although they have been used by the oil industry for decades. The most obvious application is where the area of concern (i.e., contaminant plume) is inaccessible due to above-ground structures, tankage, roads, etc. or subsurface structures such as landfills, lagoons, pits, pipelines, or wells. Horizontal wells can maintain a larger zone of influence, thus comparatively higher specific capacities than vertical wells. These advantages make horizontal wells attractive as in situ groundwater aeration (sparging), vapor recovery, pump-and-treat, and injection wells. Horizontal wells can also be used for pressure- or jet-grouting a permanent barrier under and around an affected area, bioremediation by use for delivery of microbes, nutrients, and oxygen into the affected area, or as a French drain and landfill leachate collection system. Furthermore, horizontal wells are more likely to intersect vertical or inclined fractures in comparison to vertical wells.

A small-diameter pilot hole is commonly drilled first, where inclination, azimuth, and toolface data is determined and used for directional steering. Following completion of the pilot hole, a larger-diameter hole is drilled or the pilot hole is forsaken and the larger diameter hole is implemented from the start of the drilling program. Specially designed tools and methods are subsequently used to clean the borehole and to install and develop the well to meet its intended purpose. Lateral radial wells can incorporate a vertical well of 4.5 in. or larger casing diameter and be placed from a few feet to 6800 ft below ground surface, which allows for the precise placement of horizontal radial wells at a 9- to 12-in. radius of curvature. Radial wells can extend from a central vertical well to 200 ft or more.

Monitoring Wells and Piezometers

Assessment and characterization of hydrogeologic conditions and monitoring the efficiency and effectiveness of groundwater restoration programs (and soil vapor extraction systems) are performed utilizing monitoring wells and piezometers. Wells and piezometers are installed immediately following completion of the borehole. Upon borehole completion, the well screen, with end cap and casing, is placed in the borehole. Every effort is made in proper selection of screen length and placement opposite the interval of concern while avoiding intercommunication between higher intervals of significant permeability. Several casing materials of varying strength, chemical resistance, and interference characteristics are available. These characteristics influence their use for certain site-specific hydrogeologic conditions and contaminant of concern as summarized in Table 5-2.

The sand or filter pack is emplaced around the screen in the well annulus, that region between the well casing and the borehole wall. The sand pack typically extends a minimum of 2 ft above the top of the screen. A seal is placed above the sand pack. The purpose of the seal is to isolate the screened portion of the well (the perforated casing) and prevent potential contaminants from entering the well from directly down the borehole. Because the seal is required to be impermeable, a pure bentonite, cement grout, or bentonite-grout mixture is used. The initial pure bentonite seal is first placed directly above the sand pack that surrounds the perforated portion of the casing, for the purpose of initially sealing the annulus, and preventing grout invasion in the sand (filter) pack which surrounds the well screen. The initial bentonite seal, once emplaced,

Table 5-2 Summary of Casing Material Characteristics

Material Category	Material Brand Name	Advantages	Disadvantages
Fluoropolymer	Polytetrafluoroethylene (PTFE) Tetrafluoroethylene (TFE) Fluorinated ethylene propylene (FEP)	Almost completely chemically inert Sorption of chemical Leaching of chemical constituents minimal	More expensive than PVC Difficult to handle (i.e., low coefficient of friction when wet) Low compressive strength (i.e., ductile behavior) Extreme flexibility of casing string
Metallic	Carbon steel Low-carbon steel Galvanized steel Stainless steel (304 and 316)	Strong Rigid Low temperature sensitivity	Corrosion during long-term exposure to certain geochemical environments Potentially affects water sample integrity causing changes in dissolved metals or organic compounds
Thermoplastic		Resistance to galvanic and electrochemical corrosion Lightweight High abrasive resistance High strength-to-weight ratios Durable Requires low maintenance Flexible and workable Relatively low cost	Brittleness and gradual loss of impact strength with long-term exposure to ultraviolet rays, direct sunlight, or low temperatures
Reinforced fiberglass	Fiberglass reinforced epoxy (FRE) Fiberglass reinforced plastic (FRP)	Unknown	Very little data concerning characteristics and performance

is followed by the emplacement of cement grout, bentonite-grout mixture, or in some cases, pure bentonite that extends from the top of the initial bentonite seal to the ground surface.

Initial bentonite seals are conventionally emplaced in a pellet form, which expands at a relatively fast rate when exposed to water. This occurs because bentonite possesses the ability to expand significantly when hydrated. Expansion reflects the incorporation of water molecules into the bentonite (clay) lattice (structure). Hydrated bentonite in water typically expands 10 to 15 times the volume of dry bentonite, thus providing a very tight seal between the casing and the adjacent formation materials (borehole wall). Experimental studies indicate up to a 50% height and volume increase upon hydration under ambient conditions. Therefore, an initial 2-ft-thick, dry bentonite seal would hydrate and expand up to approximately 3 ft in total length.

Guidelines for the required length of initial bentonite seals have varied, depending primarily upon the well use and function and existing subsurface geologic and hydrogeologic conditions. Recommended lengths range from approximately 2 to 5 ft for monitoring wells to a minimum of 10 ft for wells used for industrial and/or drinking water supply usages. The height of the entire seal above the top of the well screen is more important than the length of the initial bentonite seal. The initial bentonite seal is placed at a minimum length of 2 ft above the top of the well intake, where possible, to avoid potential chemical reactions between water retrieved from the well and the bentonite. A typical monitoring well construction detail is presented in Figure 5-4.

Water-table wells typically screen the water table and the head is measured directly in the well. Piezometers screen a discrete interval below the water table; however, the water level measured in the well reflects the head distribution for the entire length of the sand pack.

Most well installations fall into one of three categories: single-level, nested or group, or multiple-port completion (Figure 5-5). Single-level completions as shown in Figures 5-5 and 5-6 are usually used when the focus is on the interval immediately at and below the water table, perched zones, or a discrete zone at depth (i.e., with drinking water wells). When more than one depth is of interest, nested and multiple-port completions have merit. A nested or group of piezometers comprises several piezometers all installed in a single borehole. A nested completion is expensive and difficulties are often encountered in constructing and maintaining reliable seals.

Multiple-port piezometers have been available for about a decade, and are designed to provide data on piezometric levels and water quality from several levels within one borehole. Designed as a cost-efficient alternative to single-level completions in close proximity to each other or nested completions, they achieve increased data density by allowing depth-discrete monitoring and sampling ports which are separated by inflatable packers (Figure 5-6). Minimal purging is required prior to fluid sampling. Retrieval depth, pore pressures and vertical displacement (heave and settlement) data, and conduct of permeability testing can also be accomplished.

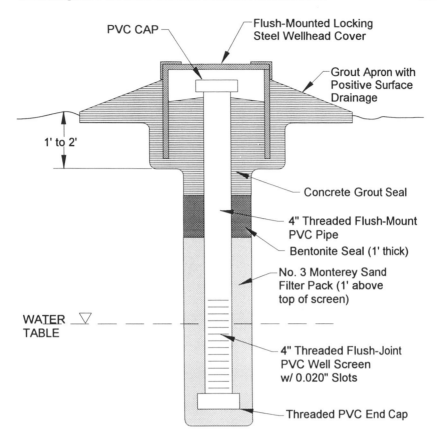

Figure 5-4   Typical groundwater monitoring well construction detail.

Horizontal and lateral radial wells in saturated unconsolidated materials perform differently in comparison to their vertical counterparts. Overall advantages include development of a large, elongate zone of influence, a horizontally continuous capture zone, relatively high specific capacity, and a long screen with relatively low screen-entrance velocities. Thus, one horizontal (lateral) well can be used in lieu of several vertical wells (Figure 5-6). Under saturated conditions, by providing an elongate concentric ellipses and continuous horizontal capture zone, groundwater sparging can be achieved by placing the well immediately below the dissolved contaminant plume. Groundwater velocities are lower than that created by pumping a vertical well, providing that the screen length is at least three times the thickness of the confining aquifer. In addition, under both confined and unconfined conditions, minimal horizontal lengths can be achieved, reflecting the greater specific capacity of horizontal wells with the exception of relatively thick and highly permeable aquifers. Under anisotropic conditions, specific capacity is also greater providing the vertical

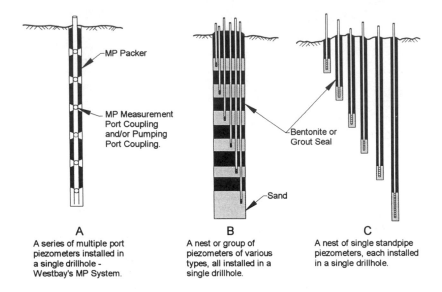

A

A series of multiple port piezometers installed in a single drillhole - Westbay's MP System.

B

A nest or group of piezometers of various types, all installed in a single drillhole.

C

A nest of single standpipe piezometers, each installed in a single drillhole.

Figure 5-5   Examples of varying types of groundwater monitoring well completions: (A) single-level, (B) nested or group, and (C) multiple-port completions.

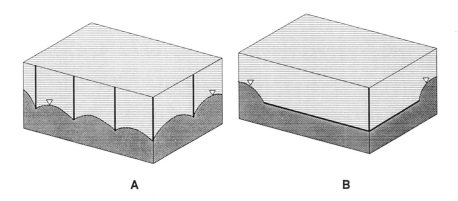

A   B

Figure 5-6   Hydraulic well-field configurations using (A) vertical and (B) lateral radial wells.

hydraulic conductivities exceeds the horizontal conductivity, as in a vertically fractured aquifer. Greater specific capacity results in lower pumping rates per foot of screen.

Vapor Wells

Monitoring wells to permit continued collection of soil vapor samples are usually constructed in a similar manner as that of standard monitoring wells. Materials of construction should be compatible with the chemicals involved. In most cases, PVC casing and screen are used. However, strong solvents such as acetone and methyl-ethyl-ketone are not compatible with PVC. Well screens should extend from below the yearly low water table to as close to the surface as practical while sealed at the surface. This type of construction allows measurement of water levels and collection of water and vapor samples. However, construction of depth-discrete screened intervals within the vadose zone opposite the affected zone(s) can also be easily constructed. Vapor wells to assess the potential for vapor presence and effective vapor recovery in fractured bedrock are constructed similar to those for hydrocarbon-affected soil evaluation. However, the occurrence of fractures as migration pathways instead of soil pores is difficult, and most fractured-rock venting is completed on a trial-and-error basis. Surface completion of the vapor wells should allow for multiple-use adaptations (i.e., an air inlet port and/or a pressure sampling point during restoration-related activities).

CONE PENETRATION TESTING

Cone penetration testing is used to delineate sediment stratigraphy by measuring the friction ratio of sleeve friction and cone-bearing capacity on a 1/2-in.-diameter metal rod which is hydraulically pushed into the ground. The values obtained are then matched to the cone-bearing capacity to determine the soil type to the nearest tenth of a foot. Recent advances in technology of cone penetrometers provide an excellent capability to determine physical soil characteristics between boreholes. Charts of penetration resistance can be calibrated to indicate stratigraphic horizons which will enable development of a more complete understanding of the subsurface at a relatively lower cost. Many contractors who provide cone penetration testing have adapted their probes to allow collection of discrete soil vapor or groundwater samples during insertion. At completion, resulting probe holes are relatively small in diameter and are easily sealed with bentonite or grout.

The soil penetrometer can also be used in conjunction with other types of sensors. Penetrometer-mounted sensors have been developed for the direct measurement of such parameters as soil pore pressure, soil radioactivity and soil electrical resistivity. Electrical resistivity and soil spectral properties, notably fluorescence (i.e., fiber optics and laser-induced fluorescence), have been used to detect and estimate the quantity of hydrocarbon-affected soils in a variety of media.

CLASSIFICATION OF SUBSURFACE MATERIALS

The description and classification of subsurface materials and conditions is an essential part of any subsurface investigation. As with most subsurface environmental studies, it is important that an intimate knowledge of material encountered be required during and immediately following the drilling of the borehole. Adequate description of soils in place is mandatory, allowing decisions to be made regarding placement of screened intervals for vapor or groundwater monitoring wells and mitigation purposes, assessment of adversely impacted soils on overall groundwater resources, and potential preferred migration pathways.

Several soil and rock classification schemes have been proposed by various agencies and other disciplines, although they do not always lend themselves to field classification for hydrogeologic or environmental engineering projects. Neither of the systems, in fact, measure such important parameters as pH, organic matter content, clay type, or predictable soil-chemical reactions.

Unconsolidated Deposits

Soils are conventionally classified and described in accordance with an approved soil description system. The three most widely used classification systems are the Unified Soil Classification System, the Burmister System, and the Comprehensive Soil Classification System as discussed below.

Unified Soil Classification System (USCS)

Most subsurface, environmental, and engineering applications use the USCS (Figure 5-7). The USCS reflects identification of soils based on textural characteristics and plasticity, and their grouping on the basis of engineering performance. The USCS also provides adequate characterization of particle size distribution and physical behavior (i.e., relative permeability). The following properties form the basis of classification:

- Percentages of gravel, sand, and fines (fraction passing the No. 200 U.S. Standard Sieve)
- Shape of the grain-size distribution curve
- Plasticity and compressibility characteristics.

The USCS is based on classification of the minus 3-in. (75 min) fraction, but all fractions should be included in the description of soil according to the grain size ranges as shown in Table 5-3.

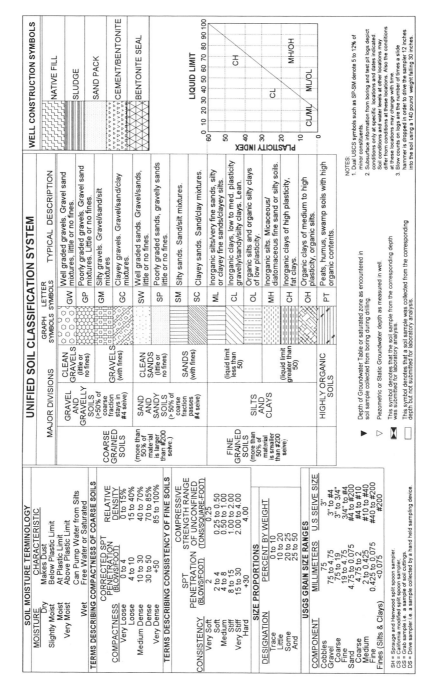

Figure 5-7   The Unified Soil Classification System and supportive Burmister System.

Table 5-3 Grain Size Scales for Sediments[a,b]

U.S. Standard Sieve Mesh #	Millimeters	Microns (μm)	Phi (φ)	Wentworth Size Class
	4096		−12	Boulder (−8 to −12φ)
	1024		−10	
Use wire	256		−8	
squares	64		−6	Cobble (−6 to −8φ)
	16		−4	Pebble (−2 to −6φ)
5	4		−2	Granule
6	3.36		−1.75	
7	2.83		−1.5	
8	2.38		−1.25	
10	2.00		−1.0	Very coarse sand
12	1.68		−0.75	
14	1.41		−0.5	
16	1.19		−0.25	
18	1.00		0.0	Coarse sand
20	0.84		0.25	
25	0.71		0.5	
30	0.59		0.75	
35	1/2 0.50	500	1.0	Medium sand
40	0.42	420	1.25	
45	0.35	350	1.5	
50	0.30	300	1.75	
60	1/4 0.25	250	2.0	Fine sand
70	0.210	210	2.25	
80	0.177	177	2.5	
100	0.149	149	2.75	
120	1/8 0.125	125	3.0	Very fine sand
140	0.105	105	3.25	
170	0.088	88	3.5	
200	0.074	74	3.75	
230	1/16 0.0625	62.5	4.0	Coarse silt
270	0.053	53	4.25	
325	0.044	44	4.5	
	0.037	37	4.75	
Analyzed	1/32 0.031	31	5.0	Medium silt
by pipette	1/64 0.0156	15.6	6.0	Fine silt
or hydro-	1/128 0.0078	7.8	7.0	Very fine silt
meter				

Table 5-3 Grain Size Scales for Sediments[a,b] (continued)

U.S. Standard Sieve Mesh #		Millimeters	Microns (μm)	Phi (φ)	Wentworth Size Class
	1/256	0.0039	3.9	8.0	Clay
		0.0020	2.0	9.0	
		0.00098	0.98	10.0	
		0.00049	0.49	11.0	
		0.00024	0.24	12.0	
		0.00012	0.12	13.0	
		0.00006	0.06	14.0	

[a] From Folk, 1968.

[b] The grade scale most commonly used for sediments in the Wentworth scale (actually first proposed by Udden), which is a logarithmic scale in that each grade limit is twice as large as the next smaller grade limit. For more detailed work, sieves have been constructed at intervals $^2\sqrt{2}$ and $^4\sqrt{2}$. The φ (phi) scale, devised by Krumbein, is a much more convenient way of presenting data than if the values are expressed in millimeters, and is used almost entirely in recent work.

Soils are divided into two primary groups: coarse grained and fine grained. Coarse-grained soils are those having 50% or more material larger than the No. 200 sieve; fine-grained soils are those with more than 50% of the material passing the No. 200 sieve. The soil groups are identified by letters and descriptive modifiers as shown in Figure 5-7. The symbols are combined to describe soil types. The first symbol indicates the primary constituent followed by the modifier. In the USCS, the symbols are all capitalized.

Coarse-grained soils are subdivided into gravels and sands depending on the major coarse-grained constituent. For purposes of identification, coarse-grained soils are classed as gravels (G) if the greater percentage of the coarse fraction (retained on No. 200 sieve) is larger than the No. 4 sieve and as sand (S) if the greater portion of the coarse fraction is finer than the No. 4 sieve. Both the gravel (G) and the sand (S) groups are further subdivided on the basis of uniformity of grading and percentage and plasticity of the fine fraction (minus No. 200) sieve. GW and SW soil groups include well-graded gravels and sands with less than 5% nonplastic fines. GP and SP soil groups are poorly graded or skip graded soils with less than 5% fines. Gravels and sands containing more than 12% fines may be classified as silty gravel (GM), silty sand (SM), clayey gravel (GC), or clayey sand (SC) depending on the plasticity of the minus No. 40 soil fraction. Silty gravels (GM) and silty sands (SM) will have liquid limits and plasticity indices which plot below the A-line of the plasticity chart also shown in Figure 5-8. Clayey gravels (GC) and clayey sand (SC) will have

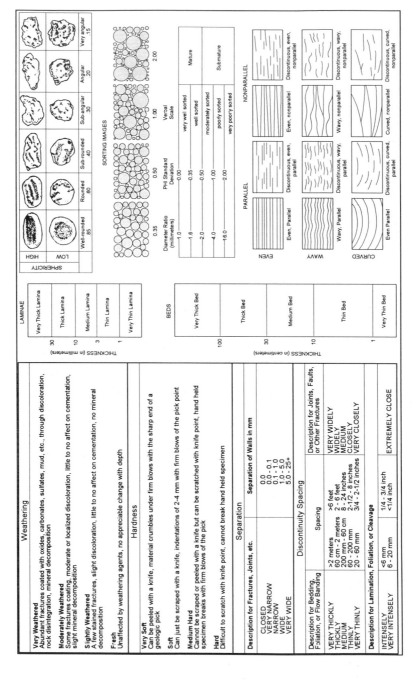

Figure 5-8   Explanation of terminology for consolidated deposits.

liquid limits and plasticity indices which plot above the A-line (grading is not a factor when the percentage of fines is greater than 12%). A dual classification symbol (i.e., GC, GP) is used when the minus No. 200 fraction is between sand 12% and the soil has characteristics intermediate between two groups.

The fine-grained soils are subdivided into silts and clays, depending on their plasticity and highly organic soils. To distinguish between silts and clays, the plasticity index vs the liquid limit is plotted as shown in Figure 5-7. Among the inorganic materials, the clays plot above the A-line and the silts below the A-line. Silts and clays are further subdivided into low (L) and high (H) plasticity based on whether the liquid limit is less than 50% (L) or greater than 50% (H).

Field techniques for differentiating between silts and clays include manual tests for dry strength, dilatancy, and toughness. Accurate identification of silts and clays is primarily a matter of experience and conscientious comparison to laboratory test results.

Highly organic soils are combined into a single classification with the symbol "Pt" and are characterized by a high organic content commonly consisting of leaves, grass, branches, and other fibrous matter, by high compressibility, and by relatively low strength. Typical samples of highly organic soils are peat, humus, and swamp soils.

Burmister Soil Identification System (BS)

The BS is commonly used in conjunction with the USCS in developing written field descriptions. The BS involves rather precise soil fraction descriptions utilizing specific terminology and symbols (Figure 5-7). The BS uses some capital letters as symbols for the components, with lower case letters to indicate proportionality or gradation. The proportionality terms "and", "some", "little", and "trace" are sometimes incorporated into written descriptions in other classification systems (Figure 5-7). The BS requires considerable laboratory and field practice before accurate identifications can be made.

Comprehensive Soil Classification System (CSCS)

The other major soil classification system used in the U.S. is the CSCS, developed by the U.S. Department of Agriculture (USDA). This system was developed to organize soils into established groups, identify their best uses, and allow for productivity estimates. The CSCS emphasizes classification for agricultural and land use, although it is sufficiently uniform for general prediction of potential chemical migration in soil.

Other Pertinent Soil Information

Fundamental to any soil description is color, moisture condition, and consistency (fine-grained soils) or compactness (coarse-grained soils). The color of the soil and natural moisture content should be recorded. Although not restricted to any specific classification scheme, for precise work soil and rock are most conveniently measured by comparison with a color chart. For soils, the Munsell System is the most widely accepted system for color identification in the U.S. The Munsell System describes all colors according to three variables, which are combined in sequence to form color designation: hue, value, and chroma. For example: yellowish-red = 5YR 5/6 (5YR = hue; 5 = value; 6 = chroma). Although the nomenclature for soil color consists of two complimentary systems, both color names and the Munsell notation should be used when describing soil color.

Soil compactness, consistency, and relative density are frequently described in the field on the basis of the Standard Penetration Test (SPT) and/or by our estimate of the unconfined compressive strength. Factors affecting penetration resistance include depth, overburden, pressure, soil type and condition, weight of drill rods, drilling fluid, and the presence of disturbed soil at the bottom of the boring. Due to the interdependence and complexity of the factors affecting the SPT, relationships of penetration-resistance to soil condition are generally only approximate. For example, the penetration resistance at shallow depths in sand is generally too low; thus, corresponding field descriptions may underestimate the compactness of the soil unless correction factors are applied to the field data. A number of additional soil properties should also be recorded, such as structure, cementation, grain shape, layering, dry strength, dilatancy, toughness, etc.

As part of any subsurface drilling program, representative soil and groundwater samples are retrieved for testing. The purpose behind these tests include further characterization of subsurface conditions, evaluation of remedial or corrective actions, or monitoring the effectiveness of a remedial program. The types of tests to be performed will vary depending upon the remedial strategies to be considered as summarized in Table 5-4.

Consolidated Deposits

Consolidated deposits encompass a wide diversity of igneous, metamorphic, and sedimentary rocks. Rocks are basically classified in accordance with grain (or crystal) size, textural relationships, and chemistry. Many sedimentary rocks, however, may physically resemble or behave in a manner similar to that of soil (i.e., clay stone, siltstones, mudstones, weakly cemented sandstones, and decomposed granite). These materials are commonly classified using a dual system, notably, incorporation of the USCS for descriptive, engineering,

Table 5-4 Summary of Site-Specific Parameter Considerations per Remedial Action

Hydro-geologic Conditions	Parameters	Remedial or Corrective Action										Pump and Treat	
		Slurry Cut of Trench/Wall	Grouting	Soil Bentonite Cap	Capping	Gas Venting	Hydraulic Fracturing	Solidification/ Stabilization	Soil Flushing Chemical Extraction	Bioremediation	Vacuum Extraction	NAPLs	No NAPLs
Vadose zone	Air-filled porosity					X					X		
	Bulk density			X	X			X					
	Chemistry												
	Pore water		X							X	X		
	Soil									X			
	Vapor										X		
	Waste	X	X	X	X	X	X	X	X	X	X	X	
	Coefficient (K_{oc})								X			X	X
	Diffusivity			X	X	X					X		
	Lithology	X	X	X	X	X	X				X		
	Marshall Index/ Cone Index							X					
	Microbial influence									X			
	Moisture content							X		X	X		
	Octonal/water coefficient (K_{ow})												
	Organic carbon distribution								X				
	Particle size distribution		X	X	X	X		X			X		
	pH		X							X			
	Pore size and shape		X			X					X		

Table 5-4 Summary of Site-Specific Parameter Considerations per Remedial Action (continued)

Remedial or Corrective Action

Hydrogeologic Conditions	Parameters	Slurry Cut of Trench/Wall	Grouting	Soil Bentonite Cap	Capping	Gas Venting	Hydraulic Fracturing	Solidification/ Stabilization	Soil Flushing Chemical Extraction	Bioremediation	Vacuum Extraction	Pump and Treat NAPLs	Pump and Treat No NAPLs
	Porosity					X	X				X		
	Retention												
	Soil organic matter												
	Soil permeability		X	X	X	X	X			X	X		
	Solubility								X		X		
	Stratigraphy	X	X	X		X	X		X	X	X		
	Temperature/gradient					X				X	X		
	Volumetric water content						X			X	X		
	Water vapor pressure												
	Boiling point												
Saturated zone	Chemistry Groundwater	X	X						X			X	X
	Soil	X	X						X			X	X
	Waste	X	X						X	X	X	X	X
	Density Groundwater												
	Depth (fluctuations)	X	X						X			X	X

Parameter					
Flow direction	X	X	X	X	X
Gradients	X	X	X	X	X
Hydraulic conductivity	X	X	X	X	X
Storativity	X	X	X	X	X
Velocity	X	X	X	X	X
Henry's Law Constant					
Lithology	X				
Molecular weight					
Octonal/water coefficient (K_{ow})			X		
Organic carbon distribution coefficient (K_{oc})			X	X	X
Organic substance vapor pressure					
pH	X	X		X	X
Porosity	X	X		X	X
Solubility	X	X	X	X	X
Viscosity				X	

and environmental purposes. The proper classification of the wide diversity of rock types is beyond the scope of this book; however, a few key references are provided in the bibliography section of this chapter.

For field purposes, it is important that certain parameters are recorded, including description of degree of weathering, hardness, separation, bedding and laminal, degree of roundness and sphericity, sorting, and maturity. Regarding fractures and joints, the degree of separation, discontinuity of spacing, and description of lamination, foliation, or cleavage should also be noted. All these features play roles in assessing preferred migration pathways in the subsurface. A summary of descriptive terms for consolidated deposits is presented in Figure 5-8.

SOIL VAPOR MONITORING

During drilling, all soil samples should be subjected to field analysis using an organic vapor analyzer or similar instrumentation to assess the potential presence of hydrocarbon vapors. Many portable organic vapor meters are available that can detect hydrocarbon compounds within a minimum operating rate of 0 to 2000 parts per million (ppm) within a minimum detectable rate of 1.0 ppm (Figure 5-9). Soil vapor monitoring provides a nondiscriminatory indication of the presence of a variety of organic compounds and can be used for relative quantification of hydrocarbon, if present, and thus serves as a useful tool in the field screening of soil samples. The conventional field procedure for assessing the potential presence of hydrocarbon vapors is summarized below.

1. The soil sample is removed from the sample tube or tip of the sampler and approximately 1 cubic in. is placed in a sealable polyethylene bag with a capacity of approximately 500 mL.
2. The sample is crushed through the walls of the bag to loosen it and provide greater surface area for vapor outgassing.
3. The hydrocarbon compounds in the sample are allowed to volatilize approximately 5 min at ambient air temperature.
4. The bag is then pierced with the probe of the analyzer and vapors are drawn out of the bag by the pump in the analyzer.
5. The readings are noted from the initial insertion when the bag is collapsed. The sustained value is recorded unless there is moisture interference. In this case, the initial high reading is recorded before any moisture interference causes the reading to diminish.
6. If soil or excessive moisture is drawn into the instrument, the sample probe is thoroughly cleaned and air passed through the system until zero or background levels are attained.
7. Readings are then tabulated with the boring number and depth of the sample as noted on the field log maintained by the on-site geologist.

Figure 5-9   Typical portable organic vapor meter for field screening of hydrocarbon-affected soil vapors.

BORING LOGS AND WELL CONSTRUCTION DETAILS

Logs are usually prepared for all types of subsurface explorations. Subsurface exploration usually is accomplished with test pits or borings. When preparing logs, the same soil classifications, descriptions, symbols, etc., should be used for all logs on the same project and preferably throughout all field exploration for the same office. In general, the same types of data are recorded on all logs. Two basic types of data are available for recording in a log: "permanent" and "fugitive" data. Permanent data are of such a nature that time of recording is not critical, whereas fugitive data, if not observed and recorded during the drilling or excavation process, are lost forever. A typical boring log is shown in Figure 5-10.

Boring Logs

The heading of each log should be completed with care and in a form that can be transcribed with a minimum of editing. The following information, as a minimum, should be included on the field log heading:

1. Project number and name.
2. Drilling dates and name(s) of person(s) recording log.

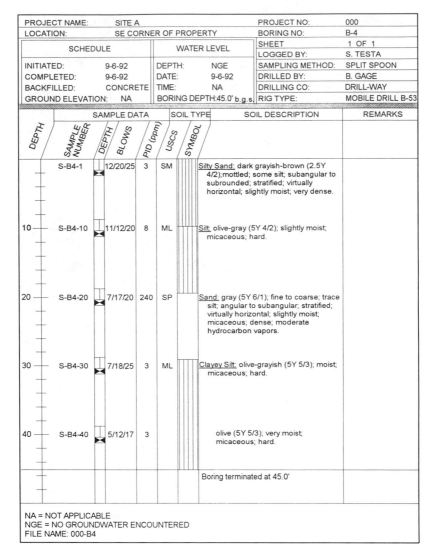

PROJECT NAME:	SITE A						PROJECT NO:	000	
LOCATION:	SE CORNER OF PROPERTY						BORING NO:	B-4	

SCHEDULE		WATER LEVEL		SHEET	1 OF 1
				LOGGED BY:	S. TESTA
INITIATED:	9-6-92	DEPTH:	NGE	SAMPLING METHOD:	SPLIT SPOON
COMPLETED:	9-6-92	DATE:	9-6-92	DRILLED BY:	B. GAGE
BACKFILLED:	CONCRETE	TIME:	NA	DRILLING CO:	DRILL-WAY
GROUND ELEVATION:	NA	BORING DEPTH: 45.0' b.g.s.		RIG TYPE:	MOBILE DRILL B-53

DEPTH	SAMPLE NUMBER	DEPTH	BLOWS	PID (ppm)	USCS	SYMBOL	SOIL DESCRIPTION	REMARKS
	S-B4-1		12/20/25	3	SM		Silty Sand: dark grayish-brown (2.5Y 4/2);mottled; some silt; subangular to subrounded; stratified; virtually horizontal; slightly moist; very dense.	
10	S-B4-10		11/12/20	8	ML		Silt: olive-gray (5Y 4/2); slightly moist; micaceous; hard.	
20	S-B4-20		7/17/20	240	SP		Sand: gray (5Y 6/1); fine to coarse; trace silt; angular to subangular; stratified; virtually horizontal; slightly moist; micaceous; dense; moderate hydrocarbon vapors.	
30	S-B4-30		7/18/25	3	ML		Clayey Silt: olive-grayish (5Y 5/3); moist; micaceous; hard.	
40	S-B4-40		5/12/17	3			olive (5Y 5/3); very moist; micaceous; hard.	
							Boring terminated at 45.0'	

NA = NOT APPLICABLE
NGE = NO GROUNDWATER ENCOUNTERED
FILE NAME: 000-B4

Figure 5-10   Typical boring log.

3. Elevation from survey reference point or from topographic map. An actual survey elevation is preferred. Elevation should be to the nearest tenth of a foot and the datum should be indicated (MSL is preferred). The elevation of borings should also be indicated on the field sketch or plan. If no survey is available, a permanent object (which will not be removed) such as the top of a curb or manhole cover edge should be used and the reference point shown on the field plan. If the elevation is approximate, so indicate.

4. Brief description of location, such as open field, next to building, etc., with plan dimensions from property line or building. If boring or test pit is not made at a pre-staked location, a simple sketch with dimensions should be prepared to define the location.

5. Groundwater information is one of the most important parts of the log. An attempt should be made to determine exactly where the water is coming from, from what stratum, and at what rate. The water level should be measured daily during the duration of the project, when possible.

6. Full name of the drilling contractor and the driller. The time the set-up was started, when drilling commenced, and when the boring was completed should be recorded. Dates should be shown, especially if drilling spans several days.

7. Description of drilling equipment and size of boring. All spaces provided in the heading for casing size, sampler used, core size and type, and tube sizes for undisturbed samples should be filled in. Additional data should be provided in the Remarks column and/or in the Daily Field Report. Make and model of drilling equipment should also be included.

8. The weight of hammer or equipment used, such as a kelly bar, inner kelly bar, and stems should be recorded when samplers are driven.

The field identification and classification of subsurface materials should be as complete and accurate as possible. The field representative should have a substantial understanding of the subsurface conditions on which to base preliminary evaluations and to recommend an effective laboratory testing program and/or screened interval, if required. A checklist to guide in the preparation of complete field soil descriptions and the recording of relevant subsurface conditions is provided below.

1. The blows required to drive each 6 in. of the standard split barrel sampler should be recorded. If there is any deviation from the specifications for the standard penetration test (SPT), the deviations should be shown on the log. On the field boring log, SPT blows are recorded for each 6 in. of drive. Any other drive test data normally would also be recorded, such as blows required to drive another type of sampler or to drive casing. When recording data from a bucket auger boring for conversion to drive energy, care must be taken to record the weight of kelly and drill stems and to identify which parts are involved in each 6-in. drive. Only the accumulated last 12 in. of drive is recorded.

2. The type and location of samples should be recorded. Bulk samples are disturbed and should be identified on logs. When disturbed samples are to be retained for alter inspection or testing, the type

container in which the sample was placed should be recorded. Disturbed samples may be retained from any sampling operation, such as an SPT sample or from wash borings, and the samples recorded. Waterproof tags or other permanent identification should be placed on all sample containers. Large bulk samples should have tags placed both outside and inside the containers. Numbering and identification of samples obtained by sampling devices may be accomplished in a number of different ways. Extreme care should be taken that the correct sample number is placed on each sample container.

3. The depth of the bottom of the sample (not sampler) is generally recorded. Measurement should be in feet with sample depth recorded to the nearest tenth.

4. The push pressure should be recorded if a sample is obtained by pushing rather than driving or rotating.

5. The percentages of gravel, sand, and fines are sometimes estimated on field logs. The assessment is visual and made on a volume basis.

6. Soil identification should be indicated in a systematic manner for each individual sample in abbreviated form as the boring progresses. Typically, information is presented sequentially with each parameter separated by a semicolon. The sequence of parameter varies throughout the industry; however, the following is commonly used: soil type (geologic significance); color; grain size; constituents; structure; moisture content; density; miscellaneous information (i.e., for hydrocarbon odors, use a term such as slight, moderate, or strong). Any additional information concerning the presence of hydrocarbon odors, visually stained soil, gravel, boulders, roots organic material, wood, or fill materials such as bricks and concrete, etc. should accompany the soil classification to provide as much information as possible concerning the materials encountered. The log should also indicate layering or stratification of the materials encountered. The term "miscellaneous fill" should be used only if a complete description of all the materials encountered in the fill is provided.

7. The color, grain size, constituents, moisture, and consistency or compactness of the soil should be described. Descriptions should also be included with other pertinent properties such as structure, plasticity, cementation, grain shape, lensing, etc.

8. Moisture content should at minimum be assessed visually and relative moisture content noted. Depth to groundwater should be noted and reference to perched, water table, or artesian conditions should also be noted.

9. The Remarks column on all logs is used to record data which do not naturally fit on other parts of the log. The Remarks column can be

the most important column on the log in that pertinent data, not otherwise logged, can be recorded.

Well Construction Details

Well construction details are typically present in a format opposite the boring log as previously described (Figure 5-11). Well construction details are finalized in the field following completion of the borehole. The need for a completed and accurate boring log is essential for proper screen and seal placement. A checklist in the preparation of well construction details is provided below.

- All dimensions should be noted, including casing and screen diameter and type, slot size, filter pack type, and grain size
- Depth of main cement bentonite seal, and depth and thickness of bentonite seals
- Well-head completion details (i.e., flush-mounted completion vs well-head monument)
- Depth of borehole vs depth of monitoring well
- Results of sieve analysis for evaluation of proper filter pack material and screen size
- Depth to water in respect to screen placement; screen length should accommodate seasonal fluctuations in water levels
- Seal placement should be opposite zones of relatively low permeability whenever possible.

SOIL VAPOR MONITORING

Soil vapor monitoring and characterization is an important aspect of any subsurface investigation where hydrocarbons are of concern. The migration of odorous and flammable vapors into excavations and basements has long been a recognized problem, but only during the past 20 years has serious scientific attention been given to the occurrence, migration, and removal of these vapors. Soil vapor monitoring and subsequent venting is one approach, under suitable conditions, that can be an effective investigative and remediation option. However, field work with volatile chemicals presents an additional set of site conditions which must be carefully evaluated to prevent personal injury or exposure to hazardous chemicals. Thorough preparation and planning should be completed prior to initiation of field work. The variety of volatile compounds and mixtures of chemicals encountered, especially in the vapor phase, can be toxic, anesthetic, or explosive.

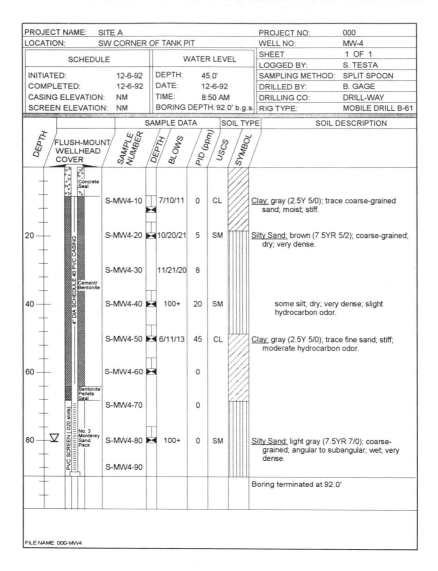

PROJECT NAME:	SITE A	PROJECT NO:	000
LOCATION:	SW CORNER OF TANK PIT	WELL NO:	MW-4

SCHEDULE		WATER LEVEL		SHEET	1 OF 1
				LOGGED BY:	S. TESTA
INITIATED:	12-6-92	DEPTH:	45.0'	SAMPLING METHOD:	SPLIT SPOON
COMPLETED:	12-6-92	DATE:	12-6-92	DRILLED BY:	B. GAGE
CASING ELEVATION:	NM	TIME:	8:50 AM	DRILLING CO:	DRILL-WAY
SCREEN ELEVATION:	NM	BORING DEPTH: 92.0' b.g.s.		RIG TYPE:	MOBILE DRILL B-61

Figure 5-11　Typical well construction detail.

Occurrence and migration of volatile contaminants in the unsaturated zone depends on many factors including the distribution of soil particle size (and mineral content), pore size and shape, percentage of saturated pores, presence of soil structure, presence of residual hydrocarbons, depth to the water table, and type of contaminant. A preliminary site investigation should include appropriate testing to quantify all of these factors along with any additional site-specific information identified during the prefield work data search.

Collection and analysis of soil gas samples as a means of locating and delineating subsurface hydrocarbon contamination has been attempted for at least 30 years. Early efforts were based on the use of explosion meters to determine the percentage of lower exposure limit (LEL) of vapor samples retrieved from soil probes. The effectiveness of these efforts was limited by the sensitivity of the instruments, since the vapor content was far below the detection limit of the instrument.

Recent advances in instrumentation, and an advanced understanding of the processes involved, have allowed development of soil vapor monitoring and surveying as a viable tool for site characterization. Two sample collection procedures are currently in common usage: static and dynamic. Static soil vapor sampling includes installation of static trapping devices on a prescribed pattern just below the land surface. A charcoal-absorbent coated wire in each is allowed to remain in place for a few days to several weeks. When removed, the collector is analyzed by Curie point desorption mass spectrometry. The resulting analysis describes the chemical characteristics of the vapor that came into contact with the collector. Dynamic vapor sampling is usually completed by inserting small-diameter probes to a predetermined depth on a planned pattern. The probes are purged by a sampling pump to assure that the vapor collected has been drawn from the surrounding soil. Both systems produce repeatable results.

Purged samples are pumped from the probes into flexible "tedlar" containers for transport to an off-site laboratory or may be pumped directly to on-site analytical instruments. Samples delivered to the laboratory are analyzed by conventional gas chromatograph (GC) or mass spectrometer (MS) procedures. On-site analysis may be completed by a wide variety of instruments which vary from true field GC units to hand-held broad spectrum photoionization detectors (PID), flame ionization detectors (FID), or similar units typically assessed for soil vapor monitoring during drilling.

Resulting data from either static or dynamic sampling is usually plotted on maps and contoured to indicate the areas of highest subsurface volatile contamination. In many cases, the resulting maps have produced very accurate descriptions of soil or groundwater contamination. Caution must be used, however, if soil vapor monitoring is the primary investigatory tool because several significant variables can affect the results of analysis. Static adsorption and Curie point analysis is an effective measure of analyzing the vapor which reached the sample point. The composition of vapor at that location may not represent the typical soil vapor composition in the soil mass because of diffusion rates, sorption, and other physical phenomena. Dynamic sampling followed by off true GC/MS analyses also represents the vapor content at the point location, but may be hindered by concentrations of carbon dioxide and oxygen which may vary from prepared standards.

Analytical laboratory procedures should be selected to represent the compounds expected or known to occur. If the product is a mixture (i.e., gasoline or technical grade paint solvents), the testing should be capable of quantifying a broad range of compounds encompassing the expected chemicals plus potential degradation products. Benzene, toluene, ethylbenzene, and total xylenes (BTEX) analysis is commonly used to indicate gasoline while it constitutes less than 25% of the fuel; thus, consideration should be given to a broader spectrum analysis such as Total Petroleum Hydrocarbon, Total Organic Carbon, and Total Organic Halogen.

Use of field FID and PID instruments to perform field screening is a rapid and economical means. At some sites, these instruments are connected directly to the soil probes and read directly. The output of these screening instruments typically is a single readout which represents an integrated response to a broad spectrum of chemical products. Each specific compound has a unique ionization response and the detection sensitivity is proportionate to the detector design. Commonly, a single composition gas in air (i.e., isobutylene at 100 ppm) and zero composition air are used to calibrate the instrument when field readings represent PID or FID "units" and not true concentrations.

Specific chemicals (or mixtures) may be outside the "optimum detection window" for the instrument used. An example is that PID instruments calibrated to isobutylene are much less sensitive to chlorinated hydrocarbons. Also, field PID response to aromatic hydrocarbons (BTEX) is diminished in the presence of elevated concentrations of carbon dioxide, high relative humidity, and in the presence of alkanes. FID hand-held field instruments rely on oxygen in the soil vapor to support combustion in the detector. Diminished oxygen content in the soil vapors or significantly elevated flammable vapor content will extinguish the detector flame response of both PID and FID units as governed by vapor flow rates. Variations in flow rate results in a lower instrument response. For this reason, more accurate data can be achieved by purging the points with a pump into a container which in turn meters the vapor into the instrument. Selection of vapor sampling probe type, depth of installation, grid spacing, detection instruments, and analytical procedures is largely a matter of judgement by the professional responsible for the project. Physical parameters such as the variety of contaminant and soil characteristics will influence the selection. Permeable soils usually are better suited for less sophisticated field testing techniques, especially if the contaminants stimulate a strong instrument response. Clayey soil or soil containing a high percentage of fine-grained materials, often do not allow the vapors to be readily transmitted throughout the soil mass. "Tight soil" tests may not actually represent the contamination present.

Soil vapor testing is a valuable qualitative tool to assist in site evaluations. When vapor concentrations are plotted on a site map and contoured, they indicate the relative intensity of the instrument response. These data are helpful to select areas for further testing or confirmation of data from soil borings or monitoring wells.

The occurrence of vapors at sampling locations is controlled by many physical and chemical variables, as discussed in Chapter 8. If all the variables were precisely defined, it would be possible to calculate the contaminant concentration in both soil and groundwater based on soil vapor analytical data. However, minor variations in the soil, such as changes in air-filled porosity, volumetric water content, soil pore shape, unsaturated zone depth, temperature, atmospheric pressure, and groundwater gradient, can alter the soil vapor content and concentrations. Actual field data (especially related to petroleum field spill studies) have resulted in a relatively high percentage of incorrect results. Compound-specific analysis (i.e., BTEX to represent gasoline) tends to yield more questionable results when benzene and toluene are used to identify gasoline contamination. False negative results (contamination present but not found) have ranged from about 14 to 37% in sands and 57 to 71% in clays. These results may represent a particular extreme situation but confirm that data should be critically evaluated. Based on this type of result, soil vapor sampling and analysis must be interpreted along with all other available data.

Suitability of a particular site for remediation by soil vapor venting is dependent on several factors:

- Average air-filled porosity of the soil
- Subsurface heterogeneities
- Areal extent of site
- Unique site properties affecting system layout
- Configuration, concentration, and variety of chemical contaminants.

If the site appears favorable, pilot testing is strongly recommended prior to design and implementation of a full-scale remedial effort.

To assess whether vapor extraction is a viable alternative for soil remediation, a pilot test is routinely conducted prior to development of a full-scale vapor extraction system. Pilot-scale stripping tests are planned to assess the soil drying rate, the vacuum required for extraction, the air flush volume, and the effective radius of influence from the extraction well. Four or five monitoring wells located at varying radii should be equipped with vacuum gauges (0.1 to 10 in. of water) to determine the shape and size of the area of influence. Shallow systems (<15 ft to bottom of extraction well) may function more efficiently if the surface is covered with an impermeable surface (asphalt, concrete or plastic sheeting) to force recharge air to enter at a perimeter location. The longer flow path causes greater exposure of fresh air to soil and thus improves mass transfer of contaminants. Pilot testing should be conducted for a period of 24 to 48 hr to assure that at least 1 pore volume has been exchanged. Procedures used are often very similar to that of groundwater pumping tests as previously discussed in Chapter 4. Initially, vapor flow is adjusted to a uniform rate and time/vacuum data is recorded for each monitoring point. Vapor quality samples are collected periodically during the test. At completion of the testing phase, data retrieved may be plotted graphically and evaluated. Design of a full-scale unit should be based on these data as well as theoretical data.

After receipt and evaluation of all available site data, the project should be reviewed to evaluate whether soil ventilation is a viable option. This objective review should be made based on the overall goals of the project. Several items of consideration are:

- Does ventilation work?
- What level of residual contamination is acceptable?
- Is time or project cost the motivating factor?
- What level of technical sophistication is required?
- What concentration of discharge is acceptable?

The very basic principle of soil venting is to create an air flow through unsaturated soil. Volatile organic compounds will be removed as they partition between the existing media and the flowing air. Operational designs range from passive venting (driven by atmospheric pressure fluctuations) to sophisticated high vacuum systems which effectively "distill" the volatile compounds from the soil. Selection of final design criteria will usually be made between these extremes.

Active systems are usually designed to fully penetrate the portion of the unsaturated zone which is to be cleaned. Venting pipes are constructed to allow maximum air entry from the designated zones, but are sealed at the surface to limit short circuiting. These 2- to 4-in.-diameter vent pipes are most often vertical (wells), but may be placed horizontally.

Most vent pipes are spaced at 15- to 40-ft intervals and are connected by a manifold to a single blower assembly to assure efficient use of equipment. Each vent head should be equipped with valve connections and gauge ports so that it may serve for flexible usage. During system operation, it may be necessary to convert an extraction vent to an inlet vent to control contaminant flow or to eliminate "stagnation" points. Header pipes are usually buried for convenience and temperature control.

Vacuum is developed by ordinary nonsparking industrial equipment which may be positive displacement, rotary, aspirator pump, or turbine. Common sizes for typical small projects are 100 to 6000 cfm with rated vacuum capacities of 5 to 30 in. of mercury. Most units are less than 10 hp. Operational costs in relation to time allotted for cleanup should be evaluated carefully. Power requirements for air movement are a function of the cube of velocity; thus, accelerated remediation can dramatically increase remediation costs.

Performance monitoring is an important factor in achieving (and verifying) a successful remediation. Sample ports, vacuum gauges and pilot tubes at all vent heads, as well as pressure and monitoring tubes, constitute a very small fraction of the overall project cost but may be a major factor in project documentation. These instruments provide the operational data necessary to fine tune the system (i.e., reduce air flow where less is needed, and increase flow where needed to reduce higher concentrations). When in doubt, measure it; when not in doubt, measure it anyway.

Vapor treatment at the exhaust port of the blower may be necessary to comply with regulatory standards. Local and state volatile organic compound (VOC) discharge limitations are highly variable. Some regulations are based on mass emissions of VOCs per day (20 to 500 lb/day and others are regulated on a compound-specific basis (benzene for example). Selection of a vapor treatment system will depend on discharge standards. Commercially available vapor treatment units include vapor phase carbon cells, catalytic oxidation units, and specially designed internal combustion engines.

BIBLIOGRAPHY

Acker, W.L., 1974, *Basic Procedures for Soil Sampling and Core Drilling:* Acker Drilling Company, Scranton, PA, 246 p.

Aller, L., et al., 1989, *Handbook of Suggested Practices for the Design and Installation of Ground-Water Monitoring Wells:* National Water Well Association, Dublin, OH, 398 p.

Barvenik, M.J. and Cadwagan, R.M., 1983, Multi Level Gas Drive Sampling of Deep Fractured Rock Aquifers in Virginia: *Ground Water Monitoring Review:* Vol. 3, No. 4, pp. 34-40.

Black, W.M., Smith, M.R. and Patton, F.D., 1986, Multiple-level Groundwater Monitoring with the MP System: In Proceedings of the National Water Well Association-American Geophysical Union Conference on Surface and Borehole Geophysical Methods and Groundwater Instrumentation: Denver, pp. 41-61.

Bohn, H., McNeal, B. and O'Connor, G., 1985, *Soil Chemistry:* John Wiley & Sons, Second Edition, New York, 341 p.

Burmister, D.M., 1951, Identification and Classification of Soils - An Appraisal and Statement of Principles: American Society of Testing Methods Special Technical Publication No. 113.

Burmister, D.M., 1970, Suggested Methods of Test for Identification of Soils: American Society of Testing Methods, Special Technical Publication No. 479.

California State Water Resources Control Board, 1989, LUFT Field Manual.

Campbell, M.D. and Lehr, J.H., 1973, *Water Well Technology:* McGraw-Hill, New York, 681 p.

Cassagrande, A., 1948, Classification and Identification of Soils: USCE Transactions, Paper No. 2351.

Cherry, J.A. and Johnson, P.E., 1982, A Multilevel Device for Monitoring in Fractured Rock: *Ground Water Monitoring Review,* Vol. 2, No. 3, pp. 41-44.

Davison, C.C., 1984, Monitoring Hydrogeological Conditions in Fractured Rock at the Site of Canada's Underground Research Laboratory: *Ground Water Monitoring Review,* Vol. 4, No. 4, pp. 95-102.

Dewitt, D.A., et al, 1987, *Soil Gas Sensing for Detection and Mapping of Volatile Organics:* National Water Well Association, Dublin, OH, 270 p.

Dickinson, W., Dickinson, R.W., Mote, P.A. and Nelson, J.S., 1987, Horizontal Radials for Geophysics and Hazardous Waste Remediation: In Proceedings of the Hazardous Materials Control Research Institute Superfund '87, pp. 371-375.

Douglas, B.J. and Olsen, R.S., 1981, Soil Classification Using Electronic Cone Penetrometer: ASCE STP, Cone Penetration Testing and Experience, Geotechnical Engineering Division, St. Louis, 57 p.

Driscoll, F.G., 1989, *Groundwater and Wells:* Johnson Filtration Systems, St. Paul, pp. 20-24.

Dyroff, G.V., 1989, *Manual on Significance of Test for Petroleum Products,* 5th Edition: American Society for Testing and Materials, Philadelphia, 169 p.

Everett, L.G., 1980, *Ground-water Monitoring:* General Electric Company, Technology Marketing Operation, Schenectady, NY, 440 p.

Exploration Logging, Inc., 1979, Field Geologist's Training Guide, pp. 4-6.

Folefoe, A.N., Archer, J.S. and Issa, R.I., 1991, Effects of Pressure Drop Along Horizontal Wellbores on Well Performance: In Proceedings of the Society of Petroleum Engineers Offshore Europe Conference, Aberdeen, pp. 549-560.

Goddard, E.N., et al., 1948, *Rock-Color Chart:* Geological Society of America, Boulder, CO.

Handman, E.H., 1983, Hydrologic and Geologic Aspects of Waste Management Disposal - A Bibliography of Publications by U.S. Geological Survey Authors, 1950-1981: U.S. Geological Survey Circular 907, 40 p.

Hanson, E.A. and Harris, A.R., 1974, A Groundwater Profile Sampler: *Water Resource Research,* Vol. 10, No. 2, pp. 375.

Hewitt, A.D., Miyares, P.H., Leggett, D.C. and Jenkins, T.P., 1992, Comparison of Analytical Methods for Determination of Volatile Organics in Soils: *Environmental Sciences and Technology,* Vol. 26, pp. 1932-1938.

Johnson, T.L., 1983, A Comparison of Well Nests Vs Single-Well Completions: *Ground Water Monitoring Review,* Vol. 3, No. 1, pp. 76-78.

Joshi, S.D., 1991, *Horizontal Well Technology:* Penhill Books, Tulsa, OK.

Keely, J.F. and Boateng, K., 1987a, Monitoring Well Installation, Purging and Sampling Techniques, Part 1 - Conceptualization: *Ground Water,* Vol. 25, No. 3, pp. 300-313.

Keely, J.F. and Boateng, K., 1987b, Monitoring Well Installation, Purging and Sampling Techniques, Part 2 - Case Histories: *Ground Water,* Vol. 25, No. 4, pp. 427-439.

Lichtenberg, J.J., Winter, J.A., Weber, C.I. and Fradkin, L., 1988, Chemical and Biological Characterization of Municipal Sludges, Sediments, Dredge Spoils, and Drilling Mud: American Society for Testing and Materials (ASTM) Special Publication No. 976, 512 p.

Lieberman, S.H., Inman, S.M. and Stromvall, E.J., 1987, Fiber Optic-Fluorescence Sensors for Remote Detection of Chemical Species in Seawater: In *Proceedings of the Symposium on Chemical Sensors,* Electrochemical Society, Pennington, NJ, pp. 467-473.

Losonsky, G. and Beljin, M.S., 1992, Horizontal Wells in Subsurface Remediation: In Proceedings of the Hazardous Materials Control Research Institute HMC-South '92 Conference, New Orleans, pp. 75-80.

Lurk, P.W., et al., 1991, Field Investigation of a Waste Oil Disposal Area Using Innovative Cone Penetrometer Technology: In Proceedings of the Hazardous Materials Control Research Institute National Research and Development Conference on the Control of Hazardous Materials, Anaheim, CA, pp. 124-126.

Marrin, D.L., 1988, Soil-Gas Sampling and Misinterpretation; *Ground Water Monitoring Review,* 8(2), pp. 51-54.

Marrin, D.L. and Kerft, H.B., 1988, Soil-Gas Surveying Techniques; *Environmental Science and Technology,* 22(7), pp. 740-745.

Munsell Color, 1988, *Munsell Soil Color Charts:* Munsell Color, Baltimore.

Rehtlane, E.A. and Patton, F.D., 1983, Multiple Port Piezometers vs. Standpipe Piezometers - An Economic Comparison: In Proceedings of the National Water Well Association Second National Symposium on Aquifer Restoration and Ground-Water Monitoring, Worthington, OH, pp. 287-295.

Senn, R.B. and Johnson, M.S., 1987, Interpretation of Gas Chromatographic Data in Subsurface Hydrocarbon Investigations; *Ground Water Monitoring Review,* 7(1), pp. 58-63.

Shacklette, H.T. and Boerngen, J.G., 1984, Element Concentrations in Soils and Other Surficial Materials of the Conterminous U.S.: U.S. Geological Survey Professional Paper No. 1270, 105 p.

Society of Petroleum Engineers, 1962, *Petroleum Production Handbook, Vol. 2, Reservoir Engineering:* (Edited by Frick, T.C. and Williams, R.W.), Society of Petroleum Engineers of American Institute of Mining Engineers, Dallas.

Soil Science Society of America, 1987, *Glossary of Soil Science Terms:* Soil Science Society of America, Madison, WI, 44 p.

Sparks, D.L. (Editor), 1986, *Soil Physical Chemistry:* CRC Press, Boca Raton, FL, 308 p.

Testa, S.M. and Winegardner, D.L., 1991, *Restoration of Petroleum-Contaminated Aquifers:* Lewis Publishers, Boca Raton, FL, 269 p.

The Asphalt Institute, 1978, Soils Manual for the Design of Asphalt Pavement Structures: Manual Series No. 10 (MS-10).

U.S. Army Corps of Engineers, 1953, The Unified Soil Classification System: USACE Technical Memorandum No. 3-357.

U.S. Environmental Protection Agency, 1981, Handbook - Stabilization Technologies for RCRA Corrective Actions: EPA/625/6-91/026, 62 p.

U.S. Environmental Protection Agency, 1991, Description and Sampling of Contaminated Soils - A Field Pocket Guide: Report No. EPA/625/12-91/002, 122 p.

Zapico, M.M., Vales, S. and Cherry, J.A., 1987, A Wireline Piston Core Barrel for Sampling Cohesionless Sand and Gravel Below the Water Table: *Ground Water Monitoring Review,* Vol. 7, No. 3, pp. 74-82.

6   GEOPHYSICAL
APPLICATIONS

". . . Obviously there are no well qualified students of the Earth, and all of us, in different degrees, dig our own small specialised holes and sit in them."

(Bullard, 1960)

INTRODUCTION

The application of geophysical techniques has been successfully used for subsurface characterization of geologic and hydrogeologic conditions, detection and delineation of certain contaminant plumes, detection of buried debris, tanks, utilities, buried waste, etc. and provides information to best situate boreholes, monitoring wells, or sampling locations. Geophysical applications are also well suited for subsurface investigation since it is nonobstructive; site disturbance may pose significant health and safety risks. Geophysical techniques are available such that no subsurface penetration of equipment or instrument is required, referred to as surface geophysics. In addition, if boreholes or wells exist, then other geophysical techniques are available and referred to as downhole geophysics.

Combining the results of several techniques can prove cost effective and provide a powerful interpretive tool for focusing and minimizing subsequent exploration and remediation activities. To allow for characterization of anomalous subsurface conditions, the appropriate technique or combination of techniques must be developed for an effective assessment strategy. Compositing of maps using several techniques in conjunction with reference to historical documentation identifying past site land-use activity, can then be used to assist in assessment-related activities and chemical testing program.

Presented in this chapter is discussion on surface and downhole geophysical techniques, respectively, as they apply to subsurface environmental assessment and characterization activities.

SURFACE GEOPHYSICAL TECHNIQUES

These techniques, their applications, and their limitations are summarized in Table 6-1 and discussed below.

Ground Penetrating Radar

Ground penetrating radar (GPR) is used to characterize shallow subsurface anomalies and to refine data generated from other systems to locate and track buried utilities. The radar energy reflected from various subsurface materials depends upon the contrast in electrical properties (which are dependent upon the mineral constituents present), density, and water content of those materials.

GPR employs radio waves that are projected, reflected, and then recovered at an antenna moving across the surface of the ground (Figure 6-1). Variations in the return signal are continuously recorded to produce a continuous cross-sectional profile of shallow subsurface conditions (Figures 6-2 and 6-3).

For favorable resolution and effective depth, sufficient differences in the electrical properties (i.e., dielectric and conductive) of the soil horizon and overlying soils must exist at the interface, which allows for the scattering or reflection of the radio waves. Overall penetration is site specific and best in dry, sandy, or rocky areas (i.e., 30 ft or less). The water table can be detected in coarse-grained materials. GPR is limited in certain clay soils reflecting substantial attenuation in highly conductive soils or in areas of saline groundwater (i.e., 3 ft or less).

Electromagnetic Induction

Electromagnetic induction (EM) is used to measure the apparent conductivity of the soil to a depth of about 20 ft, responds to electrical conductivity contrasts such as buried pipes, and assessment of variations in soil conductivity, notably in the presence and extent of certain dissolved constituents in groundwater. Inorganic constituents can increase the specific conductance of pore fluids (i.e., increase in EM), whereas organic fluids can decrease the specific conductance in pore fluids (i.e., decrease in EM). Thus, when compared to background levels, the lateral and vertical extent of contaminant plumes can be delineated, and direction of groundwater flow, gradients, and migration rates can be assessed using time series measurements. In porous media with a high moisture content, conductivity of the system will be controlled by the liquid. EM induces a current in the ground by an alternating current in a coil with magnitude and individual current phases measured by a receiving coil (Figure 6-4). Conductivity of the material reflects upon the materials, chemical makeup, porosity, permeability, and moisture content. This technique uses spatial changes in subsurface EM to infer subsurface features. Sensitivity is inferred from localized spatial differences in conductivity, not on absolute values measured. Depth is a function of the properties of the background materials, soil moisture, and relative difference in conductive properties between the target and background. Typically, areas of terrain conductivity greater than 100 millimhos/m are considered anomalous. As shown in Figure 6-5, large areas of this site were determined to fall within this category and may represent shallow layers of soil saturated with groundwater of high specific conductance or locations of buried conductive wastes and debris. General terrain conductivity conditions and the signature of a utility trench containing a buried steel pipe are shown in Figure 6-6. The geophysical signatures (EM-31 inphase and vertical magnetic gradient contour maps) of an area where several drums were buried are presented in Figure 6-8. This technique is susceptible to manmade disturbances such as power lines, fences, buried cables, rebar, etc.

Resistivity

Resistivity is used to measure the electrical resistivity, the reciprocal of conductivity (as measured using EM). Basic operating principles are thus similar to that of EM. An electrical current is injected using a pair of surface electrodes in a known geometry and the resulting voltage is measured at the surface between a second pair of electrodes. Subsurface resistivity is from the separation and geometry of the electrode positions, applied current, and measured voltage. Since most soil and minerals are electrical insulators (i.e., highly resistive), current flow is primarily through moisture-filled pore spaces. Resistivity is consequentially based on porosity, permeability, amount of pore water, and concentration of dissolved solids in pore water. Controlling variables, detection capabilities, sensitivity, and limitations are overall similar to EM.

Seismic

Seismic techniques, whether refraction or reflection, employ a seismic wave from an acoustic source and measures the subsequent travel time of seismic waves. Subsurface features can be inferred by the analysis of the travel time of the direct and refracted waves. Sources include a sledge hammer, explosives, or other method for deeper or special applications. Geophones implanted in the

Table 6-1 Summary of Surface Geophysical Techniques, Applications, and Limitations

Geophysical Technique	Physical Properties Measured	Typical Range of Values			Depth of Penetration (in feet)	Application	Limitations
		Parameter	Range	Units			
Electromagnetic induction	Electrical conductivity and resistivity	Conductivity	10^4–$10^{-2\,a}$	millimhos/m$_3$	200 to several 1000s	Soil and rock identification; Dissolved solids in groundwater	Susceptible to manmade conducting disturbances such as power lines, fences, buried cables, and rebar
Gravity	Density	Magnetic susceptibility	$10^{-6} = 10^{4\,b}$	CGS	Several 100s	Buried metal objects; Magnetic content	Very slow; Requires extensive data reduction; Sensitive to ground vibrations
Ground penetrating radar	Dielectric constant (10–1000 MHz$_2$)	Relative dielectric constant	1–100^c	E/E_o	100 (<30)	Water content; Density; Frequency; Buried objects	Substantial attenuation in highly conductingsoils; unsuitable in certain clay soils and in areas of saline groundwater
Magnetometry	Magnetic susceptibility	Density	0–3^d	g/cm^3	Several 100s	Soil and rock identification; Cavities; Porosity; Degree of saturation	Can only detect ferromagnetic substances; Susceptible to surface or near-surface clutter
Seismic reflection	Soil and rock identification	Seismic velocity and impedance	10^2–$10^{4\,e}$	Velocity	Several 1000s	Density; Degree of saturation; Cementation	Shallow surveys <100 ft critical; Sensitive to ground vibrations

Seismic refraction	Seismic velocity and impedance	Velocity	10^2–10^4 [e]	m/s	Several 100s	Density Degree of saturation Cementation	Best suited for defining layers Not capable of detecting small-scale inclusions Sensitive to ground vibration

[a] Seawater (<10^4 but >10^{-3}); clay-shale, brine saturated formations (<10^{-3} but >10^{-2}); clays ($\approx 10^{-2}$); sand and gavels (<10^1 but >10^0); massive low porosity rocks ($\approx 10^{-1}$).

[b] Sedimentary rocks (<10^{-6} but > 10^{-5}); basalt intrusions ($\approx 10^3$); magnetite(<10^{-1} but >10^0); iron alloys (> 10^1).

[c] Air, metals ($\approx^1$); ice, dry soils and rocks (>1 but <10); saturated earth materials (>10 but <50); water ≤ 100).

[d] Air-filled voids (0); water-filled voids (1); unsaturated soil (>1 but <2); saturated soil ($\approx$ 2); massive rocks (>2 but <3).

[e] Dry soil (>10^2 but < 10^3); saturated soil ($\approx 10^3$); shales (>10^3 but < 10^4); massive rocks (< 10^4).

Figure 6-1   Technician field use of GPR (photo courtesy of Spectrum, Inc.).

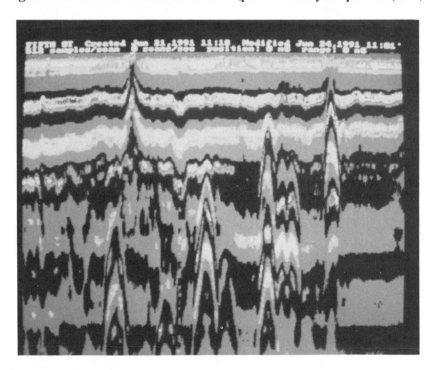

Figure 6-2   Typical computer printout of GPR profile (photo courtesy of
              Spectrum, Inc.).

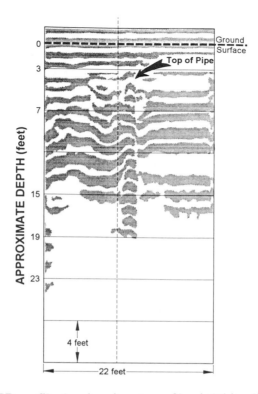

Figure 6-3   GPR profile showing signature of buried 6-in.-diameter pipe.

Figure 6-4   Technician field use of EM-31 (photo courtesy of Spectrum, Inc.).

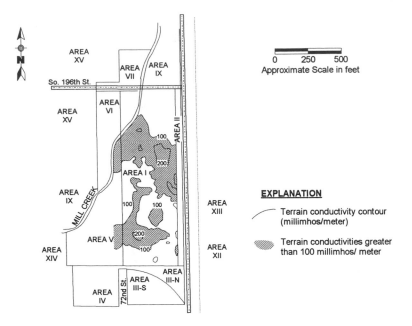

Figure 6-5   **Terrain conductivity map showing areas of anomalously high terrain conductivities which may represent groundwater contamination and/or areas of high metal concentrations in the underlying soils (after Testa, 1989).**

ground translate seismic vibrations into an electrical signal which is then displayed on the seismograph. Seismic refraction is commonly applied to investigations of up to a few hundred feet. Resolution and depth of measurements are dependent upon spacing of the geophones. Seismic reflection is effective to depths of a few thousand feet.

Information generated is similar to that of GPR. Although not commonly used for assessment of hazardous waste sites, seismic studies are very useful for investigations at deep-seated geologic repositories. The type of geologic and hydrogeologic data generated includes top of bedrock, stratigraphy, architectural structure (i.e., channeling), and structural elements (i.e., faults, fractures, cavities). Although best suited for defining layering, seismic is not suitable for detecting small-scale inclusions and is susceptible to vibration noise from other natural and cultural sources.

Gravity

Gravity surveys are not commonly employed, but can provide data on regional geologic features over hundreds of square miles and on major fractures and cavities in rocks. The Bouguer anomaly, the difference ˙ ˛tween

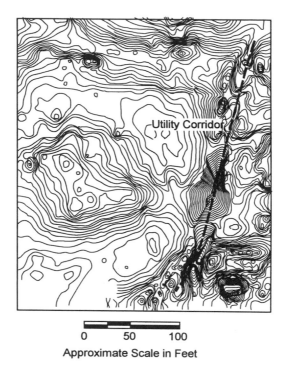

0   50   100
Approximate Scale in Feet

**Figure 6-6   General terrain conductivity contour map showing a signa-
ture of a buried utility trench containing a steel pipe (after
Testa, 1989).**

observed and theoretical gravity, is measured reflecting the changes in the
density of the soil and rock. The measurements are made up of deep-seated
effects (regional Bouguer anomaly) and shallow effects (local Bouguer anomaly).
The regional Bouguer anomaly has merit in regional studies associated with
geologic repositories, while the local Bouguer anomaly may be useful for
localized studies.

Two basic types of gravity surveys exist: standard gravity and microgravity.
A standard gravity survey uses widely spaced stations (100- to 1000-ft intervals) and a standard gravity meter to cover large distances (typically via use of
an airplane). A microgravity survey uses a microgravimeter which is capable
of measuring small differences in the gravitational field for delineating and
mapping local anomalies (i.e., bedrock channels, fractures, cavities, etc.).

Magnetometry

A magnetometry (MAG) survey, including both total field and vertical
gradient, measures variations and disturbances in the Earth's total magnetic

Figure 6-7   Technician field use of magnetometer (photo courtesy of Spectrum, Inc.).

field strength created by ferromagnetic objects (i.e., iron and steel) (Figure 6-7). The presence of subsurface regions of elevated iron content can thus be detected based on localized perturbations caused by the Earth's magnetic field. The MAG's response is proportional to the mass of the ferrous target and inversely proportional to the cube of the distance to the target. The magnetic gradiometer has greater lateral resolution than that of a total field survey; thus, buried objects which appear as a single composite total field anomaly can often be individually identified on vertical profiles or maps (Figure 6-8). This greater spatial resolution thus reduces the area of search at individual anomalies. Magnetometry has sufficient sensitivity to detect anomalies as deep as about 60 ft, but is limited to ferromagnetic substances (i.e., metal drums, pipes, etc.) or ferromagnetic-rich waste disposal units (i.e., buried trenches or waste units), and susceptible to surface and near-surface clutter.

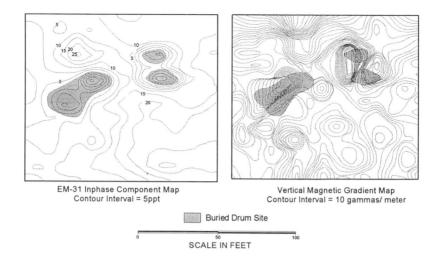

EM-31 Inphase Component Map
Contour Interval = 5ppt

Vertical Magnetic Gradient Map
Contour Interval = 10 gammas/ meter

Buried Drum Site

0   50   100
SCALE IN FEET

Figure 6-8   **Vertical magnetic gradient and inphase component maps showing anomaly signatures at an area containing multiple buried drums (after Testa, 1989).**

DOWNHOLE GEOPHYSICAL TECHNIQUES

Conventional downhole geophysical logging techniques can prove useful and, in some cases, essential to characterization of subsurface conditions at sites of environmental concern. Typically used for purposes of stratigraphic correlation and in evaluation of subsurface conditions in a well where no geologic log is available, its use as part of the characterization process required under certain regulatory programs which demand in-depth understanding of the hydrogeologic environment prior to demonstration of the adequacy of a groundwater monitoring system has been limited. Downhole techniques can be used for the purpose of identification and correlation of geologic strata, thus providing information on the uniformity of subsurface conditions and evaluate conditions when no samples are available. Zones of varying permeability (i.e., sand lenses, fractures, solution cavities and voids, etc.) can be assessed.

Identification of water table and other water-bearing zones can be ascertained, thus allowing for proper well design and screen placement. This becomes increasingly important in low-permeability formations where equilibration is slow, ranging from a few days to several weeks. This lack of immediate response may hinder identification of water-bearing zones in such environments, as well as selection of screened intervals for subsequent monitoring. Certain techniques can be used as a supportive tool to set proper screen interval(s) without the borehole remaining uncased for an undefined period of time, increasing the potential for caving, thus possibly preventing the construction of a well. Downhole terminology and relationships are illustrated in Figure 6-9.

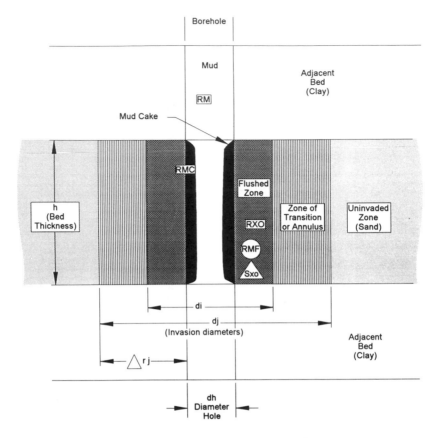

Figure 6-9   Downhole terminology and relationships.

Geophysical data generated can be digitized in the field and computer processed allowing for amplification of stratigraphic details, and thus improve resolution of geophysical responses and automatic scale factor selection. Digital processing also allows for the minimizing of operator-related errors. After computer processing and plotting, the digital records are compared with the field-copy analogs as a quality-assurance measure and identification of potential instrumentation problems. The digital logs also afford the interpreter the opportunity to plot any group of logs or combination of logs at any desired scale for analysis (Figure 6-10). These techniques and associated criteria are summarized in Table 6-2 and discussed briefly below.

Nuclear Logs

Nuclear logs include those logging techniques that either generate unstable isotopes in the vicinity of a borehole or detect the presence of such isotopes.

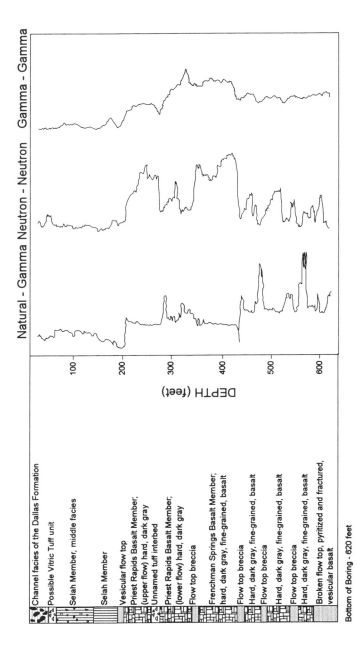

Figure 6-10 Composite downhole geophysical logs showing natural gamma, neutron-neutron, and gamma-gamma response (after Testa, 1988).

Table 6-2 Summary of Downhole Geophysical Techniques and Criteria

Logging Type	Logging Tool	Property Measured	Application	Saturated or Water-Filled Hole			Unsaturated Zone or Dry Hole			Radius of Measurement	Limitations
				Uncased	PVC	Steel	Uncased	PVC	Steel		
Acoustic	Television	Visual image	Visual image; fracture and cavity elevation; structure evaluation; casing inspection	Yes	Yes[a]	Yes[a]	Yes	Yes[a]	Yes[a]	Within borehole and surficial borehole wall	Require fluid; slow; heavy mud attenuates signal
	Velocity	Compressional wave velocity	Soil and rock identification; soil and rock porosity; fracture detection; elastic moduli determination	Yes	No	No	No	No	No	Within borehole and surficial borehole wall	Requires fluid; secondary porosity not detected; requires sophisticated analysis
Caliper	Oriented multiarm bow spring	Borehole or casing diameter	Borehole-diameter log corrections; large fracture and cavity determination	Yes	No	No	No	No	No	2–3 ft	None; crooked holes; variable resolution between tools
Electric	Dipmeter		Fracture location and orientation; bedding orientation	No	No	No	Yes	No	No	Surficial borehole wall	Requires conductive fluid
	Induction		Soil and rock identification; soil and rock porosity; pore fluid conductivity	Yes	Yes	No	Yes	Yes	No	Surficial borehole wall	Requires nonconductive fluid

Method	Measures	Applications						Location	Remarks
Normal resistivity	Formation resistivity, fluid resistivity	Soil and rock identification; soil and rock porosity; pore fluid resistivity	Yes	No	No	No	No	Surficial borehole wall	Requires conductive fluid; borehole clean water and mud affects significant
Single-point resistivity	Formation resistance, fluid resistance	Rock resistance; fluid resistance; fracture location	Yes	No	No	No	No	Surficial borehole wall	Requires conductive fluid; borehole diameter and mud affects significant
Focused resistivity	Formation resistivity, fluid resistivity	Lithology; interstitial fluid salinity	Yes	No	No	No	No	Surficial borehole wall	Requires conductive fluid
Lateral resistivity	Formation resistivity, fluid resistivity	Lithology; interstitial fluid salinity	Yes	No	No	No	No	Surficial borehole wall	Requires conductive fluid (i.e., saline muds)
Spontaneous potential (SP)	Electric potential reflecting salinity differences in interstitial borehole fluids	Soil and rock identification; electric potential caused by submitting differences	Yes	No	No	No	No		Requires conductive fluid; correct for NaCl fluids only; salinity difference required between interstitial and borehole fluids
Fluid — Conductivity	Fluid resistivity	Fluids and flow characterization; casing leak location; water quality determination	Yes	Yes[a]	Yes[a]	No	No	Within borehole	Requires fluid; variable accuracy; requires temperature correction

Table 6-2 Summary of Downhole Geophysical Techniques and Criteria (continued)

Logging Type	Logging Tool	Property Measured	Application	Saturated or Water-Filled Hole			Unsaturated Zone or Dry Hole			Radius of Measurement	Limitations
				Uncased	PVC	Steel	Uncased	PVC	Steel		
	Flow			Yes	Yes[a]	Yes[a]	No	No	No	Within borehole	Requires fluid; requires centralization; relatively high velocities required for use of spinners
	Temperature	Temperature of borehole fluid	Temperature of borehole fluid near sensor; fluid and flow characterization; casing leak location	Yes	Yes[a]	Yes[a]	Yes[a]	No	No	Within borehole	Requires fluid; variable accuracy and resolution between probes
Nuclear	Gamma gamma	Electron density	Bulk density determination; lithology; casing leak location; porosity; moisture content; depth to water table	Yes	Yes	Yes	Yes	Yes	Yes	6 in.	Optimum results in uncased hole; otherwise qualitative; severe borehole diameter effects
	Natural gamma	Gamma radition from radioisotopes	Soil and rock identification; soil and rock porosity; pore resistivity; lithology	Yes	Yes	Yes	Yes	Yes	Yes	6–12 in.	Optimum results in uncased hole; otherwise qualitative; excludes large diameter boreholes and casing strings and cement

204

Neutron neutron	Hydrogen content	Moisture content above water table; total porosity below water table; saturated porosity; unsaturated moisture content; depth to water table; activation analysis; lithology	Yes	Yes[a]	Yes[a]	Yes	Yes[a]	Yes[a]	6–12 in.	Optimum results in uncased hole; can be calibrated for casing; borehole diameter and chemical effects

[a] Limited application.

Nuclear logs are an important investigative tool in subsurface geologic, hydrogeologic, and environmental studies due to use in cased holes regardless of the type of fluid in the borehole. The major types of nuclear logging techniques are discussed below.

Natural Gamma Logs

The natural gamma (or gamma ray) logging technique is the most commonly used log for the purpose of lithology, and subsequent stratigraphic correlation and relative permeability. This technique measures naturally occurring total gamma radiation from the rocks penetrated by the borehole. The sonde used does not discriminate between gamma radiation from naturally occurring isotopes of potassium (K^{40}), thorium (Th^{232}), and uranium (U^{238}), although the record of gamma radiation depends upon the naturally occurring radioisotopes present. The average content of K^{40}, U^{238}, and Th^{232} based on 200 shale samples analyzed is on the order of 2% of the total K content in total volume, 6 ppm, and 12 ppm, respectively.

Natural gamma logging techniques can be used in either cased wells or in open boreholes containing air, water, or drilling fluid. Thus, this technique is extremely useful in logging wells where the original geologic log is either lost or destroyed.

Radioactive isotopes are encountered in numerous rock and soil types, notably shale. In general, clays and shales contain relatively high concentrations of radioactive isotopes, notably potassium, and thus emit very high levels of radiation. Sulfates (i.e., anhydrite and gypsum), coal, basic igneous rocks (i.e., basalts and gabbros), sandstones, and carbonates typically emit very low levels of radiation; however, responses in sandstones may vary depending on mineral and aggregate constituents. For example, high silica content will emit a low response; whereas, if the sandstone contains potassium feldspars, micas, glauconite, or uranium-rich water, then a relatively high response will be emitted. Conversely, silicic igneous rocks (i.e., rhyolite, dacites, granites, etc.) and potash and phosphate beds contain relatively higher amounts of radioisotopes and thus emit moderate to high levels of radiation.

Factors that must be considered in interpretation of natural gamma logs include mineral assemblages, vertical and lateral facies changes, grain size, alteration and decomposition processes, and stratification (i.e., bed thicknesses). Borehole diameter and differences in water quality (i.e., saltwater vs freshwater) has insignificant to negligible effects on interpretation of natural gamma logs.

Gamma-Spectrometry Logs

Gamma-spectrometry logs have not typically been used for environmental studies, although they provide more diagnostic lithologic information than gamma-gamma logs. This type of log can be used to identify both natural and artificial radioisotopes in groundwater. The source(s) of all three radioactive naturally occurring radioisotopes can be determined, in addition to identification of artificial radioisotopes that might contribute to groundwater contamination. Gamma spectrometry has been used to relate radioisotope concentrations to mineralogical composition, identify fractured and altered intervals in a geothermal well penetrating sedimentary rocks and in a nongeothermal well penetrating igneous rocks, and has considerable application to monitoring of radioisotopes at hazardous and radioactive waste disposal sites and uncontrolled waste sites.

Gamma-Gamma Logs

Gamma-gamma logs (or density logs) provide a record of the radiation received at a detector from a gamma source situated in the probe. However, in contrast to the natural gamma log that records naturally occurring radioisotopes, gamma-gamma logs record radioisotopes after they have been attenuated and scattered in the borehole and surrounding formation.

Gamma-gamma logging is a gamma-activated gamma geophysical logging technique which provides a porosity log that measures the electron density of a formation. A detector in the sonde counts the rate of backscattered gamma radiation induced by a medium energy, 50-μCi cobalt-60 source. The density of a rock is proportional to the electron density. Backscattering of gamma rays (Compton scattering), which refers to the interaction between incoming gamma ray particles and electrons in the formation, is also proportional to electron density. Thus, the count rate of backscattered gamma rays can be related to the bulk density of the rock.

The gamma-gamma log is sometimes referred to as a density log because density is the fundamental characteristic inferred from the log. In general, the higher the density, the lower the porosity. This type of log can also be used to calculate porosity when fluid and grain densities are known, as follows:

$$\text{Porosity } (\phi) = \frac{(\text{grain density} - \text{bulk density})}{(\text{grain density} - \text{fluid density})} \tag{6-1}$$

Bulk density is derived from a calibrated and corrected gamma-gamma log. Fluid density for rocks saturated with freshwater is on the order of 1 g/cm^3, but as high as 1.1 g/cm^3 for rocks saturated with brine. Grain and mineral density is readily attainable from mineralogy textbooks (i.e., 2.65 g/cm^3 for quartz). Gamma-gamma logs can also be used to locate cavities or unfilled annular space behind casing and the top of the cement.

Since bulk density is affected by moisture content, changes in moisture content and location of the water table can be determined. Neutron-neutron logs as discussed later are more accurate for this purpose although both being utilized is preferred. Specific yield can also be estimated by runs before and after drawdown during a pumping test.

Above the static water level, a decrease in the count rate represents an increase in density (Figure 6-10). This is because most Compton scatter will occur in loosely bound electrons such as those in water. Below the water table, an increase in the count rate represents an increase in density. Caving of the borehole may appear as an apparent decrease in density; however, observation of the natural gamma log previously discussed can help avoid such misinterpretation.

Neutron-Epithermal Neutron Logs

Neutron logs are generated by providing a source of neutrons in the probe and detectors that record the neutron interactions occurring in the vicinity of the borehole. Two different types of neutron logging techniques are available: neutron probes with a large source and long spacing used for measuring saturated porosity, and probes with a small source and short spacing used for measuring moisture content in the unsaturated zone.

The neutron-epithermal neutron logging technique is used primarily as an indication of total porosity under saturated conditions and measures the hydrogen ion (H^+) concentration in a formation. In formations where the porosity is filled with water, the neutron log measures liquid-filled porosity. This technique is used to measure the epithermal neutrons produced by the fast neutron emission from a 3-Ci AmBe (mixture of americium and beryllium) source. The epithermal neutrons are degraded in energy by collision with nuclei of atoms in wall rocks. H^+ has an atomic radius near to that of the epithermal neutrons and absorbs them much more strongly than other atomic nuclei. Consequently, this log is a direct measure of the H^+ content of the rock which is proportional to porosity and composition.

Above the static water level, epithermal neutrons are absorbed by H^+ bound in crystal lattices; consequently, the higher the count rate, the higher the percent of crystalline material and the lower the porosity. Water strongly absorbs the fast neutrons emitted by the 3-Ci AmBe source used for the neutron-epithermal neutron (porosity) logging and backscattered gamma rays from the 50 microcurie cobalt-60 source for the gamma-activated gamma (density) logging of the borings. Thus, in rocks or soils with any measurable porosity, the saturation depth (unconfined hydrostatic level) is readily apparent by a sharp decrease in the counting rate of epithermal neutrons and backscattered gamma radiation (Figure 6-10). Consequently, the amount of hydrogen in the formation is high, and thus porosity is high.

Although the primary use of neutron logs is for porosity and moisture content, lithology subsequent stratigraphic correlation can also be provided when used in conjunction with a gamma-gamma log or acoustic velocity log. Neutron logs can also provide information on specific yield of unconfined aquifers by measuring the moisture content of saturated material before and after conducting a pumping test, and porosity differences relating to well development (increase) or plugging during artificial recharge (decrease). Overall, neutron logs are more suitable for detecting small porosity changes when porosity is minimal, whereas gamma-gamma logs are more sensitive to small changes in porosity when porosity is large.

Neutron logs can be very useful in determining the depth to the water table in low permeability soil and sedimentary rocks. Wells in such settings may take days to several months as indicated in one of several wells which penetrate a heterogeneous anisotropic, tuffaceous siltstone with the consistency of a sandy silt (Figure 6-11). The slow recovery observed is a manifestation of permeability contrasts and the extremely low hydraulic conductivity of the formation as supported by numerous laboratory permeability tests conducted on undisturbed samples and conduct of several pumping tests. Evaluation of geophysical logs for wells at the site which had the longest time to equilibrate provided a fairly accurate measurement of the depth to the top of the uppermost zone of saturation as apparent on the gamma-gamma and the neutron-neutron logs (Figure 6-12). Review of their respective geophysical logs exhibited unequivocable gamma-gamma and neutron-neutron picks for the water table. Thus, it appears that the levels estimated from geophysical logging are reliable estimates and can be used for wells that have not stabilized. The identification of these levels were invaluable in development of contour maps within a timely manner and subsequent demonstration of the adequacy of these maps for development of a groundwater monitoring system.

Of significance is the excellent agreement noted between static water levels estimated from the logs and those measured in the field for all logged wells which stabilized after drilling (Table 6-3; Figure 6-13). Correlation is commonly within a few tenths of a foot. Certain wells have measured levels from 1 to 3 ft below the levels; however, these measured levels have continued to rise slowly toward the levels estimated by picks on the geophysical logs. The mean difference between the geophysical log picks and measured water levels is on the order of –0.13 ft with a standard deviation of about ±1.5 ft.

Other Nuclear Logs

Other less conventional neutron logging techniques include neutron-activation, neutron-lifetime, and nuclear magnetic resonance (NMR) logging techniques. Neutron-activation uses a neutron source and provides identification of elements present in the borehole fluid and adjacent rocks under a wide variety

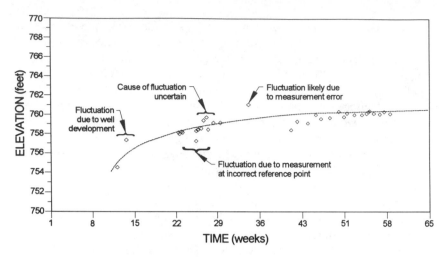

Figure 6-11   Hydrograph depicting slow response of water level in well which screens a low permeability siltstone.

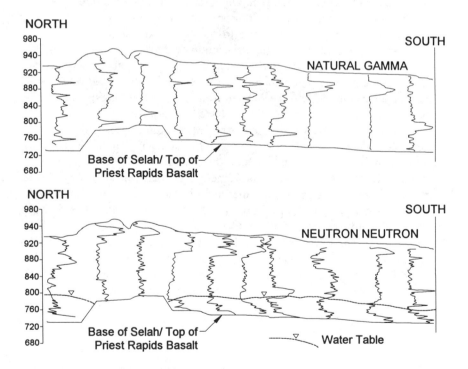

Figure 6-12   Geophysical cross section showing natural gamma and neutron-neutron profiles (after Testa, 1988).

Table 6-3 Summary of Water Level Data in Equilibrated Wells

Well No.	Geophysical Determination of Water Level Depth	Elevation	Measured Water Level Elevation	Water Level Elevation Variance	Direction of Change (+ or −)
A	146	803	803.26	0.3	+
C	135	809	811.08	2.1	−
D	215	768	767.48	0.5	+
F	201	785	783.32	1.7	+
G	206	757	755.95	1.1	+
H	111	733	734.24	1.2	−
I	164	787	787.87	0.9	−
J	147	776	778.58	2.6	−
L	216	761	760.37	0.6	+
T	76	791.5	789.03	2.5	+
MW-1	121	749	749.38	0.4	−
CC-1,2	164	794.4	796.64	2.2	−
SS-1	147	775.5	774.51	1.0	+

of borehole conditions (i.e., interpretation of ratio of carbon to oxygen and silicon to calcium in regards to lithology and in situ hydrocarbons). Neutron-lifetime (or pulsed neutron-decay) logs also use a neutron source and measure the rate of neutron population decrease near the borehole. This type of log can provide a measurement of salinity and porosity similar to resistivity logs. The lesser-known NMR (or nuclear magnetic logging, NML) can provide information on the quantity of water that is free to move into a borehole from the surrounding formation in regards to porosity and saturation data, and pore-size distribution.

Electric Logs

Electric logs include those logs where relative responses and measurements are dependent upon the flow of electric current in and adjacent to the well. Electric logs include spontaneous-potential (SP) and resistivity logging techniques. Resistivity logging, however, includes a variety of instrumentation and techniques.

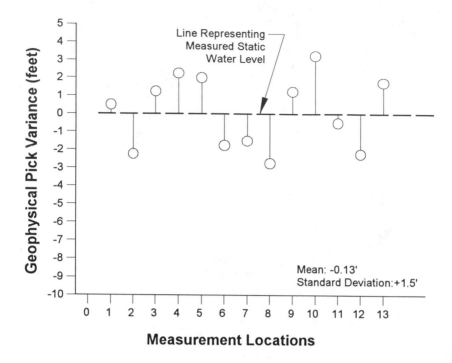

Figure 6-13   Graph of geophysical pick variance vs measurement locations.

Resistivity Logs

Resistivity logs are electric logs whose main use in the oil fields has been the determination of water-bearing vs hydrocarbon-bearing zones. Resistivity is an intrinsic property of the material being measured and is defined by

$$R = r \times S/L \tag{6-2}$$

where R = resistivity (ohm-meters), r = resistance (ohms), S = cross-sectional area normal to the flow of current (square meters), and L = length (meters).

Resistivity is based on the ability of soil or rock to transmit a current; thus, the presence of water in the pores is essential. Hydrocarbon on the other hand along with the formation material (i.e., rock matrix or grains) are nonconductive; therefore, as the hydrocarbon saturation increases, so does the formation's resistivity. Resistivity logs can also be used to distinguish permeable zones and resistivity porosity (R_w). A formation's water saturation (S_w) can also be determined using the Archie equation.

$$S_w = \frac{F \times R_w}{R_t}^{1/n}$$
(6-3)

where S_w = water saturation, F = formation factor (a/^{m}), a = tortuosity factor, m = cementation exponent, R_w = resistivity of formation water, R_t = true formation resistivity as measured by a deep reading resistivity log, and n = saturation exponent (usually 2.0).

The two basic types of log which measure resistivity are induction and electrode logs. Induction logs are the more common of the two.

Induction Logs — Induction logs include three types of curves: short normal, induction, and spontaneous potential (SP). The short-normal tool measures resistivity of the shallow, invaded zone (R_i). When compared with the deeper induction tool (R_t), invasion indicative of a permeable formation is detected. The short-normal tool performs best in conductive, high-resistivity muds (i.e., $R_{mf} > 3B_w$) vs poorly conductive salt muds (i.e., $R_{mf} \cong R_w$).

The deeper penetrating induction or more modern dual induction focused log measures conductivity, unlike the short-normal and SP logs that measure resistivity. This deeply penetrating tool is useful in formations that have been deeply invaded by mud filtrate. Since electricity is not transmitted through the drilling fluid, it can be used in air-, oil-, or foam-filled boreholes).

Electrode Logs — Electrode logs measure resistivity by placing electrodes in the borehole which are connected to a power source (i.e., generator). The current flows from the electrodes through the borehole fluid into the formation, and then to a remote reference electrode. Boreholes filled with salt-saturated drilling muds ($Rmf \cong Rw$) require electrode logs. Types of electrode tools in addition to normal include laterolog, microlaterolog, microlog, proximity log, and microspherically focused logs. The difference between these electrode resistivity measuring tools are summarized in Table 6-4.

When a porous and permeable water-bearing formation is invaded by drilling fluid, the formation water is displaced by the mud filtrate. Porosity in the water-bearing zone can be related to shallow resistivity (R_{xo}) as follows:

$$S_{xo} = F \times \frac{R_{mf}}{R_{xo}}$$
(6-4)

where S_{xo} = 1.0 (100%) in water-bearing zones.

Thus, porosity (ϕ) can be derived as follows:

Table 6-4 Summary of Electrode Logs Application

Electrode Tool Type	Application
Laterolog	Measures true formation resistivity (R_t) in boreholes filled with saltwater muds ($R_{mf} \cong R_w$).
Microlaterolog	Measures resistivity of the flushed zone (R_{xo}) with saltwater-based drilling mud
Microlog	Measures resistivity of mudcake (R_{mc}); if present, invasion has occurred; thus, the formation is permeable
Proximity log	Measures resistivity of the flushed zone (R_{xo}) with freshwater-based drilling mud where the mudcake is thicker
Microspherically focused log	Measures resistivity of the flushed zone (R_{xo}); when used with the dual laterolog, allows for correction of invasion

$$\phi = a \frac{(R_{mf}/R_{xo})^{1/m}}{(S_{xo})^2} \tag{6-5}$$

where ϕ = formation porosity, R_{mf} = resistivity of mud filtrate at formation temperature, S_{xo} = water saturation at the flush zone, R_{xo} = resistivity of flushed zone (Table 6-3 methods), a = constant (1.0 for carbonates, 0.62 for unconsolidated sands and carbonates, and 0.81 for consolidated sands), m = constant (2.0 for consolidated sands and carbonates, and 2.15 for unconsolidated sands), and F = formation factor.

All electrical logs provide information on lithology. The electrode logs can be used to attain accurate resistivity values in thinly bedded or resistive roles and provide high resolution, although these types of logs are not always readily available outside of the petroleum industry. The most important use of the normal resistivity logs is for the determination of overall water quality. The induction logs are limited due to the probe being responsive to essentially small changes in resistivity when background resistivity is minimal.

Nitrate-Fluoride Logs — A new logging tool has been discussed by Newman and Corbell (1990) that was used in conjunction with other logging tools to detect nitrate and fluoride contamination as a function of depth in existing and new wells. This tool incorporates two distinct elements: an electrode which senses the nitrate or fluoride ion activity and the electronic package which responds to the electrode voltage output. Limitations for the nitrate electrodes includes prohibited response due to temperature, ionic strength, and interfering ions which can be adjusted for in a laboratory but not during field applications. The fluoride electrodes are also affected by temperature and ionic strength in a similar manner as that for nitrate.

Spontaneous-Potential Logs

Spontaneous-potential (SP) logs record the DC voltage differences between the naturally occurring potential of moveable electrodes in the borehole and the potential of a fixed electrode at the surface. Measured in millivolts, SP logging techniques have been used widely in the oil fields but have been restricted in use in fresh groundwater environments. SP is used to identify impermeable and permeable zones such as shale and sand, respectively, as well as the boundaries of permeable beds. In addition, SP logs can also be used to determine formation water resistivity (R_w), the volume of shale (V_{sh}) in permeable beds, and detection of hydrocarbon. Suppression of the SP response can be indicative of thin beds (i.e., less than 10 ft), shaliness, and/or the presence of hydrocarbon or gas. The SP log is influenced, however, by bed thickness and resistivity, invasion, borehole diameter, shale content, and the ratio of R_{mf}/R_w.

The SP response reflects differences in salinities between the mud filtrate (R_{mf}) and formation water resistant (R_w) within permeable beds. The SP response of shale is relatively constant and thus follows a straight line referred to as a baseline. If the SP response moves toward the left or right, permeable zones are present. A response to the left indicates $R_{mf} > R_w$, whereas a response to the right indicates $R_w > R_{mf}$. In nonpermeable zones, or zones where $R_w = R_{mf}$, the SP response will not deflect from the baseline. The magnitude of the deflection reflects varying resistivities between R_{mf} and R_w, not varying permeability.

The static SP, or SSP, represents the maximum SP value that a relatively thick, highly porous, and permeable, shale-free formation can attain for a given ratio of R_{mf}/R_w. All other deflections are of a lesser magnitude. SSP can be evaluated via a convention chart or mathematical formula. V_{sh} can be calculated as follows:

$$V_{sh} = 1.0 - \frac{PSP}{SSP} \qquad (6\text{-}6)$$

where V_{sh} = volume of shale, PSP = pseudostatic spontaneous potential (SP of shaly formation), and SSP = static spontaneous potential of a thick clean sand or carbonate; also, SSP = $-K \times \log(R_{mf}/R_w)$ where K = $60 + (0.133 \times T_f)$, where T_f = borehole temperature in °F.

SP logs have been widely used for determining R_w in oil wells, but is limited in fresh groundwater systems. In NaCl-type saline water, R_w can be calculated using the following relation:

$$SP = -K \times \log (R_{mf} / R_w) \qquad (6\text{-}7)$$

where SP = log deflection in millivolts.

The SSP is the value to use in calculations although a lesser value is typically attained and difficult to correct for. Overall, the SP is favorable for quantitative work only in thick, clean, saturated sands, and not in highly resistive carbonates and shaly sands. The requirement of a nonconductive fluid (i.e., mud) in the borehole makes this logging technique restrictive for most environmental studies.

Resistivity in conjunction with SP logs have been used to support flow and containment computer modeling of a contaminant plume resulting from the injection of hazardous waste in five Class I injection wells. The injection wells are situated in the Gulf Coastal Plain, which includes coastal portions of Texas and Louisiana where most of the Class I hazardous waste injection wells are located. The injection zone is characterized as interbedded fluvial and deltaic sands and shales of Pliocene age. Data generated confirmed decreasing thickness and concentration of the contaminant plume, and assisted in defining the migration pathway.

Dipmeter Logs

Dipmeter logging includes a variety of wall-contact microresistivity probes that are used to provide data on the strike and dip of bedding planes. The more sophisticated dipmeters include four pads situated 90° apart vs the less advanced three-pad models with three probes situated 120° apart. Oriented with respect to magnetic north by a magnetometer located in the probe itself, correlation of resistivity anomalies detected by the individual pads, and the calculation of the true depth at which those anomalies occur, allows interpretation on the location and orientation of primary sedimentary structures over a wide range of borehole conditions. Fractures can also be located and orientations derived; however, limitations include the complexity of fractures which may or usually are characterized by irregularity with numerous intersections, and greater range in dip angles over a shorter depth interval.

Computer analysis and plotting of the results are of great benefit although limited when working with fractures vs bedding structures. In combination with a televiewer discussed later, a more comprehensive understanding of the relationship between complex structural elements can be achieved.

Fluid Logs

Fluid logs are used to measure certain characteristics related to the fluid column in the borehole. These characteristics include conductivity, flow, and temperature. Fluid logs are unique in that no direct signal is derived from the surrounding formation and interstitial fluids, and borehole fluid characteristics may vary rapidly with time and undergo alteration by the logging process itself.

Conductivity Logs

Logs of fluid conductivity provide data on the concentration of dissolved solids. Conductivity (or resistivity) logs provide a record of the borehole fluid which enters the probe capacity to conduct an electrical current. Conductivity is the reciprocal of fluid resistivity. The specific conductance of water samples in microsiemens per centimeter at 25°C can be converted to resistivity in ohm-meters as follows:

$$R_w = 10,000/\text{specific conductance} \tag{6-8}$$

Thus, water analyses may be converted to electrically equivalent NaCl concentrations, and resistivity values from logs can be converted to standard temperature using standard graphs (Alger, 1966). With enough data, specific conductance can be empirically related to dissolved-solids concentration and illustrated in a graph of specific conductance vs dissolved solids.

Although not informative in respect to the quality of interstitial fluids, when both fluid conductivity and temperature is known, the equivalent NaCl concentration can be determined providing corrections are made (i.e., chemical testing of water samples to determine the concentrations of various ions). Since other logs such as SP and certain resistivity logs are affected by borehole fluid salinity, conductivity logs can complement these other methods and provide a means for quantitative corrections. Conductivity logs can be important in monitoring groundwater quality beneath solid or hazardous waste operations.

Certain chemical waste or leachate can increase groundwater conductivity. Conductivity data can thus provide the basis for determination of a release from a waste unit, selection of sampling depths and necessary screened intervals, and provide insight into migration pathways. Other uses include insight into regional patterns of groundwater flow and recharge, mapping and monitoring areas of saltwater encroachment, or utilizing several wells to monitor the movement of groundwater by injecting saltwater as a tracer, providing salt concentrations used do not exceed applicable regulations in regards to overall groundwater quality. The interpretation of conductivity logs require understanding of the flow regime within the well.

Flow Logs

The measurement of flow vertically and laterally within a well and between wells can be very useful in understanding the groundwater flow regime under study and assessing the fate and transport of contaminants in groundwater. Flow characteristics can be evaluated by mechanical means such as impellers, chemical and radioactive tracers, and thermal means. Impeller flowmeters are commonly used for measuring vertical fluid movement, although limitations exist in slow-velocity regimes (i.e., less than 2 to 3 ft/min). Flowmeters of this type incorporate a three or four-headed impeller mounted on a shaft that rotates a magnet also mounted on the shaft.

Tracer methods are useful to determine much slower velocities than that capable of being measured using impellers. Salt, chemical, and radioactive tracers which require probes for detection and monitoring have been used. Salt injector-detector probe systems are centralized in the borehole for measurement of maximum velocity and to minimize boreholes wall effects. The tracer injector utilizes a liquid tracer and consists of a positive displacement piston-type pump. The quantity of tracer can be adjusted, ranging from a drop to 20 mm or greater. With radioactive-tracer logging systems, gamma or double gamma detectors are used. I131 is commonly used because it has an 8-day half-life, is water soluble, and can be detected in minute quantities. Thermal flowmeters consist of a resistance-heating element located between two thermistors within a small-diameter tube. Heat-pulse flowmeters are used to measure extremely slow velocities. Vertical velocities can range from 0.1 ft/min or less to 0.20 ft/min or more.

Once calibrated, all flowmeters can typically be used to measure vertical flow when a well or wells are open to more than one aquifer. Radioactive tracer logs can be used to identify permeable zones where flow rates are very slow. The heat-pulse flowmeter can be used to identify fractures producing and accepting water in rock.

Evaluation of the distribution of contaminants, notably chlorinated solvents, with depth in a multiperforated public water supply well was assessed via a combination of caliper logs, video survey under nonpumping conditions, and temperature, resistivity, and spinner flowmeter logs under both static and pumping conditions. The percent contribution from each producing interval to the total flow in the well was determined using spinner flowmeter logs. Depth-specific sampling with application of mass balance calculations was used to determine the concentration of contaminant being contributed to the well by each producing interval, thus allowing for qualitative assessment of depth-specific stratification of dissolved contaminants.

Temperature Logs

There are two general types of temperature logs: the temperature log *per se,* which provides a record of temperature vs depth, and the differential-temperature log, which provides a record of the rate of change in temperature vs depth and is thus more sensitive to changes on the temperature gradient.

Typically, under no flow conditions, temperature will gradually increase with depth as a function of the geothermal gradient. Geothermal gradients typically range between 0.47 and 0.6°C/100 ft of depth.

Temperature logs are very useful in groundwater studies and can provide data on groundwater movement through a borehole, location and depth intervals that either produce or accept water (thus, provide information related to permeability and hydraulic head distribution), locate cement behind casing, and correct other logs that are sensitive to temperature. Temperature logs can also be used to trace the movement of injected water or waste. Recorded temperatures may also provide information on the temperature of adjacent rocks and their contained fluids, although this use is restricted to circumstances where no fluid movement has occurred for a sufficient period of time to permit establishment of thermal equilibrium.

If the temperature of wastewater is sufficiently different from that of groundwater, temperature logs can be used to locate plumes of contaminated groundwater resulting from wastewater that had to be injected into a well or leaked from a wastewater containment structure (i.e., tank, seepage pond, etc.). In one case, identification of a wastewater plume originating from a disposal well more than 1.5 mi away was identified (Jones, 1961). The horizontal and vertical distribution of the wastewater was deciphered from a series of temperature logs in several wells, and corroborated by fluid-conductivity logs.

Caliper Logs

Caliper logs provide a continuous record of borehole diameter which reflects upon both drilling technique and lithology. Since all logs, regardless of the type of log, are affected by changes in borehole diameter, caliper logging is essential to interpretation. Obviously, an uncased or open borehole is required for caliper logging. Caliper logs also provide information pertaining to well construction and secondary porosity, notably in fractured rock or karst environments.

Caliper logging has also been used to correlate major producing aquifers (Jones, 1961), locating optimum depth and volume for cementing purposes (Keys, 1963), and assessing the impact of reactions between acidic wastes with carbonate rocks on the borehole wall, thus identifying intervals accepting most of the underground waste (Keys and Brown, 1973).

Acoustic Logs

Acoustic logging uses a transducer to transmit an acoustic wave through the fluid in the borehole and surrounding formation. Acoustic logs include velocity (or sonic or transit-time) logs, wave-form logs, cement-bond logs, and tele-viewer logs. All acoustic logs require fluid in the borehole and can be used to provide data on porosity, lithology, cement, and the location and character of fractures.

Acoustic-Velocity Logs

Acoustic-velocity logs record the time required for an acoustic wave to travel from one or more transmitters to the receivers located in the probe. Acoustic waves travel through the surrounding formation material and bore-hole at a velocity which reflects upon the matrix mineralogy and porosity. These logs are usually limited to uncased, fluid-filled boreholes. Although limited in use for environmental related studies, acoustic-velocity logs can be cross-plotted with neutron or gamma-gamma logs to identify intervals of secondary porosity, notably in carbonate rocks, fractures, and voids character-istic of karst environments.

Acoustic Wave-Form Logs

Acoustic wave-form logs are based on analysis of the various components of a received acoustic signal including amplitude changes, ratios of the veloci-ties of various components of the wave train, and frequency-dependent effects. Velocities and amplitude data generated from all parts of a recorded wave can be digitized logs allowing for quantitative analysis. In conjunction with the corrected bulk-density measurement obtained from gamma-gamma logs, elas-tic properties of rocks can be calculated from the velocities of observed compression and shear waves.

Cement-Bond Logs

Cement-bond logs are used to obtain information on the integrity of the bond between the casing and cement, and the cement and borehole wall.

Although most cement-bond logs solely measure the amplitude of the early arriving signal from the casing, accuracy can be improved with the complement of acoustic wave-form analysis.

Acoustic-Televiewer Logs

Acoustic-televiewer (ATV) logging is not a borehole television but rather provides a record of any feature in the casing or borehole wall that will alter the reflectivity of the acoustic signal. The ATV can provide high resolution data on the location and character of secondary porosity (i.e., fractures and solution cavities or openings), and structural elements (i.e., strike and dip) of planar features such as bedding planes, joints, and fractures. ATV can also provide information on the diameter and shape of the borehole, roughness of the borehole wall, and distinguish differences in rock hardness. Under ideal conditions, features as small as 1/32 in. can be ascertained. ATV can be complicated by inadequate centralization, deviation of the borehole from the vertical, and magnetic field aberrations.

BIBLIOGRAPHY

Alger, R.P., 1966, Interpretation of Electric Logs in Freshwater Wells in Unconsolidated Formations: In Society of Professional Well Log Analysts Annual Logging Symposium Transactions, Tulsa, OK, pp. CC1-CC25.

Asquith, G.B. and Gibson, G.R., 1982, *Basic Well Log Analysis for Geologists,* American Association of Professional Geologists, Tulsa, OK, 216 p.

Benson, R.C., 1988, Surface and Downhole Geophysical Techniques for Hazardous Waste Site Investigation: *Hazardous Materials Control,* March-April, pp. 9-60.

Davison, C.C., Keys, W.S. and Paillet, F.L., 1982, Use of Borehole-Geophysical Logs and Hydrologic Tests to Characterize Crystalline Rock for Nuclear-Waste Storage, Witeshell Nuclear Research Establishment, Manitoba, and Chalk River Nuclear Laboratory, Ontario, Canada: Ballette Project Management Division of Nuclear Waste Isolation Technical Report 418, 103 p. (U.S. Department of Commerce National Technical Information Service Report No. UNW1-418).

Dobrin, M.B., 1960, *Introduction to Geophysical Prospecting,* Third Edition: McGraw-Hill, New York, 630 p.

Driscoll, F.G., 1986, *Groundwater and Wells,* Johnson Division, St. Paul, 1089 p.

Henry, D.K., Mickelson, B.J. and Ohnstad, D., 1990, Well Logging and Depth Specific Sampling in a Borehole Water Supply Well as an Aid in Identifying Contaminant Stratification: In Proceedings of the Association of Ground Water Scientists and Engineers Fourth National Outdoor Action Conference on Aquifer Restorations, Ground Water Monitoring and Geophysical Methods, Las Vegas, NV, pp. 953-967.

Jones, P.H., 1961, Hydrology of Waste Disposal; National Reactor Testing Station, Idaho, Oak Ridge, Tennessee: U.S. Atomic Energy Commission Technical Information Service, IDO22042.

Keys, W.S. and Brown, R.F., 1973, Role of Borehole Geophysics in Underground Waste Storage and Artificial Recharge: In *Underground Waste Management and Artificial Recharge Symposium Transactions* (edited by Braunstein and Jules), Second Edition, New Orleans, Vol. 1, pp. 147-191.

Keys, W.S., Eggers, D.E. and Taylor, T.A., 1979, Borehole Geophysics as Applied to the Management of Radioactive Waste-Site Selection and Monitoring: In *Management of Low-Level Radioactive Waste* (edited by M.W. Carter, A.A. Moghissi, and Kahn), Pergamon Press, New York, Vol. 2, pp. 955-982.

Keys, W.S., 1982, Borehole Geophysics in Geothermal Exploration: In *Developments in Geophysical Exploration Methods* (edited by A. Fitch), Elsevier Applied Science, London, pp. 195-268.

Keys, W.S., 1989, *Borehole Geophysics Applied to Ground-Water Investigations:* National Water Well Association, Dublin, OH, 313 p.

Lawrence, T.D., 1979, Continuous Carbon/Oxygen Log Interpretation Techniques: Society of Petroleum Engineers of the American Institute of Mining, Metallurgical, and Petroleum Engineers, Paper SPE 8366, 12 p.

Mahannah, J.L., Witten, A.J. and King, W.C., 1988, Use of Geophysical Diffraction Tomography for Hazardous Waste Site Characterization: In Proceedings of the Hazardous Materials Control Research Institute Ninth National Superfund, 88 Conference, Washington, D.C., pp. 152-156.

Manchon, B., 1990, Theory-Rendering Injected Subsurface Waste Non-Hazardous as Demonstrated by Geophysical Logs: In Proceedings of the Association of Ground Water Scientists and Engineers Fourth National Outdoor Action Conference on Aquifer Restorations, Ground Water Monitoring and Geophysical Methods, Las Vegas, NV, pp. 1017-1031.

Milchie, D.W., 1968, Caliper Logging - Theory and Practice: *The Log Analyst,* Vol. 25, No. 5, pp. 16-30.

Newman, J.L. and Carbell, G.W., 1990, Continuous In-Situ Measurements of Nitrates and other Contaminants Utilizing Specialized Geophysical Logging Devices: In Proceedings of the Association of Ground Water Scientists and Engineers Fourth National Outdoor Action Conference on Aquifer Restoration, Ground Water Monitoring and Geophysical Methods, Las Vegas, NV, pp. 953-967.

Poteet, D.R., 1989, Using Terrain Conductivity to Detect Subsurface Voids and Caves in a Limestone Formation: In *Engineering and Environmental Impacts of Sinkholes and Karst* (edited by B.F. Beck), A.A. Balkema, Brookfield, VT, pp. 271-279.

Quirein, J.A., Gardner, J.S. and Watson, J.T., 1982, Combined Natural Gamma Ray Spectral/Lith-density Measurements Applied to Complex Lithologies: Society of Petroleum Engineers of the American Institute of Mining, Metallurgical, and Petroleum Engineers, Paper SPE 11143, 14 p.

Ring, G.T. and Sale, T.C., 1987, Evaluation of Well Field Contamination Using Downhole Geophysical Logs and Depth-Specific Samples: In Proceedings of the Hazardous Materials Control Research Institute Eighth National Superfund '87 Conference, Washington, D.C., pp. 320-325.

Snyder, S.W., Evans, M.W., Hine, A.C. and Compton, J.S., 1989, Seismic Expression of Solution Collapse Features from the Florida Platform: In Proceedings of the Hazardous Materials Control Research Institute Eighth National Superfund '87 Conference, Washington, D.C., pp. 281-298.

Telford, W.M., Geldart, L.P., Sheriff, R.E. and Keys, D.A., 1983, *Applied Geophysics:* Cambridge University Press, New York, 860 p.

Testa, S.M., 1988, Benefits of Downhole Geophysical Methods in Low Permeability Hydrogeologic Environments: In Proceedings of the Second National Outdoor Action Conference on Aquifer Restoration, Ground Water Monitoring and Geophysical Methods, pp. 969-985.

Testa, S.M., 1989, Overview of Subsurface Remediation at a Major Superfund Site - Western Processing Site: In *Engineering Geology in Washington* (edited by R.W. Galster), Washington Division of Geology and Earth Resources, Bulletin 78, Vol. 2, pp. 1115-1126.

7   WASTE CHARACTERIZATION

"I am, therefore I pollute."

(Descartes, modified)

INTRODUCTION

It is very difficult to get a handle on what exactly a "waste" is, let alone on the number of generators of hazardous and toxic wastes and the quantities of waste produced. No single source or data base exists on the generation and management of hazardous and toxic wastes that fulfill the requirement of being comprehensive, extensive and current. The sources that are available differ in regard to purpose, scope, content, and time frame, and include:

- USEPA's Hazardous Waste Data Management System of 1986 (HWDMS)
- USEPA National Survey of Hazardous Waste of 1981
- National Survey of Hazardous Waste Facilities under RCRA of 1983
- USEPA's Biannual Report.

A summary of these sources, including their purposes, contents, and limitations, is presented in Table 7-1.

As of April 1986, about 68,265 generators of RCRA-regulated hazardous waste were identified by the HWDMS. The largest single portion of federally regulated hazardous waste generators was in EPA's Region 5 which represents the industrial states of Illinois, Indiana, Michigan, Minnesota, Ohio, and Wisconsin. Manufacturing industries account for approximately 86% of the generators (Figure 7-1). In 1981, the chemical and petroleum industries accounted for an estimated 71% of the hazardous waste generated, and possibly up to 85% of the total quantity of waste generated. Metal-related industries generated 22% of the waste, with other industries accounting for the remaining 7%. The types of waste generated are summarized in Table 7-2.

Table 7-1 Summary of Hazardous Waste Information Sources

Source	Purpose	Content	Limitation
Hazardous Waste Data Management System (HWDMS)	Regulatory	Provides record of every facility that has applied for an EPA permit to treat, store, or dispose of hazardous waste under RCRA. Includes type of process used (i.e., surface improvement, landfill, injection well, ocean disposal, land application, incineration, and tank or container)	No current information on quantity of waste generated. Information process types available. Many current facilities were not managing waste in 1981. About two thirds of the facilities have since withdrawn their permits
Natural Survey of Hazardous Waste (1982)	Probability survey	Records statistically reliable estimate of number of facilities which generate, treat, store, or dispose of hazardous waste. Includes types of processes used. Includes characteristics of each facility	Information is dated but reliable as of 1986
Biannual report	Census	Includes facility identification waste description (DOT Code and EPA waste number), and amount of waste type for each facility to which or from which waste was shipped. Kept current every two years. Census rather than sample survey	Data collection requirements vary among states who provide information to EPA
Chemical Manufacturer's Association	Census	Includes partial estimates of quantities of hazardous waste generated by its members. Breaks out solid hazardous waste from hazardous waste waters	Solid quantities are most likely underestimated

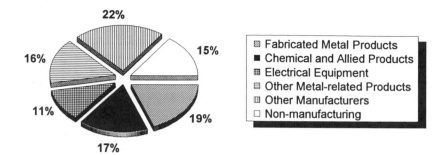

Figure 7-1   Number of hazardous waste generators per industry type.

Table 7-2 Summary of Waste Types Generated[a]

Waste Type	Waste (percent by weight)	RCRA Waste Code[b]
Spent solvents, process waste waters and sludges, and listed industry wastes	167	F or K wastes
Reactive wastes	52	D003
Corrosive wastes	35	D002
Toxic wastes	10	D004–D017
Ignitable wastes	1	D001
		U or P Wastes
Unspecified	5	Unspecified

[a] After Dietz and Burns (1989).
[b] A waste may fall in more than one EPA-listed category, and more than one characteristic category, therefore, percentages may add up to more than 100%.

   Presented in this chapter is discussion of what waste is and how a waste is characterized. Discussion of the characterization process as applied to soil, groundwater, and other materials, such as crude oil and debris, and the declassification process is also presented.

WASTE CHARACTERIZATION

   From a regulatory perspective, it is the generator's responsibility to determine whether a particular material is hazardous or toxic waste. In order for a material, including contaminated soil or groundwater, to be considered a hazardous waste, it must first be a solid waste (i.e., any discarded material), and then be proved hazardous.

   A hazardous waste is defined as any waste or combination of wastes that poses a substantial present or potential hazard to human health or living organisms because it (1) is nondegradable or persistent in nature, (2) can be biologically magnified, (3) can be lethal, or (4) can cause or tend to cause detrimental cumulative effects. A material, substance, or liquid is classified as hazardous if it is listed as a hazardous waste or meets certain specific characteristics; thus, wastes are either referred to as listed or characteristic wastes. Listed wastes may exhibit one or more of the specific characteristics, and are listed in the federal regulations (40 CFR Part 261, Subpart D). Characteristics of hazardous waste include ignitability, corrosivity, reactivity, and toxicity. These characteristics and their respective properties are summarized on Table 7-3. Although many wastes or wastestreams are derived from many industries and processes, such as spent halogenated solvents, others such as API separator sludges are generated from the one particular industry and process (i.e., petroleum-refining industry).

Other important waste types include radioactive and infectious wastes. Radioactive waste can be categorized as either high level, or transuranic (TRU) or low-level waste. High-level waste is restricted to spent fuel rods and the liquid and solid material that can be directly derived from them. TRU waste is material containing appreciable amounts of the heavy actinide elements (chiefly Np, Pu, and Am) which is derived from the production and handling of plutonium for military purposes. Low-level radioactive waste includes that derived from the mining and milling of uranium ore (i.e., tailings). Other low-level radioactive wastes include that derived from the use of radioactive materials in medical practices, research laboratories, and in industrial processes. In the U.S., as much radioactive waste is derived from miscellaneous sources as is generated via the nuclear power industry, although the half-lives of such material is usually restricted to less than a few decades.

Infectious waste makes up a minor category of potentially hazardous waste. Infectious waste refers to (1) any equipment, instruments, utensils, and fomites of a disposal nature from the rooms of patients who are suspect of having or have a communicable disease and therefore must be isolated, (2) laboratory wastes such as pathological specimens, and (3) surgical operating room pathologic specimens and disposable fomites.

The Toxicity Characteristics (TC) rule is important because it includes a significant volume of previously unregulated waste into the federal hazardous waste management system. The TC rule added 25 chemicals to the 8 metals and 6 pesticides on the previous Extraction Procedure Toxicity Characteristic (Table 7-4). The rule also replaced the Extraction Procedure (EP) with the Toxicity Characteristic Leaching Procedure (TCLP). The resultant regulatory levels for the newly added 25 chemicals are determined by the product of a health-related concentration threshold and a dilution/attenuation factor (DAF) which is derived from a newly developed groundwater transport model. Parameters and regulatory levels for the TC is presented in Table 7-4.

Certain states, such as California, maintain double standards and have set forth further regulations to determine if a material is to be classified as hazardous or nonhazardous. The decision on how the material is classified is based in part on the analytical values obtained in the lab. The regulations provide two different tables of maximum allowable threshold concentration values: the total threshold limit concentration (TTLC) and the soluble threshold limit concentration (STLC). Parameters and regulatory levels for TTLC and STLC are presented in Table 7-5. TTLC values apply to samples that have undergone a very harsh extraction protocol that is performed using concentrated nitric acid and hydrogen peroxide or hydrochloric acid (EPA Method 3050 contained in SW-846). The values obtained from this procedure can be, for all practical purposes, interpreted as the total quantity of a particular element in that sample. The second value is known as the STLC and applies to samples that have gone through a very mild leaching procedure with citric acid. This method is also referred to as the WET extraction. It supposedly replicates in vitro the leaching process that may potentially occur under a sanitary landfill environment. As water migrates through domestic garbage, picking up natural acids such as citric acid, it then continues on its migratory pathway through the waste in question, leaching into solution elements that could possibly travel down to the water table. This in vitro process consists of tumbling the sample for 18 hr in a buffered solution of pH 5 of citric acid. The ratio of sample to leaching solution is 1:10. That means that if you would multiply the result of this WET extraction by 10, you would roughly get the total value for this sample, if and only if the sample's analyte in question is completely soluble in water, which is a possibility that conservatively cannot be discarded.

Based on this 1:10 TTLC/STLC relationship, if any of the total values for a sample exceeds a value 10 times the STLC limit, it may possibly exceed the STLC values when the WET extraction test is performed. Thus, it is very practical and cost effective to schedule a set of samples to be analyzed first for totals, and if any of the values exceeds 10 times the regulatory STLC limit, then conduct of the WET extraction for only the element or elements in question is performed.

SOIL AS A HAZARDOUS WASTE

Soils are considered hazardous or toxic if the affected soil exhibits any of the characteristics of a waste (i.e., ignitability, corrosivity, reactivity, or toxicity), or contains a constituent that is a listed waste. The soil may also be considered hazardous should, upon chemical testing, a constituent or several constituents exceed an established maximum contaminant level (or regulatory level). This determination is made using the protocol previously discussed (i.e., TC, TTLC, and/or STLC).

Where no regulatory level exists for the constituent(s) of concern, remedial options are technically or economically limited (i.e., affected-soil beneath a

Table 7-3 Characteristics of Hazardous Waste

Characteristic	EPA Hazardous Waste No.	Properties
Hazardous waste Ignitability	D001	A liquid, other than an aqueous solution, containing less than 24% alcohol by volume and with a flashpoint less than 60°C (140°F); not a liquid but capable under ambient conditions of causing fire through friction, absorption of moisture, or spontaneous chemical changes and when ignited burns vigorously and persistently; it is an ignitable compressed gas as defined under 49 CFR 173.300; it is an oxidizer as defined under 49 CFR 173.151
Corrosivity	D002	Aqueous with a pH less than or equal to 2, or greater than or equal to 12.5; liquid which corrodes steel (SAE 1020) at a rate greater than 6.35 mm (0.250 in.) per year at a test temperature of 55°C (130°F)
Reactivity	D003	Normally unstable and readily undergoes violent change without detonating; reacts violently with water; forms potentially explosive mixtures with water; generates toxic gases, vapors, or fumes in sufficient quantities to present danger when mixed with water; a cyanide or sulfide-bearing waste which, when exposed to pH conditions between 2 and 12.5, can generate toxic gases, vapors, or fumes in sufficient quantities to present a danger; readily capable of detonation or explosive decomposition or reaction at ambient conditions; is a forbidden explosive as defined under 49 CFR 173.51, Class A explosive as defined under 49 CFR 173.53, or Class B explosive as defined under 49 CFR 173.88
Toxicity	D004–D043	Extract from a representative sample contains any of the contaminants listed in Table 1 at a concentration equal to or greater than the respective value provided in Table 1, using the Toxicity Characteristic Leaching Procedure (TCLP; Method 1311) or equivalent method

Acute hazardous waste[a]

Hazardous waste   Refer to hazardous waste as defined above
characteristics

Human toxicity   Fatal to humans in low doses; in absence of data on human toxicity, studies showing an
oval LD_{50} toxicity (rat) of less than 50 mg/kg, inhalation LC_{50} toxicity of less than 2 mg/l,
or a dermal LD_{50} toxicity (rabbit) of less than 200 mg/kg or significantly contributes to an
increase or irreversible or incapacitating illness

Toxic waste

Chemical toxicity   Contains any of the toxic constituents contained in Appendix VIII (40 CFR 261)[b]; capable
of posing a substantial present or potential hazard when improperly treated, stored,
transported, or disposed of, or otherwise managed

[a] Applies to a solid waste being characterized as a hazardous waste (40 CFR 261.11).
[b] Reflects 418 substances as of July 1, 1991, shown in scientific studies to have toxic, carcinogenic, mutagenic, or teratogenic effects on
humans or other life forms.

**Table 7-4 Toxicity Characteristic Leaching Procedure (TCLP)
Parameters and Regulatory Levels[a]**

Parameter	Regulatory Level (mg/L)	Parameter	Regulatory Level (mg/L)
Metals		p-Cresol	200.0[c]
Arsenic	5.0	Cresol	200.0[c]
Barium	100.0	2,4-Dinitrotoluene	0.13[b]
Cadmium	1.0	Hexachlorobenzene	0.13[b]
Chromium	5.0	Hexachlorobutadiene	0.5
Lead	5.0	Hexachloroethane	3.0
Mercury	0.2	Nitrobenzene	2.0
Selenium	1.0	Pentachlorophenol	100.0
Silver	5.0	Pyridine	5.0[b]
Volatiles		2,4,5-Trichlorophenol	400.0
Benzene	0.5	2,4,6-Trichlorophenol	2.0
Carbon tetrachloride	0.5	Organochlorine pesticides	
Chlorobenzene	100.0	Chlordane	0.03
Chloroform	6.0	Endrin	0.02
1,4-Dichlorobenzene	7.5	Heptachlor	0.008
1,2-Dichloroethane	0.5	(& its epoxide)	
1,1-Dichloroethylene	0.7	Lindane	0.4
Methyl ethyl ketone	200.0	Methoxychlor	10.0
Tetrachloroethylene	0.7	Toxaphene	0.5
Trichlorethylene	0.5	Chlorophenoxy acid	
Vinyl chloride	0.2	Herbicides	
Semivolatiles		2,4-D	10.0
o-Cresol	200.0[c]	2,4,5-TP (Silvex)	1.0
m-Cresol	200.0[c]		

[a] Used for Resource Conservation and Recovery Act (RCRA) regulated hazardous waste. Source is 40 CFR, Part 261.24 and California Code of Regulations, Title 22, Chapter 11, Article 3.

[b] Quantitation limit is greater than regulatory level. Therefore, the quantitation limit is the regulatory level.

[c] If o-, m-, and p-Cresol cannot be differentiated, total cresol concentration can be used.

major thoroughfare, active and operational facilities, etc.), the affected soil is deeply seated in low-permeability soil, the concentrations are not significantly elevated such that they pose a threat to public health, safety, and welfare, or overall groundwater resources. In such circumstances, remedial or corrective action may be limited providing that an environmental risk assessment is

Table 7-5 Total Threshold Limit Concentration (TTLC) and Soluble Threshold Limit Concentration (STLC) Parameters and Regulatory Levels[a]

Parameter	Regulatory Level	
	TTLC[b] (mg/kg)	STLC[c] (mg/L)
Inorganic compounds		
Antimony	500	15
Arsenic	500	5.0
Barium	10,000[d]	100
Beryllium	75	0.75
Cadmium	100	1.0
Chromium	2,500	5[e]
Cobalt	8,000	80
Copper	2,500	25
Lead	1,000	5
Mercury	20	0.2
Molybdenum	3,500	350
Nickel	2,000	20
Selenium	100	1
Silver	500	5
Thallium	700	7
Vanadium	2,400	24
Zinc	5,000	250
Chromium (VI)	500	5
Fluoride Salts	18,000	180
Asbestos	1%	—
Organochlorine pesticides and polychlorinated biphenyls		
Aldrin	1.4	0.14
Chlordane	2.5	0.25
DDT/DDE/DDD	1.0	0.1
Dieldrin	8.0	0.8
Endrin	0.2	0.02
Heptachlor	4.7	0.47
Kepone	21	2.1
Lindane	4.0	0.4
Methoxychlor	100	10
Mirex	21	2.1
PCBs	50	5.0
Toxaphene	5	0.5

Table 7-5 Total Threshold Limit Concentration (TTLC) and
Soluble Threshold Limit Concentration (STLC) Parameters and
Regulatory Levels[a] (continued)

| | Regulatory Level | |
	TTLC[b] (mg/kg)	STLC[c] (mg/L)
Parameter		
Volatiles		
Tricholoroethylene	2,040	204
Chlorophenoxy acid herbicides		
2,4-Dichlorophenoxyacetic acid (2,4-D)	100	10
2,4,5-Trichlorophenoxypropionic acid (Silvex)	10	1.0
Semivolatiles		
Pentachlorophenol	17	1.7
Miscellaneous		
Dioxin (2,3,7,8-TCDD)	0.01	0.001
Organic lead	13	—

[a] Used for California regulated hazardous waste. Source is California Code of Regulations, Title 22, Chapter 11, Article 3.
[b] If a substance in a waste equals or exceeds the TTLC level, it is considered a hazardous waste.
[c] If a substance is ten times (by rule of thumb) the STLC value as found on the TTLC, the WET should be used. If any substance in the waste so analyzed equals or exceeds the STLC value, it is considered a hazardous toxic waste.
[d] Excludes barium sulfate.
[e] If soluble chromium, as determined by TCLP test, is less than 5 mg/L and soluble chromium as determined by the STLC test equals or exceed 560 mg/L and waste is not otherwise identified as RCRA hazardous waste, then the waste is non-RCRA hazardous.

performed. An assessment or inventory of health and environmental risks or factors posed by site-specific conditions is an important component of any subsurface characterization per the National Contingency Plan [40 CFR Part 300.430(d)(2)]. Evaluations of site risk factors are performed to demonstrate that no significant adverse effects on human health are anticipated. Such factors include discussion of site geology and hydrogeology, meteorology, ecology, air clarification, waste characteristics, contaminant source, exposure pathways, and Applicable or Relevant and Appropriate Requirements (ARARs). The ARARs for the site include consideration of federal, state, and local guidelines and cleanup levels.

GROUNDWATER AS A HAZARDOUS WASTE

Impacted groundwater may be considered hazardous and warrant some level of remedial action should certain parameters exceed their respective promulgated TCLP levels, as previously discussed and presented in Table 7-4, or the regulatory maximum contaminant levels (MCL) for the parameter of concern. The MCL is the maximum permissible level of a contaminant in water which is delivered to the free flowing outlet of the ultimate user of a public water system — not necessarily posing an overall adverse impact to groundwater resources since the issue of beneficial use aquifer is not addressed. MCLs for groundwater exist at the federal level in two categories: National Primary Drinking Water Regulations (NPDWR) and National Secondary Drinking Water Regulations (NSDWR). The established MCLs under the NPDWR applies to the suitability of water for consumption purposes and regulates certain parameters that at elevated concentrations could be hazardous to one's health. The MCLs under the NSDWR were established to control certain contaminants in drinking water that primarily affect the aesthetic qualities of drinking water, although health concerns may exist at higher contaminations. A summary of MCLs for drinking water is presented in Table 7-6.

MCL levels should, however, be used with caution when determining appropriate cleanup levels for groundwater restoration programs since not all groundwater is suitable for human consumption. Elevated levels of certain constituents may be harmful if consumed, but may be perfectly suitable or advantageous for agricultural (irrigation) or industrial purposes. Those factors that are important and must be considered in evaluating reasonable cleanup levels include the presence of naturally degraded groundwater (i.e., saltwater intrusion), designation of beneficial vs nonbeneficial, potential hydraulic intercommunication between aquifers, natural degradation processes and rates (especially in dealing with organics at low to moderate concentrations in groundwater), and location of drinking-water supply wells within a reasonable distance from the site of concern are all important.

In the case of benzene, for example, the California state drinking water MCL is 1 ppb, which is typically targeted as a preliminary cleanup level for groundwater by the regulatory community. Low concentrations of dissolved benzene (i.e., <30 ppb) in groundwater are commonplace at sites impacted by hydrocarbon releases well after the recoverable product and majority of the dissolved hydrocarbon constituents have been removed. This presence reflects the occurrence of "residual" hydrocarbon in the saturated zone (i.e., within the zone of a fluctuating water table, capillary fringe, or cone of depression). The resultant episodic presence of very low benzene concentrations in representative groundwater samples reflects an "asymptotic" decrease or leveling off of petroleum hydrocarbon concentrations with time which is often observed at corrective action sites where petroleum hydrocarbon-impacted groundwater

Table 7-6 Maximum Contaminant Levels for Groundwater

Contaminant	Unit	Maximum Contaminant Level (MCL)
EPA National Interim Primary Drinking Water Standards		
Inorganics		
Arsenic	mg/L	0.05
Barium	mg/L	1
Cadmium	mg/L	0.010
California bacteria	1/100mL	1
Chromium	mg/L	0.05
2,4-D	mg/L	0.1
Endrine	mg/L	0.0002
Fluoride	mg/L	2.2
Gross alpha[a]	pCi/L	15
Gross beta	millirem/yr[b]	4
Lead	mg/L	0.05
Lindane	mg/L	0.004
Mercury	mg/L	0.002
Methoxychlor	mg/L	0.1
Nitrate (as N)	mg/L	10
Radium[b]	pCi/L	5
Selenium	mg/L	0.01
Silver	mg/L	0.05
2,4,5-TP Silver	mg/L	0.01
Strontium-90 (bone marrow)	pCi/L	8
Toxaphene	mg/L	0.005
Tritium (total body)	pCi/L	20,000
Turbidity	1/TU[c]	1
Organics		
Chlorinated hydrocarbons		
Endrin (1,2,3,4,10, 10-hexachloro-6, 7-epoxy-1,4,4a,5,6,7,8,81-octahydro-1,4-endo, endo-5, 8-dimethano naphthalene)	mg/L	0.0002
Lindane (1,2,3,4,5,6-hexachlorocyclohexane, gamma isomer)	mg/L	0.004
Methoxychlor (1,1,1-Trichloro-2, 2-bis [p-methoxyphenyl] ethane)	mg/L	0.1
Toxaphene ($C_{10}H_{10}Cl_4$-Technical chlorinated camphene, 67-69% chlorine	mg/L	0.005
Chlorophenoxys		
2,4-D. (2,4-Dichlorophenoxyacetic acid)	mg/L	0.1

| 2,4,5-TP Silvex (2,4,5-Trichlorophenoxypropionic acid) | mg/L | 0.01 |
| Total trihalomethanes (the sum of the concentrations of bromodichloromethane, dibromochloromethane, tribromomethane (bromoform), and trichloromethane (chloroform) | mg/L | 0.10 |

EPA National Secondary Drinking Water Standards

Aluminum	mg/L	0.05 to 0.2
Chloride	mg/L	250
Color	units	15
Copper	mg/L	1.0
Corrosivity	NA	Noncorrosive
Fluoride	mg/L	2.0
Foaming agents	mg/L	0.5
Iron	mg/L	0.3
Manganese	mg/L	0.05
Odor	Threshold odor number	3
pH	Unit ± 1	6.5–8.5
Silver	mg/L	0.1
Sulfate	mg/L	250
Total dissolved solids (TDS)	mg/L	500
Zinc	mg/L	5
Chlorinated hydrocarbons Endrin (1,2,3,4,10,10-hexachloro-6,7-epoxy-1,4,4a,5,6,7.8.81-octahydro-1,4-endo, endo-5,8-dimethano naphthalene)	mg/L	0.0002

[a] Gross alpha particle activity (including radium-226 but excluding radon and uranium).
[b] Radium-226 and radium-228.
[c] Applicable only to surface water supplies.

has been or is being remediated through pump-and-treat methods. Asymptotic concentration limits have been adopted in four states (Alabama, Florida, Nevada, and Tennessee) as acceptable cleanup levels and a basis for discontinuation of groundwater remediation activities in lieu of drinking water MCLs. Natural aerobic biodegradation of remaining aromatic hydrocarbons to concentrations below drinking water standards commonly occurs following cessation of pump-and-treat remediation, given the known aerobic nature of the impacted saturated zone, which has been demonstrated by aerobic iron bacteria biofouling of in-place hydrocarbon recovery systems. Natural aerobic degradation of

volatile aromatic hydrocarbons in unconfined, sandy aquifers is well documented, the reaction-limiting factor being availability of dissolved oxygen to
bacteria. Introduction of a suitable dissolved oxygen source into selected
monitoring wells, such as hydrogen peroxide, may also be considered as a
suitable biorestoration technique if elevated benzene concentrations are observed to persist above designated asymptotic limits. Determination of the
achievement of asymptotic limits during the course of remediation can be
verified at a significant confidence level by application of appropriate statistical procedures (ANOVA or "t-test" methods). This approach is sanctioned by
other states, such as Florida, for remediation of those sites where drinking-
water MCLs are inappropriate and unreasonable.

Selection of Maximum Contaminant Levels and Goals

The MCL is the maximum permissible level of a contaminant in water
which is delivered to any user of a public water system. The Maximum
Contaminant Level Goal (MCLG) for a contaminant differs in that it is almost
a health-based determination and defined as a nonenforceable health goal level
at which no known or anticipated adverse effects on public health occur, and
which allows for a margin of safety. The MCLs promulgated under the NPDWR
are enforceable. MCLs are usually set as close to the MCLG as feasible with
the use of best technology, treatment techniques, or other means — in other
words "best available technology" (BAT). In determining BAT, the treatment
technology must be effective for the specific contaminant and take cost into
consideration. For example, since granulated activated carbon (GAC) is explicitly defined by statute as feasible for the treatment of certain organic chemicals,
the BAT must be at least as effective as GAC in controlling synthetic organic
chemicals. In addition, such technologies are viewed as field tested if the
technologies themselves have been operated in the field and laboratory, or via
pilot-scale tests showing that the treatment technology will work for the
particular contaminant of concern. Thus, both removal efficiencies and costs
are considered in determining what constitutes BAT. Also considered is the
monitoring capabilities in determining the MCL.

MCLGs on the other hand are primarily a health-based determination and
must be set at a level at which no known or anticipated adverse effects in the
public health occur and which allows for an adequate margin of safety. Some
MCLGs for suspected carcinogens are set at zero, including viral and bacterial
contamination. The selection of MCLGs is a complex and elaborate process
which varies based on which of the three categories of contaminants being
regulated. These categories and their respective processes are summarized in
Table 7-7.

The "no-effect" level is referred to as the Reference Dose (RfD) which is
derived from the No Observed Adverse Effect Level (NOAEL) or Lower

Table 7-7 Summary of Selection Process for MCLGs

Category	Parameter	MCLG	Rationale
1	Carcinogenic	0	Also referred to as EPA Groups A, B1, and B2 substances, known and possible human carcinogens
2	Noncarcinogenic	Varies	Based upon a "no-effect" level for chronic exposure including a margin of safety; referred to as the RfD[a], it is derived from the NOAEL[b], or LOAEL[c] via animal studies; insetting the RfD, the NOAEL is reduced by an uncertainty factor ranging from 100 up to 1000[24]
3	Equivocal carcinogenic	Varies	Based upon noncarcinogenic health effects with an additional uncertainty factor; if data is sufficient, MCLG is set based on an estimated lifetime risk of developing cancer from exposure to the substance in drinking water

[a] Reference dose.
[b] No Observed Adverse Effect Level.
[c] Lower Observed Adverse Effect Level.

Observed Adverse Effect Level (LOAEL). The NOAEL and LOAEL are typically derived from animal studies. The RfD is set by reducing the NOAEL by an uncertainty factor ranging from 100 up to 1000[24]. The uncertainty factor is employed since the data must be used to estimate a no-effect level for a widely differing human population based on extrapolations from animal studies. Once the RfD is determined, a Drinking Water Equivalent Level (DWEL) is calculated assuming that a 70-kg adult drinks 2 L of water per day, and using the RfD to assist in determining the specific level in drinking water at which noncarcinogenic health effects should not occur. The DWEL is further reduced to account for other sources of human exposure to the substance. Data generated for dietary exposure from U.S. Food and Drug Administration (FDA) studies will be relied upon for inorganic substances. For organic contaminants, however, these studies are viewed by EPA as inadequate. EPA in this case assumes that only 20% of such exposure results from drinking water and thus reduces the DWEL by a factor of five to arrive at the MCLG.

CRUDE OIL

As previously mentioned in Chapter 2, the primary and most common compounds and constituents associated with oil-field properties include methane, crude oil, drilling mud, and (to a lesser degree) refined petroleum products and associated constituents including volatile organic compounds. These compounds and constituents within the regulatory framework can be considered as either a hazardous waste or material, or, in the case of California, a designated waste.

A hazardous waste or material as defined in the State of California Code of Regulations, Title 22, (Division 4, Chapter 30) (15) is ". . . a substance or combination of substances which, because of its quantity, concentrations, or physical, chemical or infectious characteristics, may either: (1) cause or significantly contribute to an increase in mortality or an increase in serious irreversible, or incapacitating reversible, illness; or (2) pose a substantial present or potential hazard to human health or environment when improperly treated, stored, transported or disposed of or otherwise managed."

Hazardous materials can further be subdivided into specific categories based on toxicity (persistent and bioaccumulative), corrosivity, ignitability, and reactivity.

A designated waste as defined in the State of California Code Regulations, Title 23 (Subchapter 15, Section 2522(a)(1)) (16) is ". . . a non-hazardous waste which consists of or contains pollutants which, under ambient environmental conditions at the waste management unit, could be released at concentrations in excess of applicable quality objectives, or which would cause degradation of waters of the State."

Crude oil is not considered a hazardous waste since it typically does not contain significant concentrations of aromatic hydrocarbons (i.e., benzene, toluene, ethylbenzene, and xylene isomers). However, crude oil may be considered a designated waste in California. Crude oil is excluded as a hazardous waste or material in California, and can also be excluded as a designated waste on the basis of analytical testing in accordance with conditions as outlined in the California Health and Safety Code (Article 13, Management of Used Oil), and if it can be shown that the substance meets certain criteria as outlined in Table 7-8.

DEBRIS RULE

Construction debris include materials such as discarded equipment, broken drums, and materials generated during the renovation or demolition of buildings. Hazardous debris is defined as a solid material (vs a process waste) with a particle size of 60 mm (about $2^3/_8$ in.) or larger, intended for land disposal, and which exhibits one of the prohibited characteristics of a hazardous waste

Table 7-8 Criteria for Determining Whether Crude Oil Is Classified as a Hazardous Waste

Parameter	Maximum Contaminant Level (ppm)
Arsenic	5
Chloride	3,000
Chromium	10
Flash point	Less than minimum standard[a]
Lead	50
Polychlorinated biphenols (PCBs)	5

[a] Minimum standard as set by the American Society for Testing and Materials for recycled products.

or is contaminated with a listed hazardous waste. Although clumps of finer-grained material such as clay clumps do not qualify as debris, mixtures of debris and other things do qualify if they are the primary material present.

Construction debris used to fall under the "mixture and derived from" rule, which required treatment of the debris to levels specified for that listed or characteristic waste prior to land disposal. With the Debris Rule effective as of May 8, 1993, the debris only needs to be treated with 1 of the 17 "best demonstrated available technologies" (BDAT). The debris may then be disposed of offsite via landfill disposal or offshore, even if specified levels have not been reached. This does exclude, however, the deliberate mixing of the debris with other wastes in order to change its textural classification. Not classified as debris are intact containers that can hold 75% of their original capacity.

DECLASSIFICATION

To reduce the overall volume of contaminated soil and other materials being sent to landfills for disposal, and minimize the costs associated with legal, administrative, and remedial programs, the current regulatory posture has been to redefine the spirit, letter, and intent of environmental regulations concerning hazardous waste. Contrary to initial impressions, the intent of the regulations is not to list all contaminated soil and water as hazardous waste, but rather to allow avenues for declassification of the waste material. This is achieved through the use, reuse, and/or recycling of the material. A variety of materials initially considered hazardous can be declassified and documented as neither hazardous nor a waste via use of the federal regulations pertaining to waste reduction and waste minimization.

The crux of declassification is the determination as to whether a material is a waste (either a solid waste or a hazardous waste). The material must thus first be proven to be a waste. To be a hazardous waste, a material must first be a solid waste which is any discarded material [40 CFR 261.2(a)(1)]. The material cannot be classified as a hazardous waste if it is not a solid waste. It is very clear in the federal regulations that materials are not considered a solid waste if it can be shown that they can be recycled, such that they are:

- Used or reused as ingredients in an industrial process to make a product, provided the materials are not being reclaimed;
- Used or reused as effective substitutes for commercial products.

The responsibility of the generators to understand and properly declassify their materials is of utmost importance in avoiding the stigma of being a generator of hazardous waste.

Due to the circuitous nature of the federal regulations, exceptions exist. One of the most misunderstood and often misinterpreted exemptions is the use of recycled materials in a manner such that they are perceived as being "used in a manner constituting disposal" [40 CFR Part 261.2(c)(1)]. "Used in a manner constituting disposal" is defined as types of recycled materials that are applied to or placed on the land or used to produce products that are applied to or placed on the land. Obviously this means that, should the material be recycled or reused, one cannot do anything with it, thus, discouraging waste reduction, waste minimization, and beneficial uses. Another exemption to this exemption also exists and is referred to as "Recyclable Materials Used in a Manner Constituting Disposal" [40 CFR Part 266, Subpart C, Section 266.20(b)]. Stated is that products produced for the general public's use that are used in a manner that constitutes disposal and that contains recyclable materials are not presently subject to regulation if the recyclable materials have undergone a chemical reaction in the course of producing the products so as to become inseparable by physical means, and if such products meet the applicable treatment standards [40 CFR, Subpart D, Part 268; "or applicable prohibition levels in Section 268.32 or RCRA Section 3004d . . ."].

A good example of the utilization of these pertinent and important regulations is in the incorporation of contaminated soil into a cold-mix asphalt

Figure 7-2 **Core of cold-mix asphalt produced by the incorporation of petroleum-affected soil via the process referred to as environmentally processed asphalt.**

product via the process referred to as environmentally processed asphalt (Figure 7-2). In lieu of generating a landfill waste, an end-product for use as road base, liner, cap, or engineered fill, among other uses, is produced. Contaminated soil becomes a recoverable resource and is within the letter, spirit, and intent of current regulations.

BIBLIOGRAPHY

Barker, J.F., Patrick, G.C. and Major, D., 1987, Natural Attenuation of Aromatic Hydrocarbons in a Shallow Sand Aquifer: *Ground Water Monitoring Review,* Vol. 7, No. 1, pp. 64-71.

Baugh, A.L. and Lovegreen, J.R., 1990, Differentiation of Crude Oil and Refined Petroleum Products in Soil: In *Petroleum Contaminated Soil, Vol. 3,* (Edited by Kostecki, P.T. and Calabese, E.J.), Lewis Publishers, Boca Raton, FL, pp. 141-163.

California Code of Regulations, 1988, Title 22, Social Security.

California Code of Regulations, 1989, Title 23, Waters.

California State Water Resources Control Board, 1989, Leaking Underground Fuel Tank Field Manual: Guidelines for Site Assessment, Cleanup, and Underground Storage Tank Closure, 62 p.

California Regional Water Quality Control Board Central Valley Region, 1991, A Compilation of Water Quality Goals.

California Regional Water Quality Control Board Central Valley Region, 1986, The Designated Level Methodology for Waste Classification and Cleanup Level Determination, updated June 1989, 79 p.

Dietz, S.K. and Burns, M.E., 1989, Quantities and Sources of Hazardous Wastes: In *Standard Handbook of Hazardous Waste Treatment and Disposal* (Edited by Freeman, H.M.), McGraw-Hill, New York, pp. 213-219.

Dyroff, G.V., 1989, *Manual on Significance of Test for Petroleum Products,* 5th Edition: American Society for Testing and Materials, Philadelphia, 169 p.

MacKay, D.M. and Cherry, J.A., 1989, Groundwater Contamination: Pump-and-Treat Remediation: *Environmental Science and Technology,* Vol. 23, No. 6, pp. 630-636.

Mercer, J.W., Skipp, D.C. and Griffin, D., 1990, Basics of Pump and Treat Groundwater Remediation Technology: U.S. Environmental Protection Agency Publication EPA-600/8-90/003, 51 p.

Noonan, D.C. and Curtis, J.T., 1990, *Groundwater Remediation and Petroleum: A Guide for Underground Storage Tanks:* Lewis Publishers, Boca Raton, FL, 140 p.

Robbins, E.A., Robertson, M. and Cesark, D.R., 1990, Chemical Nature of Crude Oil, Condensate Crude, and Natural Gas: In Proceedings of the Sixth Annual Hazardous Materials Management Conference West, Long Beach, CA, pp. 392-416.

South Coast Air Quality Management District, 1988, Ruling 1166.

Society of Petroleum Engineers (SPE), 1962, *Petroleum Production Handbook, Vol. II, Reservoir Engineering* (Edited by Frick, T.C. and Williams, R. W.): Society of Petroleum Engineers of American Institute of Mining Engineers, Dallas.

Testa, S.M., 1989, Regional Hydrogeologic Setting and its Role in Developing Aquifer Remediation Strategies: Geological Society of America Abstract, Vol. 21, No. 6, p. A96.

Testa, S.M. and Winegardner, D., 1990, *Restoration of Petroleum Contaminated Aquifers:* Lewis Publishers, Boca Raton, FL, 269 p.

Testa, S.M., Patton, D.L. and Conca, J.L., 1992, Fixation of Petroleum Contaminated Soils via Cold-Mix Asphalt as use as a Liner: In Proceedings of the Hazardous Material Control Research Institute HMC-South '92, New Orleans, pp. 30-33.

Testa, S.M. and Conca, J.L., 1992, Chemical Aspects of Environmentally Processed Asphalt: In Proceedings of the International Symposium on Asphaltene Particles in Fossil Fuel Exploration, Recovery, Refining and Production Processes, Las Vegas, NV (in press).

Testa, S.M. and Patton, D.L., 1992, Add Zinc and Lead to Pavement Recipe - Stabilize Metal-Affected Soils in Asphalt to Create Useful Paving Material: *SOILS, Analysis, Monitoring, Remediation Magazine,* pp. 22-27 and 34-35.

Testa, S.M. and Patton, D.L., 1993, The Use of Environmentally Processed Asphalt as a Contaminated Soils Remediation Method: In Proceedings of the Association for the Environmental Health of Soils (AEHS) Third Annual West Coast Conference on Hydrocarbon Contaminated Soils and Groundwater, Long Beach, CA (in press).

Travis, C.C. and Doty, C.B., 1990, Can Contaminated Aquifers at Superfund Sites be Remediated?: *Environmental Science and Technology,* Vol. 24, No. 10, pp. 1464-1466.

Tucker, W.A., 1992, UST Corrective Action: When Is It Over?: *Groundwater Monitoring Review,* Vol. 12, No. 2, pp. 5-8.

8   SUBSURFACE PROCESSES

"Methodology must be developed to predict where in the environment a chemical will be transported, the rate and extent of transformation, and its effect on organisms and environmental processes at expected ambient levels."

(Testa and Winegardner, 1991)

INTRODUCTION

Many processes affect the fate, transport, and transformation of a particular contaminant or leachate in soil and groundwater. The overall composition of leachates produced reflects upon the waste type and degree of aerobic and anaerobic decomposition. Understanding how this material can become mobilized and what happens to it once mobilized is difficult and complex. It is also difficult to approach this problem by simply employing models which accounts for convection, dispersion, adsorption, and retardation. Physical, chemical, and biological properties and processes must also be considered in conjunction with site-specific environmental factors. It is often difficult to separate physical processes from chemical and biological processes because they are often coupled within any system. Therefore, the following breakdown is necessarily artificial and is based a great deal upon historical development and our overall perceptions of these processes.

Mathematical relationships are quite useful in understanding nonintuitive processes, notably transport; thus, some basic mathematics is required. A key mathematical concept, the gradient, is defined in regards to groundwater flow in Chapter 4, and is briefly reiterated here because it is one of the most important concepts for all aspects of subsurface science as well as any study of the earth. A gradient is a change in the value of one variable with respect to another variable, like a slope. Most of our discussion will include gradients of some property with respect to distance or time, e.g., a pressure gradient across the boundary between two soils, a concentration gradient across a membrane, or a thermal gradient across a surface. Gradients are the real driving forces for change in earth systems and in the environment. Obviously, large or steep gradients can result in rapid and dramatic changes. Often, our ability to restore a contaminated site or successfully dispose of hazardous waste depends upon our ability to minimize particular gradients within the system.

A major emphasis of environmental science today also concerns an assessment of risk and performance of disposal or remediation systems. Computer programs, called predictive models, attempt to predict the behavior and migration of subsurface fluids and contaminants. Predictive models are key to this type of work, and many licensing and government regulations depend upon the results of these predictive models. However, these models are only as good as the programming and data input. Most of the relevant data is still lacking and must be chosen as educated guesses or extrapolations from less-relevant situations. The conceptual model serves as the basic paradigm of the program, but the present conceptual models are usually too simplified to be meaningful because the actual subsurface environment is very complicated, or heterogeneous. A subsurface environment exposed in the wall of a 10-m-deep disposal pit for mixed hazardous wastes is shown in Figure 8-1. The sediment is mostly sand by volume, but has significant amounts of gravel and clay layers. The conceptual model for this situation assumed that (1) the subsurface was homogeneous layers of different sands, (2) the mixed wastes migrated similarly through each material, and (3) physical flow (advection) was the only transport process acting upon the system. Obviously, the real system pictured in Figure 8-1 might behave quite differently. Refinement of our subsurface conceptual models is essential. One objective of this chapter is to develop the framework for extending our conceptual models in order to design, characterize, and manage our hazardous waste systems more responsibly.

Another term that needs to be defined before we proceed is the term "chemical species," which refers to the actual molecular form that the contaminant (or any substance) takes in the porous medium and strongly affects how that contaminant will migrate through the subsurface. As an example, technetium (Tc) is both a highly radioactive contaminant and a toxic metal contaminant. Under reducing conditions (usually occurring when molecular oxygen is absent and organic matter is present, often a condition prevailing below the water table), Tc is generally immobile as a metal. However, under oxidizing conditions, as occurs in the unsaturated zone or vadose zone, between the earth's surface and the water table, Tc forms a complex species with oxygen, called a pertechnetate ion, TcO_4^-, which is highly mobile and is not held up, or retarded, by most soils, sediments, or rocks. Strategies for waste disposal, handling, and remediation require knowledge of the speciation of the contaminant, information gained almost exclusively through experimental studies under relevant conditions.

A

Figure 8-1   Long view (A), medium view (B), and close-up (C) of the wall of a 10-m-deep disposal pit for mixed hazardous waste at the Hanford site, eastern Washington, show sedimentary structural features that may control the preferential migration pathways of fluids. (Photos courtesy of Jim Conca.)

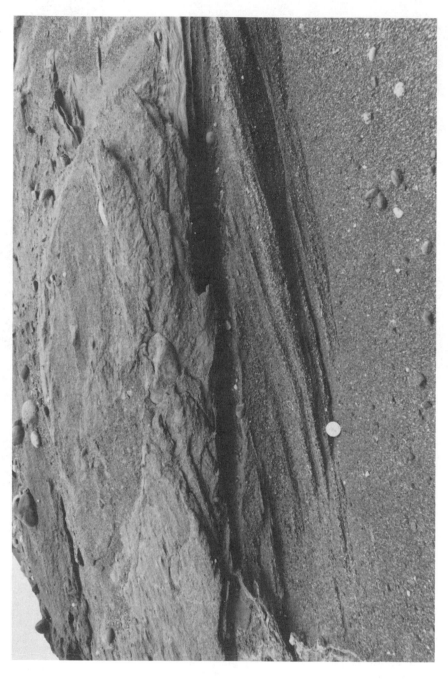

Figure 8-1B

Figure 8-1C

Presented in this chapter is discussion of the physical, biological, and chemical processes that play a role in the fate and transport of hazardous and toxic constituents in the subsurface.

PHYSICAL PROCESSES

The physical processes acting in the subsurface determine how a chemical partitions in the subsurface media, which can have a significant effect on the environmental fate of a substance. For example, if two acids of similar chemical behavior were released into the subsurface soil, one may volatilize into the air while the other may adsorb onto organic material in the soil. These processes are often grouped together as advection, dispersion, simple diffusion, volatilization, and filtration. These processes primarily describe the physical motion of substance through space and across boundaries. Note that phase changes (e.g., vaporization of water, volatilization of CCl_4, or solidification of lard oil at about room temperature) can be thought of as physical or chemical processes, and are strongly affected by the chemistry of the system.

Advection

Probably the most important subsurface process that acts in the greatest number of situations to control migration of subsurface contaminants and water is advection. Advection refers to the actual mass movement of a fluid through the porous media.

Dispersion

Hydrodynamic dispersion refers to the tendency of the fluid, or a solute or contaminant dissolved in the fluid, to spread out over time (i.e., to become dispersed in the subsurface). The mechanical component of dispersion results from the differential flow of the fluid through pore spaces that are not the same size or shape, and by different flow velocities and the fluid near the walls of the pore where the drag is greatest vs the fluid in the center of the pore (Figure 8-2).

Diffusion

Molecular diffusion also tends to disperse fluids and dissolved solutes through the random walk of individual molecules resulting from the thermal kinetic energy of each molecule and the driving force of entropy. Diffusion is important when volatile compounds are present and in situations where flow velocities are extremely low.

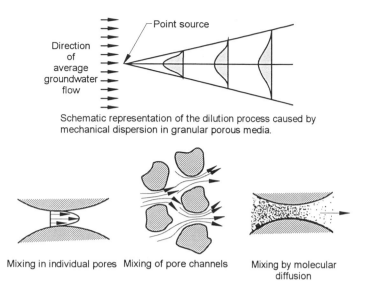

Schematic representation of the dilution process caused by
mechanical dispersion in granular porous media.

Mixing in individual pores   Mixing of pore channels   Mixing by molecular
                                                          diffusion

**Figure 8-2   Processes of dispersion in a microscopic scale (after Freeze
and Cherry, 1979).**

Volatilization

In the liquid and solid state, molecules are held together by a variety of
intermolecular forces. In going from solid to liquid, liquid to vapor, or solid to
vapor, these forces must be overcome by absorbing energy, usually thermal
energy from the environment. The volatility is the ease with which this occurs.
The more volatile, the easier it is for molecules to leave the surface. This
volatility depends upon the temperature and the chemical composition of the
compound. At any particular temperature, a volatile liquid or solid has a
characteristic vapor pressure that is the equilibrium partial pressure in the
atmosphere or air space surrounding the compound. Therefore, whether in the
pore space of a subsurface sediment at 25°C or in the atmosphere above an
open disposal trench at 25°C, if the vapor pressure of the highly volatile
compound carbon tetrachloride (CCl_4) is 0.15 atm (114 mm of Hg), then the
partial pressure of CCl_4 vapor in the pore space and in the atmosphere above
the trench will both be 0.15 atm. Vapor pressures for certain constituents
typically encountered in the subsurface are presented in Table 8-1.

If vacuum is applied momentarily to the gas in the sediment pore space to
remove the CCl_4 vapor, then more CCl_4 will volatilize to restore the partial
pressure to 0.15 atm. This is the operating principle of the vadose zone
remediation of volatile contaminants referred to as soil vapor extraction in
which vapor extraction through wells causes continuous volatilization of the
contaminant until little or no contaminant remains. Obviously, the more volatile the compound, the better this method will work. Soil vapor extraction
works well for remediating CCl_4 in the subsurface because of its high vapor

Table 8-1 Physical and Chemical Properties of Certain Common Organic Compounds

Compound	Molecular Weight	Melting Point (°C)	Boiling Point (°C)	Vapor Pressure (p, kPa)	Solubility (S, g/m³)	Henry's Law Constant kPa m³/mol[a]		
						Calc.	Exper.	Rec.
Monoaromatics at 25°C								
Benzene	78.11	5.53	80.1	12.7	1780	0.562	0.562	0.550 ± 0.025
Toluene	92.13	−95	110.6	3.80	515	0.68	0.673	0.670 ± 0.035
Ethylbenzene	106.2	−95	136.2	1.27	152	0.887	0.854	0.80 ± 0.07
p-Xylene	106.2	13.2	138	1.17	185	0.671	NA[b]	0.710 ± 0.08
m-Xylene	106.2	−47.9	139	1.10	162	0.721	NA	0.700 ± 0.10
o-Xylene	106.2	−25.2	144.4	0.882	175	0.535	NA	0.50 ± 0.06
Polynuclear Aromatics at 25°C								
Naphthalene	128.19	80.2	218	1.09×10^{-2}	34.4	0.0407	0.0489	0.0430 ± 0.004
Fluorene	166.2	116	295	8.86×10^{-5}	1.90	0.00775	0.0101	0.0085 ± 0.002
Anthracene	178.23	216.2	340	1.44×10^{-6} [c]	0.075	0.0034	0.073	0.0060 ± 0.003
Halogenated Alkanes and Alkenes at 25°C								
1,1-Dichloroethane	98.97	−96.98	57.5	30.10	5100	0.585	NA	0.58 ± 0.02
1,1,1-Trichloroethane	133.4	−30.4	74.1	16.53	720	3.06	NA	2.8 ± 0.04
1,1,2-Trichloroethane	133.4	−36.5	113.8	4.04	4420	0.122	NA	0.12 ± 0.02
1,1,2,2-Tetrachloroethane	167.85	−70.2	130.5	1.853	1100	0.283	NA	0.28 ± 0.02
Vinylchloride	62.5	−153.8	−13.4	344 (20°C)	2700 (20°C)	2.35	NA	NA

1,1-Dichloroethene	96.94	−122.1	37	79.73	400	13.32	NA	NA
1,2-Dichloroethene (CIS)	96.94	−80.5	60.3	27.46	3500	0.761	NA	NA
1,2-Dichloroethene (Trans)	46.94	−50	47.5	43.47	6300	0.669	NA	NA
Tetrachloroethene	165.83	−19	121	2.48	140	2.94	1.239	2.3 ± 0.4
Halogenated Aromatics at 25°C							
Chlorobenzene	112.56	−45.6	132	1.581	471.7	0.377	0.382	0.35 ± 0.05
d-Dichlorobenzene	147.01	−17.0	180.5	0.196	145.2	0.198	0.193	0.19 ± 0.01
1,2,3-Trichlorobenzene	181.45	53	218	0.0530[d]	16.6	0.306	0.127	NA
1,2,3,4-Tetrachlorobenzene	215.9	47.5	254	0.00876[d]	4.31	0.261	NA	NA

Note: Calc. = Calculated, Exper. = Experimental, Rec. = Recommended.

[a] Modified after MacKay and Shiv (1981).

[b] NA = not available.

[c] Calculated from the extrapolated vapor pressure with a fugacity ratio correction.

[d] Extrapolated from liquid state.

pressure of 0.15 atm, but will not work as well for the pesticide methyl parathion which has a vapor pressure of only 10^{-11} atm at 25°C. However, over many years the loss of methyl parathion through volatilization will be an important process, especially if other degradation processes are slow. Once the compound volatilizes, its vapor behaves like any other gas with respect to the physical transport processes previously described.

Filtration

Filtration occurs when fluid flows through a porous media whose pores are small enough to affect the passage of some component in the fluid, perhaps even to stop it completely. For suspended particulates and colloids, filtration occurs as a function of the pore space in the porous media. For example, if the permeability of a soil involves passage through pore spaces that are smaller than 0.1 mm, then that soil will effectively filter all particles larger than 0.1 mm and pass most particles smaller than 0.1 mm. This is not to say that every particle smaller than 0.1 mm will get through, and that an occasional particle 0.12 mm will not move a significant distance through the material.

There is a special case of filtration that occurs in the presence of a semipermeable membrane, called membrane filtration, salt filtration, or hyperfiltration, which is hypothesized to occur in the deep subsurface. The membrane is usually layers of clay minerals, whose frequently unbalanced surface charges cause a filtering of specific ions, a type of ion exclusion. The nature of the excluded ions depends upon the composition of both the water and the clays.

BIOLOGICAL PROCESSES

Biological processes in the subsurface are fascinating for many reasons, not the least being that they provide us with one of our most natural and powerful opportunities for restoring the contaminated subsurface. The range and specificity of biological reactions is astounding, and can be selected in some cases to suit different requirements. Bioengineering is currently being developed and may solve many of our mixed-waste problems that occur in heterogenous systems that cannot be treated with nonbiological techniques.

Subsurface Organisms

Macroscopic organisms exist in the subsurface generally down to the lower limit of the root zone of the native perennial plants, considered to be the base of the existing soil horizon. This depth depends upon many factors including climate, substrate type, depth of water table, type of organisms, and frequency and magnitude of tectonic or other physical processes that can affect rates of subsurface perturbations. The depth to which macroscopic organisms exist can vary from a meter or so to tens of meters for deep taprooted systems. The relict greasewood plant community at the Hanford Site in Washington State, for example, have taproots that have kept pace with a slow lowering of available subsurface water, and are now up to 80 ft long.

However, microscopic organisms can survive and thrive at lower depths. The lowest depth has not yet been determined, but organisms have been found deeper than 10,000 ft. These microorganisms are always bacteria or fungi, including the cyanobacteria, previously called the blue-green algae. As will be discussed in the subsection below, there must be an energy source and a carbon source for organisms to live. Subsurface microorganisms can be both aerobic (can utilize oxygen) and anaerobic (cannot utilize oxygen). Obligate aerobes must have oxygen and obligate anaerobes cannot grow or are killed by the presence of oxygen. Aerotolerant anaerobes are unaffected by oxygen. Facultative organisms can grow either aerobically or anaerobically. As will be discussed below, aerobes generally grow faster and are more productive than anaerobes because of the enhanced energy from using oxygen as the final electron acceptor.

Nutrition, Energy, and Cell Metabolism

A short outline of genetics, cell metabolism, and nutrition is necessary for the discussion of biological systems and processes in the subsurface because it is the metabolic processes that result in the biodegradation of contaminants.

Biological systems are constructed, or synthesized, from chemicals obtained from the environment called nutrients. This biosynthesis, as well as other cell functions, requires energy. Energy is obtained from the environment through three types of sources: light, inorganic chemicals, and organic chemicals; the organisms using these sources are called phototrophs, lithotrophs, and heterotrophs, respectively. Organisms also need a source of carbon for biosynthesis, and can obtain it from an inorganic source (CO_2) or an organic source (other organisms or products of organisms); the organisms using these sources are called autotrophs and heterotrophs, respectively. These terms are summarized in Table 8-2.

Chemical reactions are accompanied by changes in energy. The amount of energy is expressed in terms of the gain or loss during the reaction. Not all of the energy change is available to do useful work, but is lost as heat or to the disorder of the system (entropy). The available useful energy is the free energy. A negative free energy means that energy is released during the reaction (exergonic). A positive free energy means that energy is required for the reaction to proceed (endergonic). The reaction can be reversible in that the product of the reaction can re-react to form the initial reactants. In an exergonic

Table 8-2 Terms Used to Describe Various Nutritional Types in Organisms

Source Type	Subsource	Term
Energy	Light, inorganic chemicals, organic chemicals	Phototroph, lithotroph, heterotroph
Carbon	Inorganic; organic	Autotroph, heterotroph

reaction, the reaction products will build up and the reactants will be used up until a balance, or chemical equilibrium, occurs in which the rate of the forward reaction equals the rate of the backward reaction. A large negative free energy means that the reaction favors the products, and if equilibrium is reached, then little of the reactants will remain. A small negative free energy means that the reaction favors the reactants and, if equilibrium is reached, then both reactants and products will remain. This field of study which investigates chemical equilibrium is called thermodynamics.

Thermodynamics tells us only what conditions will prevail if equilibrium is achieved, but does not tell us how long it will take to achieve equilibrium. The rate of reactions is described by the field of study known as kinetics. The formation of water from gaseous oxygen and hydrogen is a good example of a kinetically controlled reaction. The reaction has a large negative free energy so that if gaseous oxygen and hydrogen are mixed and equilibrium is achieved, all of the water possible should form. However, experiments show that when they are mixed, no measurable water forms over many years. This is because the reactants require energy to first break some of their chemical bonds before the products can form. This energy is called the activation energy and can be large enough to prevent an otherwise favorable reaction from occurring. Usually the activation energy comes from thermal energy of the surrounding environment, causing most reactions to proceed faster at a higher temperature. A catalyst is any substance that lowers the activation energy of a reaction without being permanently changed by the reaction. Catalysts usually work by changing the reactants into a higher energy form that more readily reacts. The overall free energy of the reaction is not changed, only the speed of the reaction. Metals such as platinum have long been used to catalyze many organic reactions, and in automobiles to catalyze the combustion of unburned fuel.

The cell can be thought of as a chemical machine that converts energy from one form to another, breaking down molecules, building up other ones, and carrying out many types of transformations. Metabolism refers to the many chemical processes that occur within the cell that are necessary for survival of the cell. Cells are also corroding devices, translating and processing information during their lifespans and passing on information to offspring.

The components of the cell as a chemical machine are enzymes. Enzymes are protein molecules which catalyze specific chemical reactions (i.e., lower the activation energy for the reaction allowing it to occur much faster at a much lower temperature). Regulatory proteins, such as hormones, regulate enzyme functions and other reactions. The importance of this cannot be overstressed. In the absence of the appropriate enzyme, these specific reactions are usually extremely slow or completely unfavorable energetically, and would not occur to any appreciable degree. Enzymes can be so specific as to catalyze a single reaction of a specific molecule. Enzymes are named either for the reactant or substrate they bind to, or for the reaction they catalyze.

Organisms use the free energy from reactions to obtain energy for cellular activity. The process by which chemicals are broken down into simpler constituents for the release of energy is called catabolism. Respiration is catabolism that produces stored energy in the form of high-energy compounds. With the absence of sunlight below the surface, respiration is the main process of energy production in the subsurface.

The utilization of chemical energy by organisms involves oxidation-reduction reactions or redox reactions. Oxidation is defined as the removal of electrons from a substance, and reduction is the addition of electrons. In biochemistry, redox reactions often involve the transfer of whole hydrogen atoms with or without their electron. The hydrogen atom has one electron, e^-, and one proton. Therefore, a hydrogen that has lost an electron, H^+, is really a single proton. Note that oxygen need not be involved in redox reactions. Redox reactions are always coupled such that the oxidation of one compound requires the reduction of another compound. Therefore, redox reactions are often written as half-reactions. The formation of water from gaseous oxygen and hydrogen is given by the overall reaction:

$$\tfrac{1}{2}O_2 + H_2 \rightarrow H_2O$$

is the result of the two half-reactions involving the transfer of electrons and protons:

$$H_2 \rightarrow 2e^- + 2H^+$$

$$\tfrac{1}{2}O_2 + 2e^- + 2H^+ \rightarrow H_2O$$

in which hydrogen is oxidized to protons and oxygen is reduced to water.

Substances vary in their tendencies to give up electrons, and this tendency is called the reduction potential, referenced to H_2 as the standard, and is measured electrically, as E_0, in volts. Most molecules can serve as both electron donors and acceptors depending upon what other reactants are involved. Redox pairs can be arranged in order of strongest reductants at the top to strongest oxidants at the bottom, called the electron tower shown in Figure 8-3. The reduced substances at the top of the tower have the greatest potential to lose an electron and the oxidized substances at the bottom have the greatest tendency

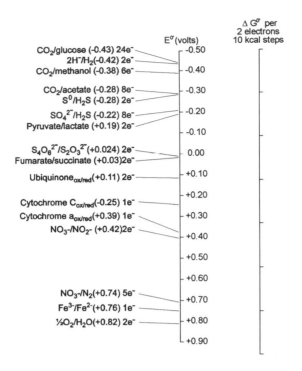

Figure 8-3   The electron tower (after Brock and Madigan, 1988).

to accept electrons. Therefore, electrons can cascade down the electron tower and be involved in many different half-reactions as they fall. Biological systems have evolved to take advantage of these redox pairs and arrange reactions within the cell to capture energy as the electrons cascade. Figure 8-3 shows why aerobic respiration provides so much energy by coupling the glucose $\rightarrow$ CO_2 to the $1/2O_2 \rightarrow H_2O$ overall reactions which occur at the extreme ends of the electron tower. Anaerobic catabolism couples the glucose$\rightarrow CO_2$ to some other redox pair, such as the $NO_3^- \rightarrow NO_2^-$.

Most biological redox reactions involve the coupling of many smaller reactions to achieve an overall redox reaction such as the glucose $\rightarrow CO_2$. The cell employs many electron carriers, or intermediates, used to control the movement of electrons during these overall redox reactions. The carriers, also called coenzymes, are divided into two groups, those that are fixed or attached to enzymes in cell membranes, and those that are freely diffusible. The most important freely diffusible coenzymes are nicotinamide-adenine dinucleotide (NAD[+]) and NAD-phosphate (NADP[+]). These redox pair catalysts, NAD[+]/ NADH used in catabolic reactions and NADP[+]/NADPH used in biosynthesis, participate in many reactions throughout the cell as schematically shown in Figure 8-4. An excellent portrayal of the three dimensionality of enzyme and coenzyme structure and specificity is reproduced in Figure 8-5. Shown in detail

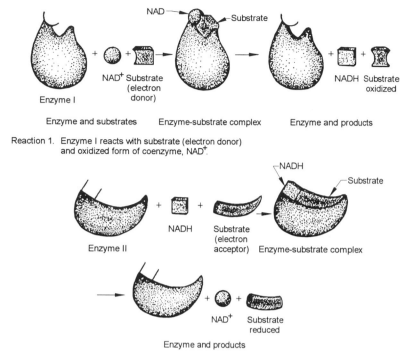

Reaction 1. Enzyme I reacts with substrate (electron donor)
and oxidized form of coenzyme, NAD⁺.

Reaction 2. Enzyme II reacts with substrate (electron acceptor)
and reduced form of coenzyme, NADH.

Figure 8-4   Schematic example of an oxidation-reduction reaction involving the coenzyme NAD+/NADH (after Brock and Madigan, 1988).

is the binding of NAD^+ to the enzyme that catalyzes the dehydrogenation of lactate and reduces NAD^+ to NADH. Note that amino acids are the building blocks of proteins, and designations such as His 195 refer to the numbered amino acids in the enzyme (i.e., the 195th amino acid is histine in the enzyme pictured in Figure 8-5).

Energy released from redox reactions must be conserved for cell functions, usually in the form of high-energy phosphorylated compounds in which a phosphate group (PO_4) is bonded with high-energy bond to some molecule. When this bond is broken, anywhere from 7 to over 14 kcal/mol is released. The most important high-energy compound in cells is adenosine triphosphate (ATP) which has two high-energy phosphate bonds, and adenosine diphosphate (ADP) which has one high-energy phosphate bond (Figure 8-6). Enzymes catalyze the phosphorylation of ADP $\rightarrow$ ATP by coupling the reaction to the breakdown of a carbon source such as the sugar (glucose).

The breakdown pathways through which these coenzymes and high-energy compounds are cycled are given different names depending upon the group of

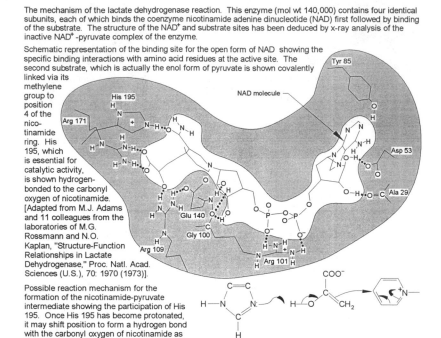

The mechanism of the lactate dehydrogenase reaction. This enzyme (mol wt 140,000) contains four identical subunits, each of which binds the coenzyme nicotinamide adenine dinucleotide (NAD) first followed by binding of the substrate. The structure of the NAD^+ and substrate sites has been deduced by x-ray analysis of the inactive NAD^+-pyruvate complex of the enzyme.

Schematic representation of the binding site for the open form of NAD showing the specific binding interactions with amino acid residues at the active site. The second substrate, which is actually the enol form of pyruvate is shown covalently linked via its methylene group to position 4 of the nicotinamide ring. His 195, which is essential for catalytic activity, is shown hydrogen-bonded to the carbonyl oxygen of nicotinamide. [Adapted from M.J. Adams and 11 colleagues from the laboratories of M.G. Rossmann and N.O. Kaplan, "Structure-Function Relationships in Lactate Dehydrogenase," Proc. Natl. Acad. Sciences (U.S.), 70: 1970 (1973)].

Possible reaction mechanism for the formation of the nicotinamide-pyruvate intermediate showing the participation of His 195. Once His 195 has become protonated, it may shift position to form a hydrogen bond with the carbonyl oxygen of nicotinamide as depicted above.

Figure 8-5   Possible portrayal of the 3-dimensionality of enzyme and co-enzyme structure and specificity (after Lehninger, 1976).

reactions. The glycolytic pathway, or glycolysis, is the pathway from glucose to pyruvate and involves eight specific reactions that, for each glucose molecule, consume two molecules of ATP and subsequently produce four molecules of ATP and two molecules of NADH. Depending upon the organism and the environment, the pyruvate then undergoes reactions to form an end product. Under anaerobic conditions (either in the deep subsurface, in a fermenting vat, or in an overworked human muscle), the pyruvate might go to ethanol or lactic acid plus CO_2. In aerobic respiration, the pyruvate enters the tricarboxylic acid cycle (the Krebs cycle) where it is completely oxidized in the presence of molecular oxygen to CO_2 and H_2O and producing 38 molecules of ATP and six molecules of NADH. This is an obvious advantage over anaerobic respiration, and the reason why aerobic biodegradation of organic contaminants is so desirable in the subsurface environment.

Enzymes, DNA, and Plasmids

Enzymes are proteins, and proteins consist of long chains of the 20 common amino acids which are connected in precise ways so that even their shape must

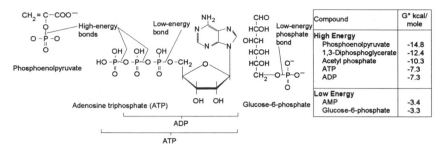

Figure 8-6 High energy phosphate bonds (after Brock and Madigan, 1988).

be correct in order to catalyze a specific reaction. The amino-acid sequence determines the protein structure and composition. How the cell determines this sequence is the role of the cell as a coding device. The code, called the genetic code, is stored in a sequence of several purine and pyrimidine bases attached to a sugar and linked by phosphate groups. In deoxyribonucleic acid (DNA), the sugar is deoxyribose. In ribonucleic acid (RNA), the sugar is ribose. The coding mechanism is the sequencing of bases in the DNA (and in RNA for some bacteria). Each amino acid is coded for by a three-base sequence on the DNA strand. Each DNA strand is bonded to another strand of complimentary DNA into a double-helix structure through the unique pairing of the purine and pyrimidine bases. When replicating, the helix opens, and complimentary DNA strands are built along each strand to form two identical sets of DNA. When translating the information contained in the base sequences, the DNA opens and the sequence is read by building a similarly complimentary RNA strand that then builds the proteins according to its newly produced coded base sequences. There are various base sequences for recognizing when to start and stop specific translations. The series of bases which code for a specific protein is called a gene. The genes are arranged into groups called chromosomes. The genetic code is, therefore, a code only for enzymes and other proteins. These proteins then control and carry out all other reactions needed by the organism.

In relation to the subsurface environment, contaminant compounds will be degraded biologically only if the appropriate enzymes are present in the cell to catalyze the specific reactions. Detoxification of a contaminant, therefore, occurs when a cell utilizes the compound in some metabolic reaction in which an enzyme catalyzes a reaction involving the compound. The reaction can be simple or complicated: for example, the cleaving of a single important functional group off a large molecule, the breaking of an organic molecule into smaller molecules which will have different chemical properties, or the complete degradation of the entire molecule into its inorganic components.

Any discussion of subsurface biological processes must include a newly recognized kind of genetic element called the plasmid. From the above discussion, it may seem as though the presence of enzymes to catalyze reactions with contaminants are random or fortuitous events based on normal cellular activity in the absence of the contaminant. However, the cell can play a more active role through the formation and use of plasmids. Plasmids are circular genetic elements that occur outside the chromosomes and reproduce independently of the chromosomes. Plasmids are not necessary for normal cell survival but have the ability to confer specific abilities to the cell, such as the biodegradation of specific compounds. This allows the cell to adapt to an environmental change, such as the presence of a new contaminant, without altering its intrinsic genetic code. Plasmids can be integrated into the chromosome to become permanent genetic material, but often disappear after the compound is no longer present. It is plasmids that have produced many of the organisms resistant to antibiotics, pesticides, etc. and it is obvious that plasmids will prove crucial to engineering subsurface microbes for biodegradation of many hazardous materials.

The Bacterial Cell and Biological Processes

The components of a bacterial cell, in this case the bacterium E. coli, is shown in Figure 8-7 and serves to synthesize what has been discussed thus far. Each species of bacteria and other subsurface organisms may have variations in some of its components. When thinking about the subsurface environment during subsequent discussions, keep in mind the cell structure shown in Figure 8-7 and what this cell may encounter in the subsurface.

Most researchers consider that subsurface microbial communities exist largely under starvation condition. When the microbial community is offered an energy source, specific populations increase in number and activity to utilize this energy. The specific population and activity may change with time as the source diminishes or certain metabolic products increase which may be utilized as a source by another population. Some enzymes are excreted from the cell or released upon cell rupture, or lysis, where they react with insoluble polymers or high molecular weight molecules to produce more usable products for the organism. This also affects the mobilities of the various compounds and their products.

The primary biological processes of interest to contaminant migration in the subsurface are biodegradation, bioassimilation, biovolatilization, bioaccumulation, and biomineralization.

Biodegradation of Xenobiotics

Xenobiotic compounds are chemically synthesized compounds that are produced artificially and have never existed naturally. Thus, organisms capable of utilizing or degrading these compounds did not previously exist. Some of the most widely distributed xenobiotics are herbicides and pesticides which can persist in the subsurface for weeks to years (Table 8-3). If a substance can be

The structural organization of prokaryotic cells.

Prokaryotes are very small, relatively simple cells having only a single membrane, the cell membrane, which is usually surrounded by a rigid cell wall. Since they have no other membranes, they contain no nucleus and no membranous organelles such as mitochondria or endoplasmic reticulum. The prokaryotes include the eubacteria, the blue-green algae, the spirochetes, the rickettsiae, and the mycoplasma or pleuro-pneumonialike organisms. They contain only one chromosome, which consists of a single molecule of double helical DNA, densely coiled to form the nuclear zone; prokaryotes reproduce largely by asexual division. Prokaryotes were the first cells to arise in biological evolution.

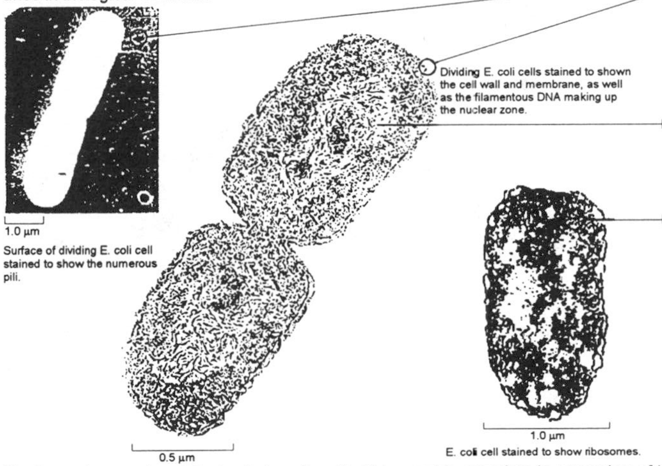

Schematic Drawing	Molecular Composition	Properties and Functions
Cell wall and membrane 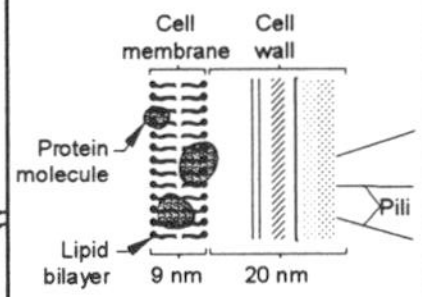 (Cell membrane, Cell wall, Pili, Protein molecule, Lipid bilayer, 9 nm, 20 nm)	The cell wall contains a rigid framework of polysaccharide chains cross-linked with short peptide chains. Its outer surface is coated with lipopolysaccharide. The pili, not found in all bacteria, are extensions of the cell wall. The cell membrane contains about 45% lipid and 55% protein; the lipids form a continuous non-polar phase. Infoldings of the cell membrane are called mesosomes.	The cell wall protects bacteria against swelling in hypotonic media. It is porous and allows most small molecules to pass. Some of the pili are hollow and serve to transfer DNA during sexual conjugation. The membrane is a selectively permeable boundary which allows water, certain nutrients, and metal ions to pass freely. Enzymes responsible for conversion of nutrient energy into ATP are located in the membrane.
Nuclear zone 	The genetic material is a single chromosome of double-helical DNA 2 mm. in diameter and about 1.2 mm. long, which is tightly coiled.	DNA is the carrier of genetic information. During cell division, each strand is replicated to yield two daughter double-helical molecules. From one strand of DNA the genetic message is transcribed to form messenger RNA.
Ribosomes 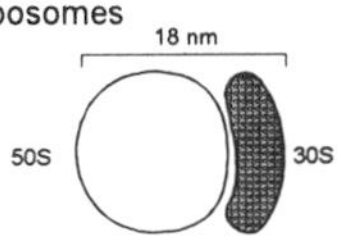 (30S, 18 nm, 50S)	Each E. coli cell contains about 15,000 ribosomes. Each ribosome has a large and a small subunit. Each subunit contains about 65% RNA and 35% protein.	Ribosomes are the sites of protein synthesis. Messenger RNA binds in the groove between the subunits and specifies the sequence of amino acids in the growing polypeptide chains.
Storage granules 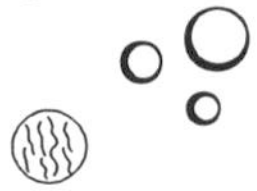	E. coli and many other bacteria contain storage granules that are polymers of sugars. Some bacteria contain granules of poly-β-hydroxybutyric acid.	When needed as fuel, these polymers are enzymatically degraded to yield free glucose or free-β-hydroxybutyric acid.
Cytosol 	The soluble portion of the cytoplasm is highly viscous; the protein concentration is very high, exceeding 20%.	Most of the proteins of the cytosol are enzymes required in metabolism. The cytosol also contains metabolic intermediates and inorganic salts.

Dividing E. coli cells stained to shown the cell wall and membrane, as well as the filamentous DNA making up the nuclear zone.

1.0 μm

E. coli cell stained to show ribosomes.

0.5 μm

Surface of dividing E. coli cell stained to show the numerous pili.

1.0 μm

Electron micrographs of the bacterium E. coli. This aerobic organism is a member of the coliform group of bacteria, typically found in the human intestinal tract. The mature cell is a cylindrical rod about 2 μm long and 1 μm in diameter; it weighs about 2 pg. E. coli cells multiply rapidly on a simple medium containing glucose as carbon source and ammonium ions as nitrogen source; the division time may be as short as 20 min. at 37°C. Most of our knowledge of the molecular basis of genetics has arisen from the study of various strains and mutants of E. coli and E. coli bacteriophages. although more is known about the biochemistry and genetics of E. coli than for any other cell, we are still very far from a complete molecular description.

Figure 8-7 Schematic showing various components of a bacterial cell (after Lehninger, 1976).

utilized by microorganisms, it will disappear from the subsurface faster. Of course, loss can also occur by volatilization, leaching, and chemical or thermal breakdown. Organisms that are able to degrade xenobiotics are fairly diverse and include genera of both bacteria and fungi. Some xenobiotics serve as both carbon and energy sources and are metabolized completely to CO_2. Others may be degraded partially or totally provided some other organic is present as the primary energy source, a phenomenon known as cometabolism. Sometimes partial cometabolic breakdown of a contaminant produces an even more toxic compound, as in the case of methanogenic biodegradation of trichloroethylene (TCE), which produces a highly toxic vinyl chloride as a metabolic product.

Because metabolic reactions are so specific, compounds that are closely related can exhibit surprisingly different degradation rates. For example, 2,4-dichlorophenoxyacetic acid (2,4-D) and 2,4,5-trichlorophenoxyacetic acid (2,4,5-T), two chlorinated herbicides that differ by only a single chlorine atom, show vastly different degradation rates. Ordinarily 2,4-D degrades rapidly after a few days while 2,4,5-T is hardly degraded at all. Recently, common *Pseudomonas* bacteria have been isolated that can utilize 2,4,5-T as the sole carbon and energy source, a metabolic development thought to have recently arisen from genetic mutation with the plasmid that codes for the degradation of 2,4-D. This shows the rapidity of evolutionary developments and the possibility that plasmids can be developed and transferred to organisms to confer specific degradation capabilities for hazardous compounds.

Biovolatilization

Organisms can also remove contaminant species from a solution by other methods. Bacteria, molds, and fungi can volatilize metals by attaching a methyl group, $-CH_3$, called methylation, that makes the methylated metal very volatile. Examples have been seen for mercury, selenium, tellurium, arsenic, and tin. However, some of the methylated metals are highly toxic themselves and very mobile.

Bioassimilation and Bioaccumulation

Bioassimilation refers to the process of the uptake of a contaminant species into the cell. The accumulation of a contaminant species by the organism may be accomplished either inside the cell or external to the cell. Except for biodegradation of organic compounds, much of the study on biological interactions with contaminants has focused on bacteria-metal interactions.

Precipitation of metals outside the cell occurs when the organism produces metabolic products which are excreted and result in the immobilization of the metal. Sulfate-reducing bacteria in the anaerobic subsurface excrete hydrogen sulfide (H_2S), which can react with metals in solution to produce metal sulfide minerals such as covellite (CuS), pyrite (FeS_2), sphalerite (ZnS), and galena

Table 8-3 Persistence of Herbicides and Insecticides in Soil

Substance	Time for 75–100% Disappearance
Chlorinated insecticides	
DDT [1,1,1-trichloro-2,2-bis(p-chlorophenyl)ethane]	4 years
Aldrin	3 years
Chlordane	5 years
Heptachlor	2 years
Lindane (hexachloro-cyclohexane)	3 years
Organophosphate insecticides	
Diazinon	12 weeks
Malathion	1 week
Parathion	1 week
Herbicides	
2,4-D(2,4-dichloro-phenoxyacetic acid)	4 weeks
2,4,5-T(2,4,5-trichloro-phenoxyacetic acid)	20 weeks
Dalapin	8 weeks
Atrazine	40 weeks
Simazine	48 weeks
Propazine	1.5 years

(PbS). This process has been used to treat metal-contaminated lakes and other waters.

Some microorganisms generate chemicals that have a high binding efficiency for metals. Siderophores are chelating agents excreted by the cell that bind iron and catalyze its uptake into the cell. Siderophores have been modified to bind Cd, Cr, Cu, Pb, Hg, Ni, Zn, Co, Cs, Sr, Th, and U. Extracellular polymers also play an important role in metal uptake into cells and are often polysaccharides which adsorb the metal ions.

Organisms can also accumulate metals by binding or precipitating them directly onto their cell walls and cell membranes. Cell walls and cell membranes tend to be negatively charged owing to the presence of OH^-, $HCOO^-$, and HPO_3^- groups of the cell components, and so positively charged ions are readily sorbed and bound to their surfaces. As with most cellular processes, the pH of the environment is crucial.

Biomineralization

Biomineralization refers to the precipitation of minerals by organisms, either as internal hard parts or as external structures. Mostly, biomineralization occurs in larger organisms in the form of shells, reefs, skeletal systems, etc. and many of these structures can sequester high levels of contaminant species, especially metal species. However, macroscopic organisms generally do not occur in the subsurface below the active biological horizons. Some bacteria do form internal mineral parts, such as the marine magnetotactic bacteria that precipitate the magnetic mineral magnetite to align themselves along the earth's magnetic field as a way of orienting themselves during movement.

CHEMICAL PROCESSES

Although a comprehensive treatment of chemical processes acting in the subsurface is not possible, some fundamentals of aqueous chemistry and thermodynamics do provide a great deal of understanding of these processes and problems. More comprehensive treatments can be found in Stumm and Morgan (1981) and Hem (1989).

Solute Concentrations in Natural Waters

Any meaningful discussion of aqueous solutions must define the concentration units of dissolved constituents. The solute is the substance that is dissolved in the solvent. Acetone is the solvent in nail polish remover and the dissolved nail polish is the solute. In salt water, the water is the solvent and the salts are the solutes. Suspended particles or droplets of other liquids in the solution are not solutes because they are not dissolved in the solvent.

Constituents are often parenthetically designated with (g), (l), and (s) to differentiate between the gaseous, liquid, and solid states, respectively. The designation (aq) means that the constituent is dissolved in water. As examples, CO_2 (g) means carbon dioxide gas; CO_2 (aq) means carbon dioxide gas dissolved in water; CCl_4 (l) means liquid carbon tetrachloride; and $CaCO_3$ (s) means solid calcium carbonate.

A mole of any substance is the formula weight of the substance in grams. For example, the atomic weight of sodium (Na) is 22.99, so 1 mol of Na weighs 22.99 g. The atomic weight of chlorine (Cl) is 35.45, so 1 mol of Cl weighs 35.45 g. The molecular weight of common salt (NaCl) is 22.99 + 35.45 = 58.44, so 1 mol of NaCl weighs 58.44 g and contains 1 mol of Na and 1 mol of Cl. The atomic weight of oxygen (O) is 16.0, so 1 mol of O weighs 16.0 g; however, 1 mol of molecular oxygen (O_2), which has a molecular weight of 32.0, weighs 32.0 grams and contains 2 mol of atomic oxygen. Nitrate ion (NO_3^-) has 1 mol of nitrogen (14.0 g), thus, 1 mol of NO_3^- weighs 62 g. Note that electrons are essentially weightless, so the charge does not matter as far as weight is concerned.

Molarity is the number of moles in 1 L of solution designated mol/L or just M. Thus, a 1 M aqueous NaCl solution means that 58.44 g of NaCl are in 1 L of water. A 0.1 M aqueous NaCl solution means that 5.844 g of NaCl are in

1 L of water. For convenience, we shall assume solutions are aqueous unless otherwise stated (i.e., the solvent is water).

The designations of parts per million (ppm), or parts per billion (ppb), are mass concentration units, such that a 1 ppm NaCl solution means that there is 1 one-millionth of 1 g of NaCl for every gram of water, or 1 lb of NaCl for every 1 million lb of water. Unlike molarity, however, the weight is not equally split between Na and Cl but must be calculated from their relative molecular weights. Because 1 L of water weighs 1000 g, 1 ppm is equivalent to 1 mg/L of water, and thus these units are used interchangeably.

When discussing chemical reactions, we need to know the molarity of solutions because a reaction written as $A + 2B \rightarrow C$ means 1 mol of A plus 2 mol of B produces 1 mol of C. However, often we only know the total weight of the solutes dissolved in a solution, or the total dissolved solids (TDS). The TDS of any water, usually given in mg/L, is determined by weighing the solid residue obtained by evaporation of a filtered sample.

Two other concentration units are defined here for completeness but will not be used in this review. Molality (m) is the number of moles of solute dissolved in 1 kg of solvent. Because 1 L of pure water weighs 1 kg, molality is essentially the same as molarity for dilute solutions (i.e., less than about 0.001 M). Most or our discussions about the subsurface will assume dilute solutions. Equivalents per liter (epL) is a concentration unit that involves the charge, or valence, of an ion or charged species, and is equal to the number of moles of the solute times its charge number per liter of solution. As an example, a 1 M NaCl solution, in which the solutes occur as the dissolved ionic species Na^+ and Cl^-, is a 2 epL solution because it has 1 mol each of two singly charged species. However, a 1 M $CaSO_4$ solution in which the solutes occur as the dissolved ionic species Ca^{2+} and SO_4^{2-}, is a 4 epL solution because it has 1 mol each of two doubly charged species. In situations where the charge is important, one can think of Ca^{2+} as being twice as effective as Na^+, or as being equivalent to two Na^+ ions.

Groundwater contains a wide variety of dissolved inorganic constituents in various concentrations as a result of chemical and biochemical interactions with the geologic materials through which it has flowed, and to a lesser extent the surface water source from which it came. The general categories and types of natural waters are presented in Table 8-4.

Although the detailed composition of each water is different, the major constituents in most natural waters, as discussed in Chapter 4, are sodium, magnesium, calcium, chloride, bicarbonate, and sulfate (Na^+, Mg^{2+}, Ca^{2+}, Cl^-, HCO_3^-, and SO_4^{2-}) which normally constitute more than 90% of the TDS and occur in the >10-mg/L concentration range. Important minor constituents in most natural waters are potassium, silica, iron, strontium, carbonic acid, boron, nitrate, carbonate, and fluoride (K^+, SiO_2, Fe^{2+}, Sr^{2+}, H_2CO_3, B, NO_3^{2-}, and F^-) which normally occur in the 0.01 to 10 mg/L concentration range. The rest of the elements occur as trace constituents in the <0.0001 to 0.01 mg/ L concentration range.

Table 8-4 Simple Water Classification Based on Total Dissolved Solids

Category	Total Dissolved Solids[a,b] (mg/L or ppm)
Freshwater	<1,000
Brackish water	100–10,000
Saline water	10,000–100,000
Brine water	>100,000

Type	Average range in TDS
Rain/snow water	1–50
River/lake water	10–200
Estuaries	500–10,000
Sea water	~35,000
Groundwaters	
Immature	100–1,000
Mature	1,000–5,000

[a] After Holland (1984), Hein (1989), Drever (1988), Cherry (1979), and Strumm and Morgan (1981).
[b] Water having a TDS greater than 2,000 mg/L is generally too salty to drink.

Notice that these constituents are represented as simple ionic species. The actual species in solution depends upon many conditions. Most simple ions, if not part of a complex species, are hydrated by the surrounding water molecules, usually about six, to form a hydrated complex (i.e., $[Na \cdot 6H_2O]^+$).

Organic constituents are ubiquitous in natural waters, although the concentrations are much lower than the inorganic constituents. Little is known about the chemistry and speciation of dissolved organic matter but the total concentration is an important part of groundwater investigations and is designated as dissolved organic carbon (DOC). DOCs are commonly in the 0.1 to 10 mg/L concentration range.

Chemical Equilibria

To develop the basis for understanding chemical reactions, especially in the subsurface, discussion of chemical equilibria is warranted. It has long been known that the driving force for a chemical reaction is related to the concentrations of the reactants and the concentration of the products. Consider the reaction:

$$aA + bB \rightleftharpoons cC + dD \tag{8-1}$$

where a, b, c, and d are the number of moles of the chemical constituents A, B, C, and D, respectively. The double half-arrows ($\rightleftharpoons$) mean that the reaction is reversible and will go in both directions. The Law of Mass Action provides the relationship between the reactants and the products at equilibrium and is expressed as

$$K = \frac{[C]^c[D]^d}{[A]^a[B]^b} \tag{8-2}$$

where [A] is read as "the activity of the species A, raised to the power of its mole number (a)," and K is the thermodynamic equilibrium constant or stability constant. The activity of a species is the effective concentration of the species, and is about equal to the actual concentration for dilute or ideal solutions. For more concentrated solution, the ions in solution begin to interfere with each other's ideal behavior and the activities must be corrected for these interactions.

Note that in the mass action expression of Equation 8-2 the products are in the numerator, so that the larger the value of K, the more the reaction favors the products. For example, if K is 1^{-20} then there will be almost no reactants left at equilibrium, whereas if K is 10^{-10} then there will be almost no products formed. If K is neither very large nor very small then significant amounts of both reactants and products will exist in solution at equilibrium. K is highly dependent upon temperature, pressure, and chemistry within the system. Note also that there is no rate, or kinetic, information in the mass action expression.

For example, consider the reaction that occurs when groundwater flows through a limestone aquifer (Freeze and Cherry, 1979). A limestone is mostly composed of the mineral, calcite (calcium carbonate):

$$CaCO_3 \rightleftharpoons Ca^{2+} + CO_{3^{2-}} \tag{8-3}$$

This reaction will proceed to the right (mineral dissolution) or to the left (mineral precipitation) until the mass-action equilibrium is reached. It may take years, or even thousands of years, to reach equilibrium. If the system is disturbed (i.e., reactants or products are added or removed), or the temperature changes, the system will proceed until equilibrium is once again achieved. If disturbances are frequent enough, equilibrium will never be achieved.

We can compare different solutions by knowing their ionic strength (I) which is half the sum of the charge concentrations of all the ions in solution, or

$$I = 1/2 \sum C_i z_{i^2} \tag{8-4}$$

an expression which looks more formidable than it is. C_i is the molar concentration of each ion (in M units, *not* the mass units ppm or mg/L) and z_i is the charge number of each ion. Because most groundwaters contain only six to ten ionic species in any appreciable concentration, as mentioned above, the ionic strength can usually be calculated easily. As an example, consider a porewater that exists in the vadose zone at the Hanford Site with the composition as presented in Table 8-5.

The ionic strength is

$$I = 1/2 \{ (0.00080)(1) + (0.00010)(1) \\ + (0.00043)(4) + (0.00031)(4) \\ + (0.00074)(1) + (0.00004)(1) \\ + (0.00067)(1) + (0.00017)(1) \\ + (0.00039)(4)$$

or

$$I = -0.0035$$

Note that for Ca^{2+}, Mg^{2+}, and SO_4^{2-}, $z_{i2} = 4$ because of the double charges, making their relative contributions to the ionic strength much higher than the monovalent ions with only a single charge. This reflects the fact that the higher the charge on an ion, the more "effect" it has on surrounding molecules.

A dilute solution is defined as having an ionic strength less than about 10^{-3} M or 0.001 M. The vadose zone water is close to the limit for considering it a dilute solution. Most groundwaters are between about 0.001 and 0.1 M. Because sea water has an ionic strength of about 0.7 M, it is not dilute and calculations or reactions need to include the correct activity for each species, not actual concentrations.

Henry's Law and Dissolved Gases

When water is exposed to gas or vapor, an equilibrium is established between the gas and the liquid by exchange of molecules across the liquid-gas interface. This equilibrium can be written as a reaction and has a mass action expression similar to any reaction. The equilibrium is controlled by the partial pressure of the gas in the atmosphere. Henry's Law states that in dilute solutions, the concentration of a solute in solution is proportional to its partial pressure in the gas phase or atmosphere. The constant of proportionality is referred to as Henry's Law constant (K_H). Consider water in contact with air containing carbon dioxide gas, CO_2 (g). The overall reactions that control the interaction between the CO_2 (g) and the water is

Table 8-5 Summary of Chemical Analysis of Porewater in Vadose Zone at the Hanford Site

Species	Concentration[a] (mol/L or M)
Na$^+$	0.00080
K$^+$	0.00010
Ca^{2+}	0.00043
Mg^{2+}	0.00031
HCO$_3$-	0.00074
F-	0.00004
Cl-	0.00067
NO$_3$-	0.00017
SO$_4^{2-}$	0.00039

[a] These concentrations would probably be given in millimoles per liter (mmol/L or mM) to make them easier to read (i.e., 0.80 mM instead of 0.00080 M).

$$CO_2(g) + H_2O \rightleftharpoons CO_2(aq) \qquad (8\text{-}5)$$

$$CO_2(g) + H_2O \rightleftharpoons H_2CO_3(aq) \qquad (8\text{-}6)$$

Although most of the CO_2 that enters the water remains as gas and only a small fraction forms carbonic acid (H_2CO_3), it is customary to denote all of the dissolved CO_2 as H_2CO_3(aq) as long as the subsequent mass action expressions include this generality. The Henry's Law expression for Reaction 8-9 is

$$K_{H_{CO_2}} = \frac{[H_2CO_3]}{[H_2O][CO_2(g)]} \qquad (8\text{-}7)$$

We shall now introduce some very important conversions when dealing with mass action expressions and reactions. First, in dilute solutions and most natural waters, the activity of H_2O is taken as unity, or $[H_2O] = 1$. Second, ideal gas behavior for the gases is assumed such that their activity is equal to their partial pressure in atm, or $[CO_2 (g)] = P_{CO_2}$. Third, the activity of solids are taken as unity, or $[CaCO_3] = 1$. Therefore, the mass action expression for Reaction 8-6 simplifies to

$$K_{HCO_2} = \frac{[H_2CO_3]}{P_{co_2}} \qquad (8\text{-}8)$$

At 25°C at sea level, the atmospheric partial pressure of $CO_2(g)$ is 3.2×10^{-4} atm, also denoted as $10^{-3.5}$ atm, and the Henry's Law constant for CO_2 (g) is $10^{-1.5}$. Therefore, $[H_2CO_3] = 10^{-3.5} \times 10^{-1.5} = 10^{-5.0}$ M. Each gas has its own Henry's Law constant that varies with temperature.

Dissociation

Dissociation is the process whereby a molecule splits into smaller components, often in response to entering a solution or to changing conditions, and can be represented as an equilibrium reaction with a mass action expression similar to any reaction. A most important dissociation reaction is that of water:

$$H_2O \rightleftharpoons H^+ + OH^- \qquad (8\text{-}9)$$

Its mass action expression is

$$K_w = \frac{[H^+][OH^-]}{[H_2O]} \qquad (8\text{-}10)$$

Taking $[H_2O]$ as unity gives

$$K_w = \rightleftharpoons [H^+][OH^-] \qquad (8\text{-}11)$$

At 25°C at sea level, $K_w = 10^{-14}$ and is the basis for the pH scale. K_w is called the dissociation constant for water. Because it is convenient to multiply and divide numbers by addition and subtraction of their exponents, numbers are represented by their negative logarithms (i.e., their base-ten exponents, called a p-scale). Therefore, if the concentration of H^+ is 10^{-5} M, then $[H^+] = 10^{-5}$ and pH = 5. Because it is the negative log, the lower the p-value, the higher the actual value. Likewise, $K_w = 10^{-14}$, so $pK_w = 14$. If a solution at 25°C at sea level has $[H^+] = 10^{-5}$, then pH = 5, $pK_w = 14$, pOH = 14 − 5 = 9, or $[OH^-] = 10^{-9}$. This convention makes it easier to determine the solution concentrations from mass action expressions.

Two very important dissociations in natural waters are the two dissociations of carbonic acid (H_2CO_3), each involving the loss of H^+. The first dislocation is

$$H_2CO_3 \rightleftharpoons H^+ + HCO_{3^-} \qquad (8\text{-}12)$$

The equilibrium constant (K_1), called the first dislocation constant for carbonic acid, is

$$K_1 = \frac{[H^+][HCO_{3^-}]}{[H_2CO_3]} \qquad (8\text{-}13)$$

The second dissociation is

$$HCO_{3^-} \rightleftharpoons H^+ + CO_{3^{2-}} \qquad (8\text{-}14)$$

The equilibrium constant, K_2, is called the second dissociation constant for carbonic acid, and is given by

$$K_2 = \frac{[H^+][CO_{3^{2-}}]}{[HCO_{3^-}]} \qquad (8\text{-}15)$$

At 25°C at sea level, $K_1 = 10^{-6.3}$ and $K_2 = 10^{-10.25}$.

Acid/Base Reactions

A reaction that involves the movement of electrons and protons is called an acid/base reaction. The Lowry-Bronsted definition states that an acid is a species that tends to lose or donate protons, and a base is a species that tends to gain or accept protons, e.g.,

$$HCl + H_2O \rightleftharpoons H_3O^+ + Cl^-$$
$$\text{acid base} \qquad\qquad (8\text{-}16)$$

The Lewis definition states that an acid is a species that tends to accept electrons, and a base is a species that tends to donate electrons, e.g.,

$$BF_3 + F- \rightleftharpoons BF_{4^-}$$
$$\text{acid base} \qquad\qquad (8\text{-}17)$$

the concentration of hydrogen ion (H^+) in aqueous solutions is extremely important because H^+ occurs in so many reaction of interest, and because the dissociation of water produces H^+. The pH is used as an indicator of the acidity of a solution, and pOH as the basicity of a solution. Because $K_w = 10^{-14}$, neutrality is $[H^+] = [OH^-] = 10^{-7}$, and pH = pOH = 7. Therefore, the pH can be thought of as a master variable when working with aqueous solutions, and most of the species will act as Lowry-Bronsted acids and bases.

Dissolution/Precipitation

When water contacts minerals and other solids, dissolution of the solid begins and continues until equilibrium concentrations are reached in the water for the dissolved species, or until all of the solid is consumed. The solubility of a mineral is defined as the mass of material that will dissolve in a unit volume of water under specified conditions. Likewise, if the solution is saturated with respect to some species, a solid may precipitate from solution until equilibrium is reached. A dissolution/precipitation reaction with its mass action relationship can be written as for any reaction. The equilibrium or stability constant in this case, K_s, is often called a solubility constant. Consider dissolution of the mineral calcite:

$$CaCO_3 \rightleftharpoons Ca^{2+} + CO_3^{2-} \qquad (8\text{-}18)$$

$$K_S = \frac{[Ca^{2+}][CO_3^{2-}]}{[CaCO_3]} \qquad (8\text{-}19)$$

Because the activity of solids are taken as unity, $[CaCO_3] = 1$ and

$$K_S = [Ca^{2+}][CO_3^{2-}] \qquad (8\text{-}20)$$

At 25°C at sea level, $K_s = 10^{-8.4}$.

There are two categories of dissolution of solids with respect to the mechanism of dissolution. Congruent dissolution means that the dissolution products are all dissolved species (i.e., no solids left behind). Table salt and calcite are two minerals that dissolve congruently:

$$NaCl \rightleftharpoons Na^+ + Cl^- \qquad (8\text{-}21)$$

$$CaCO_3 \rightleftharpoons Ca^{2+} + CO_3^{2-} \qquad (8\text{-}22)$$

Incongruent dissolution means that the dissolution products are not all dissolved species but contain new minerals or amorphous solids as a residue. Most rock-forming silicate minerals dissolve incongruently, For example, the feldspar mineral albite can dissolve to form the clay mineral kaolinite plus some solutes:

$$2NaAlSi_3O_8(s) + 11H_2O \rightarrow 2Na^+ + 2OH^- + 4H_4SiO_4 + Al_2Si_2O_5(OH)_4(s)$$

$$(8\text{-}23)$$

albite   kaolinite

The single arrow, $\rightarrow$, means that the reaction is not reversible and will only proceed in one direction toward kaolinite.

Broadly speaking, a buffer is any solid, species, reaction, etc. that tends to reduce or prevent a change within the system. Buffer is often used with respect to acid/base systems in which an acid/base reaction is controlled or buffered by some aspect of the system. Antacid tablets use the dissolution of various solids, such as $CaCO_3$, $MgOH$, or $Al(OH)_3$, to buffer the acidity of the stomach. $CO_3{}^{2-}$ or OH^- from the dissolution reactions combine with H^+, removing excess H^+ as a new equilibrium is reached in the stomach. Nonchemical buffers also occur, exemplified by the melting of ice which buffers the temperature in a glass of water so that it remains at $0°C$ as long as ice remains in contact with the liquid.

Hydrolysis

Hydrolysis is a reaction involving the addition of water to a reacting species. Because water is the most prevalent fluid on the earth and is so reactive, hydrolysis has traditionally been treated as a separate phenomenon. However, hydrolysis is an aspect of acid/base equilibria requiring no concepts beyond what has already been discussed. An example is the hydrolysis of the ammonia chloride salt:

$$NHO_{4+}H_2O \rightleftharpoons H^+ + NH_3 \qquad (8\text{-}24)$$

$$K_h = \frac{[NH_3][H^+]}{[NH_{4+}]} \qquad (8\text{-}25)$$

Oxidation/Reduction (Redox)

Redox reactions were discussed previously under the subsection Nutrition, Energy, and Cell Metabolism. To reiterate, oxidation is defined as the removal of electrons from a substance, and reduction is the addition of electrons. Uncontaminated surface waters are usually oxidizing because of mixing with atmospheric oxygen. However, the subsurface's isolation from the atmosphere results in subsurface waters becoming oxygen depleted, and finally reducing, as oxygen is used up in reactions with organic matter and in other hydrochemical reactions such as sulfide oxidation, iron oxidation, nitrogen oxidation (nitrification), and manganese oxidation. Examples of some overall reactions are:

$$O_2 + CH_2O = CO_2 + H_2O$$
$$O_2 + 1/2HS^- = 1/2SO_4^{2-} + 1/2H^+$$
$$1/4O_2 + Fe^{2+} + H^+ = Fe^{3+} + 1/2H_2O$$
$$O_2 + 1/2NH_{4+} = 1/2NH_{4+} + H^+ + 1/2H_2O$$
$$O_2 + 2Mn^{2+} + 2H_2O = 2MnO_2(s) + 4H^+$$

Overall, aromatic compounds such as ethylbenzene, naphthalene, and phenol can be oxidized easily by free radicals, whereas chlorinated compounds such as dioxins and PCBs are not easily oxidized.

Redox reactions also have equilibrium constants and mass action expressions associated with them. However, the equilibrium constant, K, associated with most redox reactions are extremely large (i.e., as high as 10^{50}). This means that if equilibrium is reached, the reactions always go to completion until one or more of the reactants are gone. On the other hand, redox reactions are often kinetically controlled with high activation energies, as previously demonstrated for H_2 and O^2 gas, so that the reactions may proceed very slowly, unless catalyzed. Because redox reactions involve the movement of electrons and changes in the oxidation state of the element, the master variable in these reactions is the concentration of electrons ($[\epsilon]$), which can be expressed as pϵ, analogous to pH.

Complexation

Complexation or chelation is the process by which metal ions and organic or other nonmetallic molecules, referred to as ligands, can combine, forming stable metal-ligand complexes. A complex consists of a central atom closely surrounded by a number of other atoms or molecules that donate electrons to the central atom. The surrounding atoms or molecules are called ligands. Complexes are very important to the migration of contaminants because many metals readily form complexes that are extremely mobile. Certain ligands are specially used for their complexing ability, the most common being ethylene diamine tetra-acetic acid (EDTA) shown in Figure 8-8 complexing a central atom of iron. Once complexed, the metal is generally prevented from undergoing other reactions or interactions that the free metal cation would undergo. In most natural waters, common ligands are small inorganic anionic groups, such as SO_4-, CO_3-, etc.

The Carbonate System

The above information can now be synthesized to describe the chemistry of a solution. The simple carbonate system ($CO_2 - H_2O - CaCO_3$) is probably the

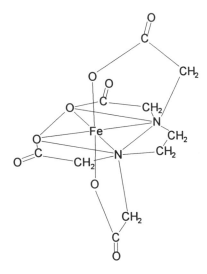

The ethylene diamine tetra-acetic complex of iron. The oxygen and nitrogen atoms occupy the corners of an octahedron with the iron atom at the center.

Figure 8-8   The ethylene diamine tetra-acetic complex of iron. The oxygen and nitrogen atoms occupy the corners of an octahedron with the iron atom at the center (after Mahan, 1975).

single most important set of reactions that control natural water chemistry and it is also the best known. The following is a review of the reactions and mass balance expressions for the carbonate system as an example of equilibrium thermodynamics, remembering that every other element in solution has a similar set of reactions, making solution chemistry a complicated result of many individual systems.

Consider a small pond composed of pure water with some calcite rocks on the bottom of the pond. The atmosphere over the pond has the normal composition at the earth's surface with a partial pressure of CO_2 (g) equal to $10^{-3.5}$ atm. The relationships and equilibrium constants required to fully characterize this system have already been provided in Equations 8-6, 8-8, 8-9, 8-11 to 8-15, 8-18, and 8-20. These equations contain all of the species that will occur in solution, i.e., CO_2 (g), H_2O, H^+, OH^-, H_2CO_3, H_2CO_3., $CO_{3^{2-}}$ and Ca^{2+}, and the mass action relationships contain all of the relationships between their concentrations. Therefore, we can combine them all to give the concentrations of each species as a function of $[H^+]$ which can be easily measured. For instance, combining and rearranging Equations 8-8 and 8-13 gives

$$\frac{K_1 K_{H_{CO_2}} P_{CO_2}}{[H^+]} = [HCO_{3^-}]$$

Combining this expression with Equation 8-15 gives

$$\frac{K_1 K_2 K_{HCO_2} P_{CO_2}}{[H^+]^2} = [CO_{3^{2-}}]$$

At 25°C at sea level these become

$$\frac{10^{-6.3}10^{-1.5}10^{-3.5}}{[H^+]} = [HCO_{3^-}]$$

$$\frac{10^{-6.3}10^{-10.25}10^{-1.5}10^{-3.5}}{[H^+]^2} = [CO_{3^{2-}}]$$

or

From Equation 8-8 we had calculated
From Equation 8-11 we have

$$\frac{10^{-11.3}}{[H^+]} = [HCO_{3^-}] \qquad (8-26)$$

$$\frac{10^{-21.8}}{[H^+]^2} = [CO_{3^{2-}}] \qquad (8-27)$$

$$10^{-5.0} = [H_2CO_3]$$

$$\frac{10^{-14}}{[H^+]} = [OH^-]$$

From Equations 8-20 and 8-26 we have

$$\frac{10^{-8.4}[H^+]^2}{10^{-21.8}} = 10^{-13.4}[H^+]^2 = [Ca^{2+}]$$

We see that the concentration of all species are now given in terms of [H⁺], the concentration of H⁺, which means that a single pH measurement would allow calculation of all these concentrations.

Geochemical computer codes are predictive models that utilize as many sets of reactions and mass action expressions for as many elements in solution as possible. These codes are used to predict what species and concentrations will occur in solution over time as the system achieves equilibrium based upon what solution chemistries and solids are present. Also, these codes are used to observe what changes occur if the system is perturbed (i.e., if a new solid is introduced to the system or if the temperature changes). Obviously, these codes are best if the system corresponds to the reactions for which we have information and does not contain any unknown reactions or species, a case that is not yet common for most real situations.

Speciation

The species in which an element will occur in solution is determined by all of the reactions with other species in that solution in which it can be involved. This is why it is so important to know the chemical composition of the solution and the solids in contact with the solution. As we saw above for the $CO_2 - H_2O$ system, every element has a host of reactions that control the concentration of its various species in solution. Fortunately, relatively few reactions and species dominate in natural waters simply because the materials and conditions of the earth's surface buffer the solutions with respect to many possible reactions. The most probable species that exist in natural waters under aerobic conditions are presented in Table 8-6. Of course, these will depend upon what solids are present, what unusual chemistries may be occurring, and what contaminants might be present that can change the various equilibria relationships. The aqueous species of importance to radionuclide migration under different redox conditions are presented in Table 8-7. When viewing Tables 8-6 and 8-7, remember that the major species in solution, such as HCO_3^-, SO_4^{2-}, NO_3^-, Cl^-, Na^+, or Ca^{2+}, determine many of the species and complexes of other elements that will occur in solution simply because they are the most abundant species and are available for reaction.

There is a special class of species, called polymers, that are chemically very important. A polymer is a long-chain molecule composed of identical small molecular units or groups that are attached together over indefinite lengths. Organic compounds can form many varieties of polymers that have different properties than the individual subunits (i.e., single sugars can polymerize into long polysaccharides).

Thermal Degradation

Many organic compounds and some inorganic compounds degrade upon heating, a process known as thermal degradation. Depending upon the species, the temperature can be low as for some large organic compounds, or may need to be quite high as for small stable compounds. In the subsurface, thermal degradation is a slow but constant process for most organic compounds and in the absence of other more active processes, is the primary degradation process in the subsurface.

Table 8-6 Probable Main Dissolved Inorganic Species in Natural Waters Under Aerobic Conditions[a]

Element	Probable Main Species	Element	Probable Main Species
H	H_2O	Rb	Rb^+
Li	Li^+	Sr	Sr^{2+}
Be	$BeOH^+$	Y	$Y(OH)_3$
B	H_3BO_3, $B(OH)_4^-$	Zr	$Zr(OH)_n^{4-n}$
C	HCO_3^-	Mo	MoO_4^{2-}
N	N_2, NO_3^-	Cd	$CdCl_2^0$ (S), Cd^{2+} (F), $CdOH^+$ (F)
O	H_2O	Sn	$SnO(OH)_3^-$
F	F^-, MgF^+ (S)	Sb	$Sb(OH)_6^-$
Na	Na^+	Te	$HTeO_3^-$
Mg	Mg^{2+}	I	IO_3^-, I^-
Al	$Al(OH)_4^-$	Cs	Cs^+
Si	$Si(OH)_4$, MgH_3SiO_4 (S)	Ba	Ba^{2+}
P	HPO_4^{2-}, $MgPO_4^-$ (S)	La	La^{3+}, $LaOH^{2+}$
S	SO_4^{2-}, $NaSO_4^+$ (S)	Ce	Ce^{3+}, $CeOH^{2+}$
Cl	Cl^-	Pr	Pr^{3+}, $PrOH^{2+}$
K	K^+	Nd	Nd^{3+}, $NdOH^{2+}$
Ca	Ca^{2+}	Other	
Sc	$Sc(OH)_3^0$	rare	Me^{3+}, $MeOH^{2+}$
Ti	$Ti(OH)_4^0$	earths	
V	$H_2VO_4^-$, HVO_4^{2-}	Lu	$LuOH^{2+}$
Cr	$Cr(OH)_3^0$, CrO_4^{2-}	W	WO_4^{2-}
Mn	Mn^{2+}, $MnCl^+$ (S)	Re	ReO_4^-
Fe	$Fe(OH)_2^+$	Au	$AuCl_2^-$ (S), $Au(OH)_3^0$ (F)
Co	Co^{2+}, $CoCO_3^0$ (?)	Hg	$HgCl_4^{2-}$ (S), $Hg(OH)_2^0$ (F) $HgOHCl$ (F)
Ni	Ni^{2+}, $NiCO_3^0$ (?)		
Cu	$CuCO_3^0$, $CuOH^+$	Tl	$TlCl^0$ (S), Tl^+ (F)
Zn	$ZnOH^+$, Zn^{2+}, $ZnCO_3^0$	Pb	$PbCO_3^0$, $Pb(CO_3)_2^{2-}$
Ga	$Ga(OH)_4^-$	Bi	BiO^+, $Bi(OH)_2^+$
Ge	$Ge(OH)_4^0$	Ra	Ra^{2+}
As	$HAsO_4^{2-}$, H_2AsO	Th	$Th(OH)_n^{4-n}$, $Th(CO_3)_n^{4-2n}$ (?)
Se	SeO_3^{2-}	U	$UO_2(CO_3)_3^{4-}$
Br	Br^-		

[a] S, species prevalent in seawater only; F, species prevalent in fresh water.

Table 8-7 Predominant Solution Species of Elements[a]

Elements	Little Affected by Oxidation-Reduction	In an Oxidizing Environment	In a Reducing Environment
Am	Am^{3+}, $AmSO_4^+$, $AM(OH)^{2+}$		
Sb		$HSbO_2^o$, $Sb(OH)_3^o$, $SbOF^o$, $Sb(OH)_4^-$	SbO^+
Ce	Ce^{3+}, $CeSO_4^+$		
Cs	Cs^+		
Co	Co^{2+}, $Co(OH)_2^+$		
Cm	Cm^{3+}, $CmOH^{2+}$, $Cm(OH)_2^+$		
Eu	Eu^{3+}, $EuSO_4^+$, $Eu_2P_2O_7^{2+}$		
I	I^-, IO_3^-		
Np		NpO_2^+, $NpO_2HPO_4^-$, NpO_2HCO_3	$NpOH^{3+}$, Np^{4+}
Pu		PuO_2^{2+}, $PuO_2(CO_3)(OH)_2^{2-}$, PuO_2^+	$PuOH^{2+}$, Pu^{3+}
Pm	Pm^{3+}		
Ra	Ra^{2+}		
Ru		$Ru(OH)_2^{2+}$, RuO_4^-, RuO_4^{2-}	RuO_4^-
Sr	Sr^{2+}		
Tc		TcO_4^-	Tc^{2+}
Th	ThF^{3+}, $Th(OH)_3^+$		
3H	H^+, ^3H-O-H		
U		UO_2^{2+}, UO_2F^+, $UO_2(OH_2)^o$, $UO_2(CO_3)_3^{4-}$	UO_2^{2+}, UOH^{3+}, UO_2^+, $UO_2(CO_3)_3^{4-}$
Zr	$Zr(OH)_4^o$, $Zr(OH)_5^-$, ZrF^{3+}		

[a] In a solution characterized by pH 4 to 9, pO2 0.68 to 80, pCO2 1.52 to 3.52, pCl- = pNO3- = pSO4^{2-} = 3.0, pF- 4.5 and pH2PO4 = 5.0 environment without organic ligands.

Photochemical Reactions

So far, the reactions discussed have proceeded due to the thermal kinetic energy of the reacting molecules. However, there is another source of energy, that associated with light, that can drive reactions. These are referred to as photochemical reactions. The energy associated with light (E) is a function of the frequency (v) of the light times a constant (h) known as Planck's constant:

$$E = h\,v \tag{8-28}$$

Each quantum of light (the term for an individual energy packet) can be deposited in molecular bonds, either dissociating the molecule or creating excited molecular states that more readily react. An important set of photochemical reactions comprise the NO–NO_2–O_3 system in the creation of ozone:

$$NO_2 + hv \rightarrow NO + O \tag{8-29}$$

$$O + O_2 \rightarrow O_3 \tag{8-30}$$

Photochemical reactions are obviously most important in the atmosphere, in the upper hundred feet or so of water, and on the surface of exposed soil and rock. However, the reaction products are important to the subsurface if they enter the subsurface or influence conditions in the subsurface. Organic chemicals on exposed soil surfaces can undergo photochemical reactions, as shown in Figure 8-9, in which flumetralin and disulfoton are significantly degraded in the upper few millimeters of the soil.

Ion Exchange and Sorption

Sorption refers to the process by which a species leaves the solution and is adsorbed into the surface of a solid resulting in an increase in the concentration of the contaminant in the soil in lieu of relatively lower concentration in the pore water or within the saturated zone. Ion exchange is the process in which a species in solution replaces a species from the solid phase (i.e., exchange places). Both of these processes can be treated as reactions and have mass action expressions associated with them. In natural environments, mineral surfaces usually have a net negative charge because of exposed anions such as oxygen. These surface groups usually hydrate upon wetting to form surface hydroxide groups, OH^-. To neutralize the charge, cations such as Na^+, Ca^{2+}, K^+, etc. sorb onto the surface. The magnitude of the charge, and hence the sorption properties of the surface, is strongly dependent upon the pH, e.g., at low pH, which is high $[H^+]$, there will be lots of H^+ in solution to compete for sorption onto the negatively charged sites.

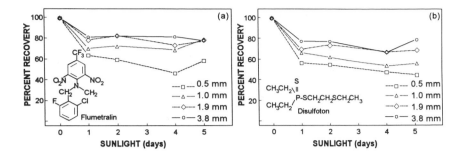

Figure 8-9   Sunlight photolysis (May) of flumetralin (a) and disulfoton (b) on a Truckee silt loam soil at various soil depths (after Miller et al., 1989).

Ion exchange is important in clay minerals, which are composed of layered sheets that can have interlayer species such as Na^+, Ca^{2+}, K^+, and H_2O. These can exchange with species in solution such as strontium (Sr), or any of a number of metal ions. Because sorption and ion exchange are processes that tend to remove specific species from solution, they are important subsurface processes for retarding the migration of many contaminants.

Retardation Factors

The transport parameter that describes the overall chemical transport behavior of a species through porous media is called the retardation factor (R_f). R_f for a particular species is the ratio of the solution velocity and the species velocity (or ratio between the rate of groundwater movement and rate of contaminant movement) and can be determined in column flow experiments in an analogous way to what occurs and are estimated in the field as shown in Figure 8-10. R_f is a bulk property of the system, combining all the effects of solution chemistry, porous media properties, chemical reactions, etc. and is a synthesis of all the subsurface processes that we have discussed thus far.

Consider a single pore space or channel between two mineral grains in a saturated, compacted layer of the clay, bentonite, in the subsurface (Figure 8-11). A groundwater solution begins to infiltrate this pore space and contains a variety of contaminants, including uranium, neptunium, technetium, and iodine (U, Np, Tc, and I). However, the actual species in solution might be the uranyl ion, the neptunyl ion, the pertechnetate ion, and the iodine ion (UO_2^{2+}, NpO_2^{2+}, TcO_4^- and I^-, respectively). Furthermore, these will be hydrated, or in other words, have hydration shells around them as shown in Figure 8-11. The top half of Figure 8-11 shows the initial infiltration. As we have discussed, different species have different tendencies to adsorb onto the mineral surfaces. The anions TcO_4^- and I^- are not significantly sorbed and diffuse on through at

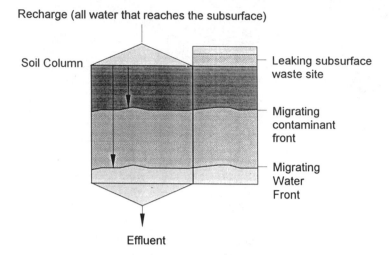

Recharge (all water that reaches the subsurface)

Soil Column

Leaking subsurface waste site

Migrating contaminant front

Migrating Water Front

Effluent

Figure 8-10   Schematic showing determination of a retardation factor.

some relatively fast rate close to the diffusion rate of the water itself, designated as D_s. However, UO_2^{2+} and NpO_2^{2+} can be highly sorbed onto the mineral surface and may even displace other sorbed species such as Na^+, that are not as strongly sorbing. Therefore, the apparent diffusion rate of U and Np will be much less, and the migration of U and Np is said to be retarded by the bentonite. However, at some later time, shown in the bottom half of Figure 8-11, the sorption sites on the clay surfaces will be filled so that no more UO_2^{2+} and NpO_2^{2+} sorption will occur, and the diffusion rate for these species will jump up to the relatively fast rate of the other nonretarded species. This may take many years depending upon the amount of contaminant, the number of sorption sites, and concentration of all species in solution. Competition among the different species for different sorption sites will occur.

The retardation factor for any particular chemical species is given by

$$R_f = V_{gw} / V_{sp} = 1 + \rho K_d / n \qquad (8\text{-}31)$$

where V_{gw} is the velocity of groundwater, V_{sp} is the velocity of the species, p is the dry bulk density of the porous media, K_d is the distribution coefficient (the ratio of the concentration of the species on the soil particles to the concentration of the species in solution), and n is the porosity. If none of a particular species is sorbed or lost to solid phases, then $K_d = 0$ and $R_f = 1$ for that species and no retardation will occur. High R_f indicates that the contaminant is strongly adsorbed, whereas low R_f indicates that the contaminant is weakly adsorbed. In column experiments, a breakthrough curve is obtained for the particular species and R_f is determined as the pore volume at which C/C_o = 0.5. It is generally assumed that retardation does not depend on water content,

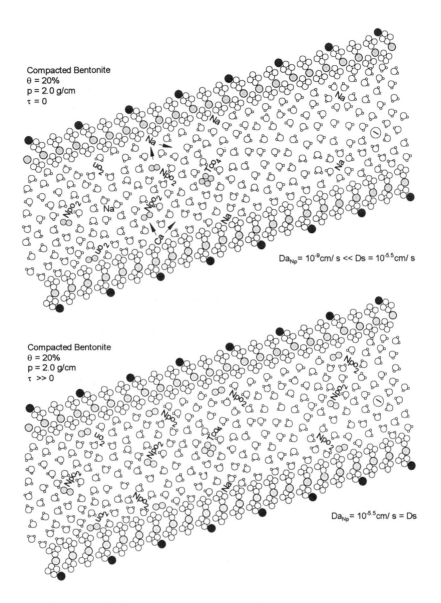

Figure 8-11   Chemical schematic showing single porespace or channel between two mineral grains in a saturated, compacted layer of bentonite.

but only on the total volume of water that has passed through the system, making $n = \theta_s$, the saturated volumetric water content. However, there is evidence that $n = \theta$.

As an example, the retardation factor for a Hanford Site soil with respect to uranium was determined using Hanford groundwater contaminated with approximately 28 ppb of U. The Hanford soil was characterized as a coarse to medium-grained fluvio-glacial sediment of a dominantly volcanic source lithology, having approximately 85 to 90% sand, 5 to 10% silt, and 3 to 6% clay by weight for the <2 mm size fraction. U exhibited a surprisingly low retardation factor of 2.3 (Figure 8-12), probably because of its existence as the mobile negatively charged carbonate species of the uranyl ion, $[UO_2 \cdot (CO_3)_2]^{2-}$. As previously mentioned, under different conditions the uranyl ion (UO_2^{2+}) is unusually highly retarded by adsorption to soil particle surfaces because of its positive charge. The retardation behavior exhibited in Figure 8-12 will most likely be the case for many of the transuranics at arid western sites where alkaline, carbonate-containing soils are abundant.

Because of its breakthrough characteristics in these soils under field conditions, uranium was used to compare retardation behavior at two different volumetric water contents: 27.9 vol% (77% of saturation) and 9.8 vol% (26% of saturation). To isolate the water content as the only variable, the experiments were run with the experimental conditions and parameters as identical as possible, e.g., the same sample volume and density, the same solution chemistry, and same sample holders, dispersion systems, and effluent collectors, and similar average residence times of the solution in each sample. When the breakthrough curves for U at the two different water contents are plotted using the total porosity, $n = \theta_s$ as the pore volume, the agreement is poor. However, the agreement is excellent when the volumetric water content, θ, is used as the pore volume (Figure 8-12). This suggests that $n = \theta$ in Equation 8-31, which allows R_f and K_d data obtained from saturated experiments to be extrapolated to unsaturated conditions.

R_f can also be derived using field generated data where the distribution coefficient (K_d) is determined using actual chemical concentrations measured in groundwater and the adjacent soil. K_d is simply determined by the relationship

$$K_d = \frac{C_s}{C_w} \tag{8-32}$$

where K_d (mL/g) = the distribution coefficient, C_s = mass of solute on the solid phase per unit mass of solid phase ($\mu g/g$), and C_w = concentration of solute in solution (mg/L). C_s should technically be based on analysis of a dry sample reflecting only the chemical adsorbed on the soil and not in the soil water being considered. However, since soil samples submitted to the laboratory are not dried, moisture content on a portion of the sample should be conducted and then corrected for the chemical constituent actually in the water fraction of the soil. In lieu of this, the concentration of the chemical in the soil moisture is typically assumed to be equal to the chemical concentration in the water, thus providing K_d results that may be underrepresented.

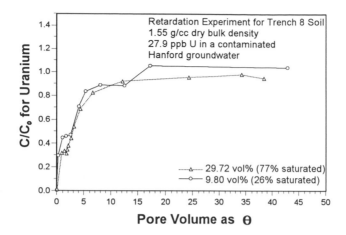

Figure 8-12   Graph showing C/C₀ for uranium vs pore volume as O for retardation experiment for trench 8 soil (after Conca and Wright, 1992).

R_f can be calculated by assuming bulk density and effective porosity values based on soil type using Equation 8-31. R_f results via this method may vary significantly reflecting heterogeneities of the soil sample, and sampling interval of the soil (i.e., 1.5 ft or less) in respect to the relatively large interval (i.e., 10 ft or more) representative of the groundwater sample and collection of samples from low transmissive zones. K_d can also be determined via several other means including methods based on water solubility, molecular structure, surface area, octanol-water partition coefficient (K_{ow}), and laboratory methods.

Because retardation factors depend upon so many factors, it is difficult and dangerous to make generalizations about a particular situation. Experiments need to be done for each type of situation and each contaminant of interest. About all one can say is that positively charged species, cations, are generally highly retarded in most soils, sediments, and rocks; whereas, negatively charged species, anions, are generally poorly retarded. However, changes in the redox state of the solution and high organic carbon contents can dramatically change these situations.

BIBLIOGRAPHY

Alloway, B.J., 1990, *Heavy Metals in Soils:* John Wiley & Sons, New York.
Amarantos, S.G. and Petropoulos, J.H., 1981, DEMO 81/2; Certain Aspects of Leaching Kinetics of Solidified Radioactive Wastes - Laboratory Studies: Greek Atomic Energy Commission, Athens.

Atlas, R.M., 1981, Microbiological Reviews: Microbial Degradation of Petroleum Hydrocarbons: an Environmental Perspective: *Microbiological Reviews,* Vol. 45, pp. 180-209.

Benedetto, A.T, Lottman, R.P., Cratin, P.D. and Ensley, E.K., 1970, Asphalt Adhesion and Interfacial Phenomena: Highway Research Record No. 340, National Research Council Highway Research Board, Washington, D.C.

Berner, E.K. and Berner, R.A., 1987, *The Global Water Cycle: Geochemistry and Environment:* Prentice-Hall, Englewood Cliffs, N.J.

Beveridge, T.J. and Doyle, R.J., 1989, *Metal Ions and Bacteria:* John Wiley & Sons, New York.

Bogue, R.H., 1955, *The Chemistry of Portland Cement, Second Edition:* New York.

Bonazountas, M. and Wagner, J., 1984, Modeling Mobilization and Fate of Leachates Below Uncontrolled Hazardous Waste Sites: In Proceedings of the HMCRI 5th National Conference on Management of Uncontrolled Hazardous Waste Sites, Washington, D.C., pp. 7-102.

Bras, R.L., 1990, *Hydrology;* Addison-Wesley Publishing Co., Reading, MA.

Brock, T.D. and Madigan, M.T., 1988, B*iology of Microorganisms, 5th Edition:* Prentice-Hall, Englewood Cliffs, N.J.

Brown, J.R., 1987, Soil Testing: Sampling, *Correlation, and Interpretation:* Soil Science Society of America, Madison, WI.

Brulé, B., Ramond, G. and Such, C., 1986, Relationships Between Composition, Structure and Properties of Road Asphalts: State of Research at the French Public Works Central Laboratory: Transportation Research Record 1096, TRB, National Research Council, Washington, D.C., pp. 22-34.

Buelt, J.L., 1983, *Liner Evaluation for Uranium Mill Tailings: Final Report,* PNL-4842: Pacific Northwest Laboratory, Richland, WA., 136 p.

Burk, M., Derham, R. and Lubowitz, H., 1974, Recommended Methods of Reduction, Neutralization, Recovery, or Disposal of Hazardous Wastes, Vol. 14, TRW Systems Group.

Carter, J.G., 1990, *Skeletal Biomineralization: Patterns, Processes and Evolutionary Trends, 1:* Van Nostrand Reinhold, New York.

Cheng, H.H., 1990, *Pesticides in the Soil Environment: Processes, Impacts, and Modeling:* Soil Science of America, Madison, WI.

Cheremisinoff, P.N. and Holcomb, W.F., 1976, Management of Hazardous and Toxic Wastes: *Pollution Engineering,* Vol. 8, pp. 24-28.

Cloud, P., 1988, *Oasis in Space: Earth History from the Beginning:* W.W. Norton, New York.

Conca, J.L. and Wright, J.V., 1991, Aqueous Diffusion Coefficients in Unsaturated Materials: In Proceedings on Scientific Basis for Nuclear Waste Management XIV, *Materials Research Society Symposium Proceedings,* Vol. 212, pp. 879-884.

Conca, J.L. and Wright, J.V., 1992, Diffusion and Flow in Gravel, Soil, and Whole Rock: *Applied Hydrology,* January, pp. 5-24.

Cook, T.D., 1972, *Underground Waste Management and Environmental Implications:* American Association of Petroleum Geologists, Tulsa, OK.

Corbitt, R.A., 1990, *Standard Handbook of Environmental Engineering:* McGraw-Hill, New York.

Corey, A.T., 1986, *Mechanics of Immiscible Fluids in Porous Media:* Water Resource Publications, Littleton, CO, 259 p.

Corson, W.H., 1990, *The Global Ecology Handbook: What You Can Do About the Environmental Crisis,* Beacon Press, Boston.

Daiev, Ch.T. and Vassilev, G.P., 1985, On the Diffusion of 90Sr from Radioactive Waste Bituminized by the Mould Method: *Journal of Nuclear Materials,* Vol. 127, pp. 132-136.

Daugherty, D.R., Pietrzak, R.F., Fuhrmann, M. and Columbo, P., 1988, An Experimental Survey of the Factors that Affect Leaching from Low-Level Radioactive Waste Forms, BNL-52125; Brookhaven National Library, Upton, N.Y.

Dawson, G.W. and Mercer, B.W., 1986, *Hazardous Waste Management:* John Wiley & Sons, New York.

Devinny, J.S., Everett, L.A., Lu, J.C.S. and Stollar, R.L., 1990, *Subsurface Migration of Hazardous Wastes:* Van Nostrand Reinhold, New York, 387 p.

Devitt, D.A., et al., 1987, *Soil Gas Sensing for Detection and Mapping of Volatile Organics:* National Water Well Association, Dublin, OH, 270 p.

Dragun, J., 1988, *The Soil Chemistry of Hazardous Materials:* Hazardous Materials Control Research Institute, Silver Spring, MD, 458 p.

Drever, J.I., 1988, *The Geochemistry of Natural Waters, Second Edition:* Prentice Hall, Englewood Cliffs, NJ.

Elliott, L.F. and Stevenson, F.J., 1977, *Soils for Management of Organic Wastes and Waste Waters:* Soil Science Society of America, Madison, WI.

Eschrich, H., 1980, Properties and Long-Term Behavior of Bitumen and Radioactive Waste-Bitumen Mixtures, Swedish Nuclear Fuel and Waste Management Company, Stockholm, SKBF KBS Technical Report No. 80-14.

Fayer, M.J., Rockhold, M.L. and Campbell, M.D., 1992, Hydrologic Modeling of Protective Barriers: Comparison of Field Data and Simulation Results: *Soil Science Society of America Journal,* Vol. 56, pp. 690-700.

Fawcett, H.H., 1984, *Hazardous and Toxic Materials:* John Wiley and Sons, New York.

Fish, W. and Elovitz, M.S., 1990, CR(VI) Reduction by Phenols in Immiscible Two-Phase Systems: Implications for Subsurface Chromate Transport, *Transactions of the American Geophysical Union,* Vol. 71, pp. 1719.

Flint, R.F. and Skinner, B.J., 1974, *Physical Geology:* John Wiley and Sons, New York.

Freeman, H.M., 1989, *Standard Handbook of Hazardous Waste Treatment and Disposal;* McGraw-Hill, New York.

Freeze, R.A. and Cherry, J.A., 1979, *Groundwater:* Prentice-Hall, Englewood Cliffs, NJ, 604 p.

Gulf Coast Hazardous Substance Research Center, 1990, Solidification/Stabilization Mechanisms and Applications: Beaumont, TX.

Harris, J.O., 1958, Preliminary Studies on the Effect of Micro-organisms on the Physical Properties of Asphalt: *Transactions of the Kansas Academy of Science,* Vol. 61, pp. 110-113.

Harrison, R.M., de Mora, S.J., Rapsomanikis, S. and Johnson, W.R., 1991, *Introductory Chemistry for the Environmental Sciences:* Cambridge University Press, New York.

Haxo, H.E., 1976, *Assessing Synthetic and Admixed Materials for Liner Landfills; Gas and Leachate from Landfills; Formation, Collection and Treatment,* (edited by E.J. Genetelli and J. Circello); Report EPA 600/9-76-004, U.S. Environmental Protection Agency, Cincinnati, OH, NTIS Report PB 251161, pp. 130-158.

Hazardous Materials Control Research Institute, 1990, Hazardous Waste Treatment by Genetically Engineered or Adapted Organisms; Silver Springs, MD.

Heichel, G.H., 1990, *Impact of Carbon Dioxide, Trace Gases, and Climate Change on Global Agriculture:* Soil Science Society of America, Madison, WI.

Henstock, M.E., 1983, *Disposal and Recovery of Municipal Solid Waste;* The Butterworth Group, London.

Hickle, R.D., August, 1976, Impermeable Asphalt Concrete Pond Liner; *Civil Engineering,* pp. 56-59.

Hillel, D., 1980, *Fundamentals of Soil Physics;* Academic Press, Orlando, FL.

Hillel, D., 1980, *Applications of Soil Physics;* Academic Press, New York.

Hillel, D., 1991, *Out of the Earth: Civilization and the Life of the Soil;* The Free Press, New York.

Holcomb, W.F. and Goldberg, S.M., 1976, Available Methods of Solidification for Low-Level Radioactive Wastes in the United States; U.S. Environmental Protection Agency Technical Note OPR/TAD 76-4.

Holloway, J.R. and Wood, B.J., 1988, *Simulating the Earth: Experimental Geochemistry:* Unwin Hyman, London.

Howard, P.H., Boethling, R.S., Jarvis, W.F., Meylan, W.M. and Michalenko, E.M., 1991, *Handbook of Environmental Degradation Rates;* Lewis Publishers, Boca Raton, FL, 725 p.

Istok, J., 1989, *Groundwater Modeling by the Finite Element Method:* American Geophysical Union, Washington, D.C.

Jacobs, L.W., 1989, *Selenium in Agriculture and the Environment:* Soil Science Society of America, Madison, WI.

Jones, T.K., 1965, Effects of Bacteria and Fungi on Asphalt: *Material Protection,* Vol. 4, pp. 39.

Kaufmann, W.J. III, 1977, *Relativity and Cosmology, Second Edition:* Harper and Row, New York.

Klute, A., 1982, *Methods of Soil Analysis, Part 1, Second Edition:* American Society of Agronomy, Madison, WI.

Klute, A., 1982, *Methods of Soil Analysis, Part 2, Second Edition:* American Society of Agronomy, Madison, WI.

Kostecki, P.T. and Calabrese, E.J., 1989, *Petroleum Contaminated Soils, Vol. 3;* Lewis Publishers, Chelsea, MI.

Kral, D.M., 1983, *Chemical Mobility and Reactivity in Soil Systems:* Soil Science Society of America, Madison, WI.

Kriech, A.J., 1990, Evaluation of Hot Mix Asphalt for Leachability, Heritage Research Group Report HRG No. 3959AOM3, Indianapolis.

Last, G.V. et al, 1992, Characteristics of the Volatile Organic Compounds-Arid Integrated Demonstration Site, PNL-7866 UC-630; Pacific Northwest Laboratories, Richland, WA.

Leo, A.C., Hansch, and Elkins, D., 1971, Partition Coefficients and Their Uses: *Chemical Review,* No. 71, pp. 525-616.

Lerman, A., 1979, Geochemical Processes: *Water and Sediment Environments;* John Wiley and Sons, New York.

Lindren, G.F., 1990, *Managing Industrial Hazardous Waste:* Lewis Publishers, Boca Raton, FL.

Mackay, D. and Shiv, W.Y., 1981, Henry's Law Constants for Organic Compounds: *Journal of Physical Chemistry,* Vol. 10, No. 4.

Mercer, J.W. and Cohen, R.M., 1990, A Review of Immiscible Fluids in the Subsurface: Properties, Models, Characterization and Remediation; *Journal of Contaminant Hydrology*, pp. 107-163.

Miller, G.T. Jr., 1990, *Living in the Environment: An Introduction to Environmental Science, 6th Edition;* Wadsworth Publishing, Belmont, CA.

Ming, D.W., and Henniger, D.L., 1989, Lunar Base Agriculture: Soil Science Society of America, Madison, WI.

Moll, G. and Ebenreck, S, 1989, *Shading Our Cities:* Island Press, Washington, D.C.

Moore, J.G., Godbee, H.W., and Kibbey, A.H., 1977, Leach Behavior of Hydrofracture Grout Incorporating Radioactive Wastes: *Nuclear Technology*, Vol. 32, pp. 39-52.

Nahon, D.B., 1991, *Introduction to the Petrology of Soils and Chemical Weathering:* John Wiley and Sons, New York.

Newsom, J.M., 1985, Transport of Organic Compounds Dissolved in Ground Water: *Ground Water Monitoring Review, Spring Issue,* Vol. 5, No. 2, pp. 28-36.

Pojasek, R.B., 1978, Stabilization, Solidification of Hazardous Wastes: *Environmental Science Technology*, Vol. 12, pp. 382.

Pojasek, R.B., 1980, *Toxic and Hazardous Waste Disposal 1:* Ann Arbor Science , Ann Arbor, MI.

Pojasek, R.B., 1980, *Toxic and Hazardous Waste Disposal 2:* Ann Arbor Science, Ann Arbor, MI.

Pojasek, R.B., 1980, *Toxic and Hazardous Waste Disposal 4:* Ann Arbor Science, Ann Arbor, MI.

Press, F. and Siever, R., 1974, *Earth:* W.H. Freeman and Co., San Francisco.

Rankama, K., and Sahama, T.G., 1950, *Geochemistry:* University of Chicago Press, Chicago.

Riley, R.G. and Zachara, J.M., 1992, Chemical Contaminants on DOE Lands and Selection of Contaminant Mixtures for Subsurface Science Research: U.S. Department of Energy Office of Energy Research Subsurface Science Program.

Rowe, W.D. and Holcomb, H.I., 1974, The Hidden Commitment of Nuclear Wastes: *Nuclear Technology*, Vol. 24, pp. 286-289.

Sawhney, B.L. and Brown, K., 1989, *Reactions and Movement of Organic Chemical in Soils:* Soil Science Society of America, Madison, WI.

Schwille, F., 1988, *Dense Chlorinated Solvents in Porous and Fractured Media Model Experiments:* Lewis Publishers, Chelsea, MI, 146 p.

Stelly, M., 1981, *Chemistry in the Soil Environment:* Soil Science Society of America, Madison, WI.

Soil Science Society of America, 1987, Glossary of Soil Science Terms: Madison, WI.

Stumm, W. and Morgan, J.J., 1981, *Aquatic Chemistry: An Introduction Emphasizing Chemical Equilibria in Natural Waters, Second Edition;* John Wiley & Sons, New York.

Testa, S.M. and Winegardner, D.L., 1991, *Restoration of Petroleum-Contaminated Aquifers;* Lewis Publishers, Boca Raton, FL, 269 p.

U.S. Environmental Protection Agency, 1978, June, Criteria, Identification, Methods and Listing of Hazardous Wastes: Draft Proposal Rules issued under Section 3001 RCRA Legislation.

U.S. Environmental Protection Agency, 1978, June, Owners and Operators of Hazardous Waste Treatment, Storage and Disposal Facilities: Draft Proposal Rules issued under Section 3004 RCRA Legislation.

U.S. Environmental Protection Agency, 1974, Disposal of Hazardous Wastes: Publication SW-115.

Volesky, B., 1990, *Biosorption of Heavy Metals:* CRC Press, Boca Raton, FL.

Wetzel, R.G., 1983, *Limnology:* Saunders College Publishing, Philadelphia.

Weber, P.M. Jr., McGinley, W.J. and Katz, L.E., 1991, Sorption Phenomena in Subsurface Systems: Concepts, Models, and Effects on Contaminant Fate and Transport: *Water Resources,* Vol. 25, pp. 499-528.

9 NONAQUEOUS PHASE LIQUIDS (NAPLS)

"The most dangerous place for me to live, as a barrel of oil, would be in the U.S. where there are thousands of hunters and each has a different weapon."

(A.I. Levorsen)

INTRODUCTION

Groundwater contamination as a result of the subsurface presence of nonaqueous phase liquids (NAPLs) is ubiquitous. Sources of NAPLs include the release of crude oil and refined petroleum-related products from above-ground and underground storage tanks, pipeline corridors, dry wells, and accidental releases. Over the past decade, much focus has been placed upon underground storage tanks, although over the past few years emphasis has shifted to other industries and operations such as petroleum refining, bulk liquid storage terminals, major pipeline networks, gas production, steel industry coking, and wood treating. NAPLs, referred to by the federal regulations as "free product," can occur in the subsurface as both lighter than water (LNAPLs) and denser than water (DNAPLs) as shown in Figure 9-1. Once groundwater is encountered, a LNAPL will tend to form a pool overlying the capillary fringe and water table. DNAPLs behave much the same as LNAPLs in the vadose zone; however, once groundwater is encountered, DNAPLs will tend to continue to migrate vertically downward through the water column until a significant permeability contrast is encountered, providing enough of a volume of DNAPL was released. The subsurface presence of NAPLs for the most part reflects unnoticed small or episodic releases over long periods of time, with a subordinate amount due to a one-time release as experienced with a significant breach of a pipeline or tankage which are typically responded to promptly (Figure 9-2).

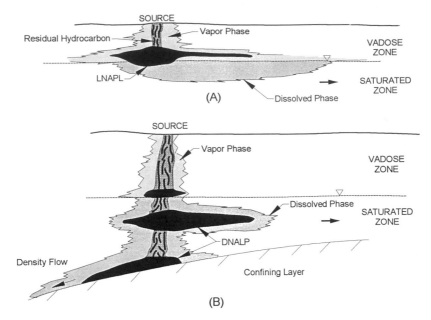

Figure 9-1   Schematic showing generalized distribution of LNAPLs (A) and DNAPLs (B) in the subsurface; arrow denotes general direction of groundwater flow.

Although the processes affecting water movement in the subsurface have been stressed in this book (Chapters 4 and 8), certain aspects of NAPL behavior need to be understood with regard to the many organic contaminants that may adversely affect the subsurface environment. Although there are nonorganic, nonaqueous phase liquids, such as mercury, the focus will be on the organic NAPLs. Free-phase NAPL refers to the NAPL existing as an independent phase, not as a dissolved component in the pore water or pore atmosphere. The environmental concerns associated with sites affected with NAPLs revolve around hydrocarbon-affected soil (residual hydrocarbon), the NAPL itself (which can serve as a continued source of soil and groundwater contamination), dissolved hydrocarbon constituents in groundwater, and hydrocarbon vapors (Figure 9-3).

Presented in this chapter is discussion of the more important properties associated with NAPLs migration, with subsequent emphasis on the occurrence and detection of LNAPLs and DNAPLs, respectively.

PROPERTIES OF NAPLs

Migration of NAPLs in the subsurface is governed by numerous properties including density, viscosity, surface tension, interfacial tension, immiscibility,

Figure 9-2   A significant breach of a valve results in the release of prod-
             uct into this bermed but unlined bulk storage hydrocarbon
             tank storage area.

Figure 9-3   Hydrocarbon vapors emitting from a LNAPL pool overlying
             a shallow water table at a refinery site.

capillary pressure, wettability, saturation, residual saturation, relative permeability, solubility, and volatilization. The two most important properties of liquids which control their flow behavior are density and viscosity. Density is the mass of a material per unit volume, which is the ratio of a substance's density to that of some standard, notably water. Water has a density of about 1 g/cm^3, whereas carbon tetrachloride (CCl_4), an important DNAPL contaminant, has a density of about 1.58 g/cm^3 and will tend to sink through the water table. The viscosity is the ability of a fluid to resist deformation or flow, and is a measure of the tendency of a fluid to flow, e.g., molasses has a high viscosity relative to water. Viscosity is highly temperature dependent and has common units of centipoise. Water has a viscosity of 1.00 centipoise at 20°C, whereas carbon tetrachloride has a viscosity of 0.97 centipoise at 20°C. Therefore, the two fluids will physically flow about the same. However, with respect to flow through porous media, surface tension is extremely important. Surface tension is responsible for capillary effects and spreading of the NAPL over the water table. At about 20°C, water has a surface tension of 73.05 dynes/cm; whereas, CCl_4 has a surface tension of only 26.95 dynes/cm. Therefore, water will be held in an unsaturated porous media by surface tension to a much greater degree relative to carbon tetrachloride (i.e., the permeability of porous media will be different with respect to each liquid). The ramifications will be important for contaminant transport of mixed wastes.

Other NAPLs often present in the subsurface are petroleum, oils, tars, and biological fluids. When more than one fluid is present, there is a need to describe how well they mix, referred to as their miscibility. Water and vegetable oil are immiscible fluids. Many of the NAPLs are immiscible with water and will occur as separate fluid bodies, droplets, zones, etc. in the subsurface environment.

It should be noted that the properties of any liquid can change as its composition changes. Saltwater is much denser than freshwater, and they will not readily mix without agitation. Likewise, a specific oil composition will determine its density, viscosity, and surface tension, as well as its other properties as further discussed under chemical processes (Chapter 8).

Free-phase NAPLs move in response to the organic liquid pressure gradients, just as water moves in response to the hydraulic pressure gradients. As a dissolved component in water, NAPLs will move with the water and also by diffusion through the water. NAPLs are generally less easily adsorbed than water because of their lower surface tensions and lower dipole moments, and will preferentially move in the larger channels, pores, or fractures. This may cause irregular and discontinuous flow patterns relative to the flow of water through the same material. However, some organic NAPLs, such as alkyl phosphates, which are used in many extraction processes, are highly polar and will behave differently.

Saturation (s) is the volume fraction of the total void volume occupied by a specific fluid at a point. Saturation values can vary from zero to one with the

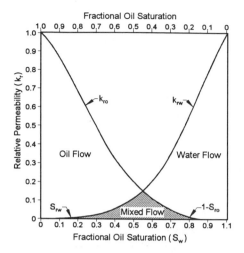

Figure 9-4   Relative permeability relationships controlling the flow of two immiscible fluids (after Leverett, 1939).

saturation of all fluids equal to one. Residual saturation (S_r) is the saturation at which the NAPL becomes discontinuous and immobile due to capillary forces. Residual saturation is dependent upon many factors including pore size distribution, wettability, fluid viscosity and density ratios, interfacial surface tension, gravity and buoyancy forces, and hydraulic gradients.

Relative permeability is the reduction of mobility between more than one fluid flowing through a porous media, and is the ratio of the effective permeability of a fluid at a fixed saturation to the intrinsic permeability. Relative permeability varies from zero to one and can be represented as a function of saturation (Figure 9-4). Neither water nor oil is effectively mobile until the S_r is in the range of 20 to 30% or 5 to 10%, respectively, and even then the relative permeability of the lesser component is approximately 2%. Oil accumulation below this range is for all practical purposes immobile (and thus not recoverable). Where the curves cross (i.e., at a S_{rw} of 56% and $1-S_{ro}$ of 44%), the relative permeability is the same for both fluids. With increasing saturation, water flows more easily relative to oil. As $1-S_{ro}$ approaches 10%, the oil becomes immobile, allowing only water to flow.

Wettability refers to the preferential spreading of one fluid over solid surfaces in a two-fluid system and is dependent upon the interfacial tension. The wetting fluid (water) will tend to coat the surface of grains and occupy the smaller spaces or pore throats; whereas, the nonwetting fluid (oil or NAPL) will tend to occupy the largest openings. The contact angle (ϕ), as measured in the water, determines whether the porous media will be preferentially wetted by water or NAPL varying from 0 to 180°. The system is water wet if ϕ is less than approximately 70°, and NAPL-wet if ϕ is greater than 110°; neutral conditions persist when ϕ is between 70 and 110° (Figure 9-5).

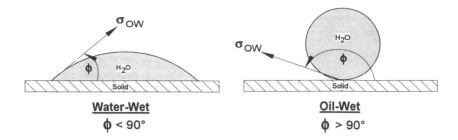

Figure 9-5   Configurations of wettability.

As a NAPL moves through the subsurface within unsaturated zone, it can migrate by several mechanisms depending upon the vapor pressure of the liquid, the density of the liquid, the solubility of the liquid (how much dissolves in water at equilibrium), and the polar nature of the NAPL. The NAPL will move from the liquid phase into the vapor phase until the vapor pressure is reached for that liquid. NAPL will move from the liquid phase into the water phase until the solubility is reached. Also, NAPL will move from the gas phase into any water that is not saturated with respect to that NAPL. Because hydraulic conductivities can be so low under highly unsaturated conditions (Figure 9-6), the gas phase may move much more rapidly than either of the liquid phases, and NAPLs can be transported to wetter zones where the NAPL can then move from the gas phase to a previously uncontaminated water phase. In order to understand and model these multiphase systems, the characteristic behavior and the diffusion coefficients for each phase must be known for each sediment or type of porous media, an incredible amount of information, much of which is presently lacking.

LIGHT NONAQUEOUS PHASE LIQUIDS (LNAPLs)

The problems associated with LNAPLs are well documented in the literature, ranging from small releases where just enough LNAPL is present to be a nuisance, to pools ranging up to millions of barrels of LNAPL and encompassing hundreds of acres in lateral extent. Subsurface migration of LNAPL (and DNAPL) are affected by several factors, including volume of release, time duration, area of infiltration, physical properties of the NAPL, properties of the media, and subsurface flow dynamics.

The migration of LNAPL in the subsurface can be divided into three phases: seepage through the vadose zone, spreading over the water table with development of a pancake layer, and accumulation stability within the water capillary zone (Figure 9-7). During seepage through the vadose zone, downward migration of the hydrocarbon can occur as a bulk product zone of affected soil or by

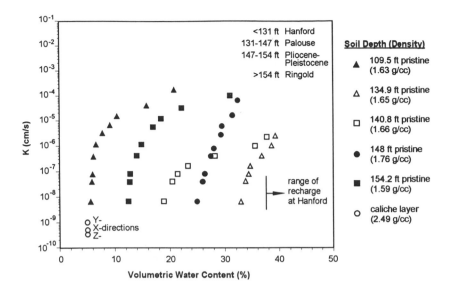

Figure 9-6 Representative hydraulic conductivities under highly unsaturated conditions.

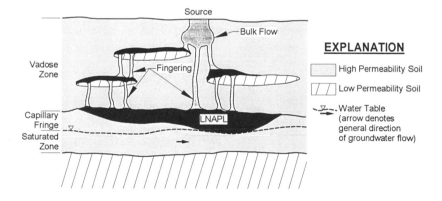

Figure 9-7 Schematic showing distribution of LNAPL in the subsurface.

fingering as illustrated in Figure 9-7. Fingering tends to occur under slow velocity conditions such that

$$q < \frac{kpg}{\mu} \tag{9-1}$$

where q = hydrocarbon velocity, k = permeability, p = LNAPL density, g = gravity, and μ = viscosity.

Large capillary forces as anticipated with tight, porous, and nonhomogeneous media will also contribute to fingering. Recognition of conditions favorable for development of fingering is important in adequately assessing subsurface presence, and lateral and vertical extent of hydrocarbon-affected soil. If fingering occurs, the LNAPL (or DNAPL) may not occupy the entire cross-sectional area through which it passes allowing water to flow through and increasing dissolution. Fingering also results in deeper penetration of the NAPL relative to that of bulk hydrocarbon migration. If conditions are such that

$$q > \frac{kpg}{\mu} \qquad (9\text{-}2)$$

then fingering will not occur and the hydrocarbon will migrate vertically downward in bulk. If the soil is stratified or contains significant fine-grained materials, some horizontal spreading may occur. To complicate matters, migration directions may vary due to facies architecture, stratigraphic controls imposed by depositional environment (i.e., channeling), bedrock orientation, and fractures.

The residual saturation (S_r) capacity of a soil is the saturation at which the LNAPL becomes discontinuous and is immobilized by capillary forces, and is generally about one third that of their respective water-holding capacity. Immobilization of a given hydrocarbon mass is dependent upon the media porosity and physical characteristics of the LNAPL. Estimated soil volumes required to immobilize a given LNAPL volume migration vertically downward through the subsurface can be calculated via several means as follows:

$$Vs = \frac{0.2\,V_{hc}}{\phi(RS)} \qquad (9\text{-}3)$$

where Vs = cubic yards of soil required to attain residual saturation, V_{hc} = volume of discharged hydrocarbon (in barrels where 42 gal = 1bbl), ϕ = soil porosity, and RS = residual saturation capacity.

Less rigorous approaches are:

$$D = \frac{Vs}{A}, \qquad (9\text{-}4)$$

$$D = \frac{KV_{hc}}{A}, \qquad (9\text{-}5)$$

Table 9-1 Typical Soil Retention Capacity Values for Soils[a]

Soil Type	Soil Retention Capacity		
	Gasoline	Kerosene	Light Gas-Oil
Stone to coarse gravel	400	200	100
Gravel to coarse sand	250	125	62
Coarse to medium sand	130	66	33
Medium to fine sand	80	40	20
Fine sand silt	50	25	18

[a] Modified after Dietz (1971) and de Pastrovich (1979).

or

$$D = \frac{1000V_{hc}}{AR_f C} \tag{9-6}$$

where D = maximum depth of hydrocarbon penetration in the vadose zone, A = area of infiltration, K = constant based on soil retention capacity for LNAPL and LNAPL viscosity (Table 9-1), C = approximate correction factor based on LNAPL viscosity (i.e., 0.5 for gasoline to 2.0 for light fuel oil), and R_f = soil retention capacity (Table 9-2).

If the release is sufficiently large and enough time has elapsed, the LNAPL will eventually approach saturated conditions (i.e., perched or unconfined water table conditions). The capillary fringe is initially encountered. The capillary fringe rises above the water table to a height dependent upon grain-size distribution as discussed in Chapter 4 (Table 4-6). Essentially, the height of the capillary fringe increases with decreasing grain size. As the LNAPL enters the water capillary zone, pores not occupied by capillary or residual water begins to fill with little mixing. Additional LNAPL accumulation causes the development of a pool; the lateral spreading that occurs has been referred to in the literature as a pancake layer. The maximum amount of lateral spreading(s) that is anticipated can be estimated as follows:

$$S = \left(\frac{1000}{F}\right) V - \left[\frac{Ad}{K}\right] \tag{9-7}$$

Table 9-2 Typical Residual Saturation Data for Various LNAPL Types[a]

Residual Fluid Type	Hydrogeologic Conditions[b]	Media	Residual Saturation (S_r) or Retention Factor (R_p, in L/m³)	
			S_r	R_f
Water	Unsaturated	Sand	0.10	
		Silt	0.07	
		Sandy clay	0.26	
		Silty clay	0.19	
		Clay	0.18	
LNAPL				
Gasoline	Unsaturated	Coarse gravel		2.5
		Coarse sand and gravel		4.0
		Medium to coarse sand		7.5
		Fine to medium sand		12.5
		Silt to fine sand		20
		Coarse sand	0.15–0.19	
		Medium sand	0.12–0.27	
		Fine-coarse sand	0.19–0.60	
		Stone, coarse sand	0.46–0.59	
Kerosene		Gravel, coarse sand		5
		Coarse to medium sand		8
		Fine to medium sand		15
		Fine sand and silt		25

Product	Zone[b]	Material		
Diesel and light fuel oil	Soil		40	
Light oil and gasoline	Soil		0.18	
Lube and heavy fuel oils	Soil		10.18	
Middle distillates	Unsaturated		0.15	
		Coarse gravel		5.0
		Coarse sand and gravel		8.0
		Medium to coarse sand		15
		Fine to medium sand		25
		Silt to fine sand		40
Fuel oils	Unsaturated	Coarse gravel		10
		Coarse sand and gravel		16
		Medium to coarse sand		30
		Fine to medium sand		50
		Silt and fine sand		80
Crude oils	Saturated	Sandstone	0.35–0.43	
		Sandstone	0.16–0.47	
		Sandstone	0.26–0.43	
DNAPL				
Benzene	Saturated	Sand (92% sand, 5% silt, 3% clay)	0.24	
Tetrachloroethane	Saturated	Fracture (0.2 apperture)		0.05 1/m²
Tetrachloroethene	Unsaturated	Coarse Ottawa sand	0.15–0.25	
1,1,1-Trichloroethane	Saturated	Coarse Ottawa sand	0.15–0.40	
Trichloroethene	Unsaturated	Medium sand	0.20	
		Fine sand	0.19	
		Fine sand	0.15–0.20	

a Modified after Mercer and Cohen (1992) and Testa and Winegardner (1990).

b Unsaturated refers to NAPL-water-air or water-air systems; saturated refers to NAPL-water system.

where S = maximum spread of the pancake (M^2), F = thickness of the pancake (mm), V = volume of infiltrating bulk hydrocarbons (m^3), A = area of infiltration (M^2), d = depth to groundwater (m), and K = constant dependent upon the soil retention capacity for the LNAPL based on viscosity (Table 9-1).

K values generally decrease with decreasing grain size. Sufficiently large seepage rates can produce a hydraulic mound which permits lateral spreading in directions inconsistent with the gradient of the water table, and may be stratigraphically controlled. The initial stage of lateral spreading is dominated by gravity forces. However, as the gravitational potential diminishes, capillary forces tend to control the rate of lateral spreading. Large LNAPL accumulations may also eventually depress the capillary fringe, thus being in direct contrast with the saturated zone.

Once formed, the LNAPL is referred to as a pool which is an areally continuous accumulation of LNAPL. Subpools are individual accumulations of relatively uniform free-product types based on geochemistry (API gravity, specific constituents, isotope ratios, etc.) that have coalesced, reflecting multiple-source accumulations with time.

The LNAPL pool, once formed, also maintains a capillary fringe. The anticipated capillary rise of LNAPL can be calculated as follows:

$$h = \frac{2T \cos \theta}{rpg} \qquad (9\text{-}8)$$

where h = rise (in cm), T = surface tension (in dynes), ϕ = wall-liquid interface angle, p = density of fluid, g = acceleration of gravity, and r = radius of tube in centimeters. Because the surface tension of LNAPL is characteristically less than water, the height of the LNAPL capillary rise is less than water.

LNAPL Apparent vs Actual Thickness

The subsurface presence of LNAPL can occur under both perched and water-table conditions and under unconfined and occasionally confined conditions. LNAPL product in the subsurface is typically delineated and measured by the utilization of groundwater monitoring wells. While monitoring wells have provided some insight as to the extent and general geometry of the LNAPL pool, as well as the direction of groundwater flow, difficulties persist in determining the actual thickness and, therefore, the volume and ultimately the duration of recovery and remediation. One of the more difficult aspects in dealing with the subsurface presence of hydrocarbons is that accumulations measured in monitoring wells do not directly correspond to the actual thickness in the formation. The thickness of both LNAPL (and DNAPL) as measured in a monitoring well is thus an apparent thickness rather than an actual formation thickness (Figure 9-8). This relationship is schematically illustrated in Figure 9-9.

This discrepancy can be a result of one or a combination of factors or phenomena. Some of the more common factors or phenomena are schematically illustrated in Figure 9-10 and include:

- Grain size differences reflected in varying heights of the capillary fringe;
- Actual formation thickness of the mobile LNAPL hydrocarbon;
- Height of perching layers, if present;
- Seasonal or induced fluctuations in the level of the water table;
- Product types and respective specific gravities; and
- Confining conditions.

The capillary fringe height is grain-size dependent, as discussed in Chapter 4 and shown in Figure 9-10. As grain size decreases, the capillary height increases. Coarse-grained formations contain large pore spaces that greatly reduce the height of the capillary rise. Fine-grained formations have much smaller pore spaces which allow a greater capillary height.

Since the LNAPL occurs within and above the capillary fringe, once the bore hole or monitoring well penetrates and destroys this capillary fringe, LNAPL migrates into the well bore. The free-water surface that stabilizes in the well will be lower than the top of the surrounding capillary fringe in the formation; thus, mobile hydrocarbons will flow into the well from this elevated position. LNAPL will continue to flow into the well and depress the water surface until a density equilibrium is established. To maintain equilibrium, the weight of the column of hydrocarbon will depress the water level in the well. Therefore, a greater apparent thickness is measured than actually exists in the formation; thus, the measured or "apparent" LNAPL thickness is greater for fine-grained formations and less for coarser-grained formations, which may be more representative of the true thickness.

The measured or "apparent" hydrocarbon thickness is not only dependent upon the capillary fringe, but also on the actual hydrocarbon thickness in the formation (Figure 9-10). In areas of relatively thin LNAPL accumulations, the error between the apparent well thickness and the actual formation thickness can be more pronounced than in areas of thicker accumulations. The larger error reflects the relative difference between the thin layer of LNAPL in the formation and the height it is perched above the water table. The perched height is constant for thick and thin accumulations; however, a thick accumulation can depress and even destroy the capillary fringe, as illustrated in Figure 9-7.

The thickness measured in a monitoring well with LNAPL situated on a perched layer at some elevation above the water table, can produce an even larger associated thickness error (Figure 9-10). This commonly occurs when the well penetrates the perched layer and is screened from the perching layer

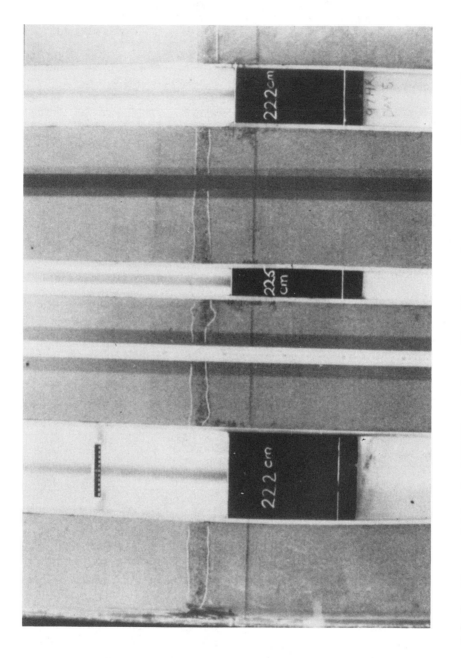

Figure 9-8   Sand-box model showing LNAPL overlying capillary fringe, and apparent vs actual LNAPL thickness. Saturated condition (water table) is represented by the straight horizontal line.

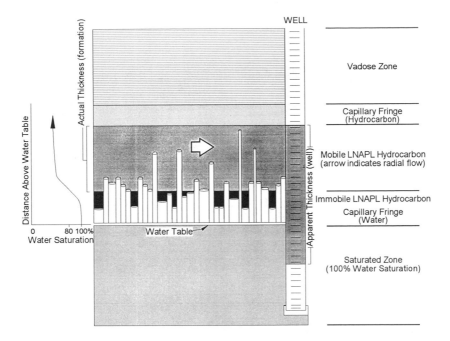

Figure 9-9 **Schematic showing generalized relationship of actual vs apparent LNAPL thickness in a well and adjacent formation (after Testa and Winegardner, 1991).**

to the water table. LNAPL then flows into the well from the higher or perched elevation. The accumulated apparent thickness is a direct result of the difference in their respective heights. If a situation such as this exists, the difference in the respective heights and weights of the column of hydrocarbon should be accounted for in determining the actual thickness.

Additionally, vertical fluctuations in the water table due to recovery operations or seasonal variations in precipitation have a direct effect upon the apparent or measured LNAPL thickness (Figure 9-10). As the water table elevation declines gradually due to seasonal variations, for instance, an exaggerated apparent thickness occurs, reflecting the additional hydrocarbon that accumulated in the monitoring well. The same is true for an area undergoing recovery operations where the groundwater elevation is lowered through pumping, and thicker apparent thicknesses may be observed.

The reverse of this effect has also been documented at recovery sites (Figure 9-10). When sufficient recharge to the groundwater system through seasonal precipitation events or cessation of recovery well pumping occurs with the water table at a slightly higher elevation, thinner LNAPL thicknesses may be observed. During this situation a compression of the capillary zone occurs, lessening the elevation difference between the water table and the free hydro-

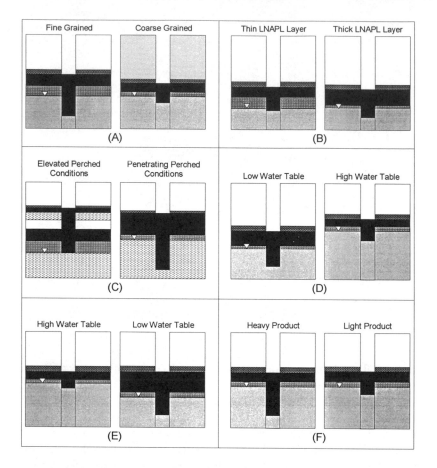

Figure 9-10   Causes for discrepancies in apparent LNAPL thickness as measured in a monitoring well in comparison with the actual thickness in the formation (after Testa and Winegardner, 1991).

carbon, which reduces the apparent thickness.

Differences in product types, and thus in API gravities, can account for variations in the apparent thickness as measured in a monitoring well (Figure 9-10). Heavier product types of relatively low API gravities will tend to depress the LNAPL hydrocarbon/water level interface more than would be anticipated from a product of higher API gravity.

In addition, monitoring wells screened across LNAPL within confined aquifers will exhibit an exaggerated thickness. This exaggerated thickness reflects relatively high confining pressures, which force the relatively lower density fluid upward within the borehole. Thus, the measured thickness is a

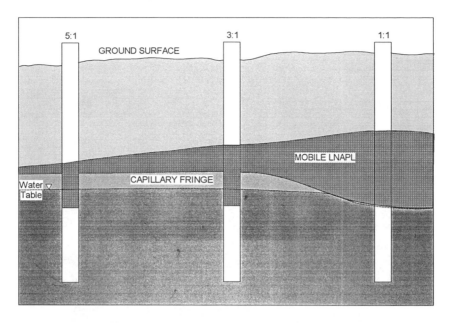

5:1

3:1

1:1

GROUND SURFACE

MOBILE LNAPL

CAPILLARY FRINGE

Water ▽
Table

**Figure 9-11   Cross section of a LNAPL pool showing well to formation
LNAPL thickness ratios (after Testa and Winegardner,
1991).**

function of the hydrostatic head and not the capillary fringe, which has been
destroyed by the confining pressures.

It is often assumed that apparent thicknesses are greater near the edge of a
LNAPL pool and smaller toward the center. The relationship between apparent
thicknesses as measured in wells and formation thickness across a LNAPL
pool is shown in Figure 9-11. Ratios are essentially smaller near the center of
the pool, where the thickness of the LNAPL is sufficient to displace water from
the capillary fringe. Conversely, where the LNAPL thickness is less, toward
the edge of the pool, ratios increase.

Apparent/actual LNAPL thickness ratios can be very high at the perimeter
of LNAPL pools, notably under low permeability conditions. Once a well is
installed, it can take several months before the LNAPL migrates from the
formation into the well reflecting the presence of low permeability soils in the
zone of LNAPL occurrence. As clearly shown in Figure 9-12, a well screened
at the perimeter of a known LNAPL pool initially had no detectable LNAPL
until 4 months after installation, whereas upon detection the apparent thickness
slowly increased with time up to 15.71 ft.

In developing water table contour maps when LNAPL is present, since the
water level (water table) as measured in the well is depressed by the weight of
the hydrocarbon, a corrected depth to water is calculated as follows:

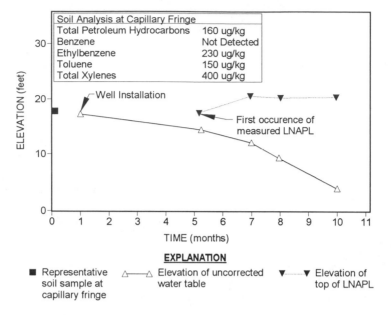

Figure 9-12   Graph showing gradual presence of LNAPL in well over time in low permeability soil.

$$DTW_c = \text{Static } DTW_m - (LNAPL_{ap} \times G) \qquad (9\text{-}9)$$

where DTW_c = corrected depth to water, DTW_m = measured depth to water, $LNAPL_{ap}$ = measured apparent LNAPL thickness in well, and G = specific gravity at 60°F. The API gravity data for LNAPL can easily be converted to specific gravity as follows:

$$API = \frac{145}{G} - 131.5 \qquad (9\text{-}10)$$

or

$$G = \frac{145}{API + 131.5} \qquad (9\text{-}11)$$

The resulting LNAPL thickness is conservative in that it incorporates both the actual thickness of LNAPL in the adjacent formation and the height of the capillary fringe. For most practical purposes, this level of accuracy is sufficient, although the more complex and extensive the site conditions are, the need for more sophisticated approaches may be warranted.

LNAPL Recovery Considerations

Recoverability of LNAPLs from the subsurface refers to the amount of mobile LNAPL available. Hydrocarbon that is retained in the subsurface media is typically not recoverable by conventional means. Recovery of LNAPL can be accomplished via a variety of strategies including linear interception systems (i.e., ditches, trenches, and hydraulic troughs), one- and two-well pumping systems and skimming systems. Enhancement techniques include vacuum assist and sparging.

With any "oil" recovery operation, retrieval of about 30% of the oil that resides in the subsurface is considered by most as successful. Even though environmental concerns dictate more stringent requirements when it comes to the subsurface presence of LNAPLs, complete recovery is not technically feasible in most cases. Factors that affect recoverability include relative permeability, viscosity, amount of residual hydrocarbon, areal distribution of the pool, coproduced water-handling capabilities, and site-specific constraints (i.e., presence of above- and underground structures).

Relative permeability ratios constantly change during recovery operations, and dictate that at certain ratios the LNAPL will be immobile and be retained as residual hydrocarbon. Residual hydrocarbon losses are much higher under saturated conditions than under unsaturated conditions, which also results in relatively thinner LNAPL thickness during a rise in the water table, and a thickening during a decline in the water table. Recoverability also varies in regards to pool size and extent. Recovery of a relatively small pool of limited areal extent and concentrated thicknesses will be more effective than a thin pool which encompasses a large areal extent. The approach to most, if not all, LNAPL recovery operations is to maximize the volume of LNAPL being recovered while minimizing the volume of coproduced water. In evaluating the effectiveness of moderate-to-large-scale LNAPL recovery programs, the coproduced water, which can range up to and possibly exceed 1000 gal/min for major large-scale recovery programs (i.e., long-term recovery programs implemented at many major refineries), is often the most limiting factor. Finally, site-specific factors may restrict the placement of recovery or reinjection wells in optimum locations.

DENSE NONAQUEOUS PHASE LIQUIDS (DNAPLs)

The potential for extensive subsurface contamination by DNAPLs is very high, but of less certainty in regards to the full occurrence and magnitude of this problem. The high degree of sensitivity to DNAPLs reflects their widespread production and use by a wide range of industries and high frequency of handling. In the late 1970s, dissolved chlorinated solvent compounds were found to be ubiquitous and substantial in many major groundwater systems throughout the U.S. at concentrations exceeding 10 times that of drinking water standards. However, only dissolved phases were reported, seldom immiscible nonaqueous phases. It was initially and erroneously assumed that the relatively high volatility in the vadose zone and high solubility in the saturated zone accounted for the absence of the immiscible phase.

DNAPLs are broadly classified on the basis of certain physical properties: density, viscosity, and solubility. Chlorinated solvents (i.e., trichloroethylene, TCE; tetrachloroethylene, PCE; and trichloroethane, TCA), creosote, and coal tar (i.e., polyaromatic hydrocarbons, PAHs) are the most common. Although many chlorinated solvents are characterized by relatively high density and low viscosity, creosote and coal tar compounds have relatively lower densities and high viscosities. In comparison to water, the chlorinated solvents are characterized by relatively high density, low viscosity, and significantly higher specific gravities. Creosote and coal tar compounds are characterized by viscosities of only 10 or 20 times greater than that of water, and specific gravities only slightly greater than water. Some of the more prevalent DNAPLs reported at CERCLA sites and their respective physical properties are presented in Table 9-3.

Based on these differences, the behavior of both the immiscible phase and dissolved phase of DNAPL in the subsurface is different. For example, immiscible chlorinated solvents are relatively mobile and strongly influenced by gravity while the dissolved phases are also relatively mobile reflecting sorption properties and significant solubilities of some constituents. The dissolved constituents can thus migrate large distances. For example, immiscible creosote and coal tar compounds are less influenced by gravity, thus, less mobile. In addition, lower solubilities and concomitant stronger sorption result in lower mobility.

DNAPLs' mobility is influenced by their respective density, viscosity, and interfacial tension with water. Mobility within the soil matrix is influenced by small-scale features such as soil type and intrinsic permeability, mineralogy, pore size, pore geometry, macropores, and large-scale features such as heterogeneities and isotropic conditions, and structural and stratigraphic features (Figure 9-13).

Once released into the subsurface, DNAPLs migrate vertically downward through the vadose zone with some lateral spreading where significant permeability contrasts are encountered. In the vadose zone, DNAPL residual hydrocarbon and manifested as trapped in pore space via surface tension, dissolved constituents into residual soil water, or in a vapor form. When sufficient DNAPL reaches the saturated zone (i.e., water table), the DNAPL continues to migrate downward by the influence of gravity until a significant permeability contrast is encountered. Depending on the volume of DNAPL under saturated condition, residual DNAPL occurs in pore spaces and as dissolved constituents.

Should a significant release occur and the capillary or entry pressures are overcome, the DNAPL will eventually reach the saturated zone (water table).

Table 9-3 Most Prevalent Chemical Compounds at U.S. Superfund Sites with a Specific Gravity Greater Than One[a]

Compound	Density (g/cc)	Dynamic Viscosity (cp)[b]	Kinematic Viscosity (cs)[c]	Water Solubility (mg/L)	Henry's Law Constant (atm-m^3/mol)	Vapor Pressure (mm Hg)
Halogenated Semivolatiles						
1,4-Dichlorobenzene	1.2475	1.2580	1.008	8.0 E + 01	1.58 E − 03	6 E − 01
1,2-Dichlorobenzene	1.3060	1.3020	0.997	1.0 E + 02	1.88 E − 03	9.6 E − 01
Aroclor 1242	1.3850			4.5 E − 01	3.4 E − 04	4.06 E − 04
Aroclor 1260	1.4400			2.7 E − 03	3.4 E − 04	4.05 E − 05
Aroclor 1254	1.5380			1.2 E − 02	2.8 E − 04	7.71 E − 05
Chlordane	1.6	1.1040	0.69	5.6 E − 02	2.2 E − 04	1 E − 05
Dieldrin	1.7500			1.86 E − 01	9.7 E − 06	1.78 E − 07
2,3,4,6-Tetrachlorophenol	1.8390			1.0 E + 03		
Pentachlorophenol	1.9780			1.4 E + 01	2.8 E − 06	1.1 E − 04
Halogenated Volatiles						
Chlorobenzene	1.1060	0.7560	0.683	4.9 E + 02	3.46 E − 03	8.8 E + 00
1,2-Dichloropropane	1.1580	0.8400	0.72	2.7 E + 03	3.6 E − 03	3.95 E + 01
1,1-Dichloroethane	1.1750	0.3770	0.321	5.5 E + 03	5.45 E − 04	1.82 E + 02
1,1-Dichloroethylene	1.2140	0.3300	0.27	4.0 E + 02	1.49 E − 03	5 E + 02
1,2-Dichloroethane	1.2530	0.8400	0.67	8.69 E + 03	1.1 E − 03	6.37 E + 01
Trans-1,2-Dichloroethylene	1.2570	0.4040	0.321	6.3 E + 03	5.32 E − 03	2.65 E + 02
Cis-1,2-Dichloroethylene	1.2480	0.4670	0.364	3.5 E + 03	7.5 E − 03	2 E + 02
1,1,1-Trichloroethane	1.3250	0.8580	0.647	9.5 E + 02	4.08 E − 03	1 E + 02

Table 9-3 Most Prevalent Chemical Compounds at U.S. Superfund Sites with a Specific Gravity Greater Than One[a] (continued)

Compound	Density (g/cc)	Dynamic Viscosity (cp)[b]	Kinematic Viscosity (cs)[c]	Water Solubility (mg/L)	Henry's Law Constant (atm-m³/mol)	Vapor Pressure (mm Hg)
Methylene Chloride	1.3250	0.4300	0.324	1.32 E + 04	2.57 E − 03	3.5 E + 02
1,1,2-Trichloroethane	1.4436	0.1190	0.824	4.5 E + 03	1.17 E − 03	1.88 E + 01
Trichloroethylene	1.4620	0.5700	0.390	1.0 E + 03	8.92 E − 03	5.87 E + 01
Chloroform	1.4850	0.5630	0.379	8.22 E + 03	3.75 E − 03	1.6 E + 02
Carbon Tetrachloride	1.5947	0.9650	0.605	8.0 E + 02	2.0 E − 02	9.13 E + 01
1,1,2,2-Tetrachloroethane	1.6	1.7700	1.10	2.9 E + 03	5.0 E − 04	4.9 E + 00
Tetrachloroethylene	1.6250	0.8900	0.54	1.5 E + 02	2.27 E − 02	1.4 E + 01
Ethylene Dibromide	2.1720	1.6760	0.79	3.4 E + 03	3.18 E − 04	1.1 E + 01
Nonhalogenated Semivolatiles						
2-Methyl Napthalene	1.0058			2.54 E + 01	5.06 E − 02	6.80 E − 02
o-Cresol	1.0273			3.1 E + 04	4.7 E − 05	2.45 E − 01
p-Cresol	1.0347			2.4 E + 04	3.5 E − 04	1.08 E − 01
2,4-Dimethylphenol	1.0360			6.2 E + 03	2.5 E − 06	9.8 E − 02
m-Cresol	1.0380	21.0	20	2.35 E + 04	3.8 E − 05	1.53 E − 01
Phenol	1.0576		3.87	8.4 E + 04	7.8 E − 07	5.293E − 01
Naphthalene	1.1620			3.1 E + 01	1.27 E − 03	2.336E − 01
Benzo(a)Anthracene	1.1740			1.4 E − 02	4.5 E − 06	1.16 E − 09
Fluorene	1.2030			1.9 E + 00	7.65 E − 05	6.67 E − 09
Acenaphthene	1.2250			3.88 E + 00	1.2 E − 03	2.31 E − 02

Anthracene	1.2500		7.5 E – 02	3.38 E – 05	1.08 E – 05
Dibenz(a,hh)Anthracene	1.2520		2.5 E – 03	7.33 E – 08	1 E – 10
Fluoranthene	1.2520		2.65 E – 01	6.5 E – 06	E – 02 E – 06
Pyrene	1.2710		1.48 E – 01	1.2 E – 05	6.67 E – 06
Chrysene	1.2740		6.0 E – 03	1.05 E – 06	6.3 E – 09
2,4-Dinitrophenol	1.6800		6.0 E + 03	6.45 E – 10	1.49 E – 05
Miscellaneous					
Coal tar	1.028[d]	18.98[d]			
Creosote	1.05	1.08[e]			

[a] After Huling and Weaver (1991).
[b] centipoise (cp), water has a dynamic viscosity of 1 cp at 20°C.
[c] centistokes (cs).
[d] 45°F (70).

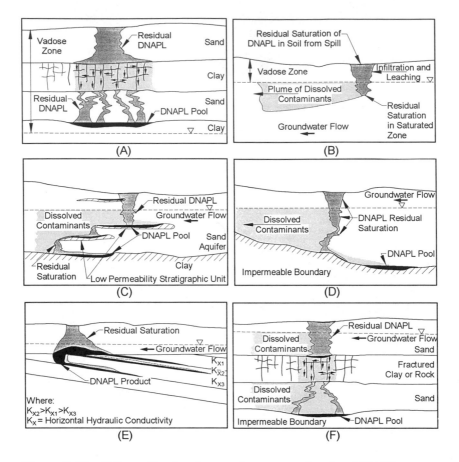

Figure 9-13   Schematic showing various scenarios for DNAPL occurrence.

The pressure head required for penetration increases as the grain size within this interval decreases. Once penetrated, gravity forces will continue to dominate in conjunction with capillary and viscous forces and density contrasts. Movement of the DNAPL continues until a significant permeability contrast is encountered (i.e., low permeability confining layer, bedrock, bedding plane, etc.). Once encountered, the DNAPL is often gradient controlled following the dip of the interface between the bottom of the aquifer and top of the lower confining layer, regardless of groundwater flow directions.

The DNAPL will tend to fill in surficial lows along this interface, or migrate along preferential permeable pathways such as fractures, bedding planes, rock holes, and coarse-grained zones. DNAPL, migration are also influenced by small-scale technical changes (i.e., openings less than 20 μm). Thus, clayey strata considered as relatively impermeable may not be so reliable in dealing with DNAPLs. If the overall thickness of the DNAPL pool which accumulates is uniform, the lower portion of the aquifer is containing most of the pool. Variable pool thickness reflects unevenness of or an inclination of the surface of the top of the confining layer at the lower portion of the saturated zone.

Once a pool is formed, continued contribution of DNAPL will result in enlargement of the pool, penetration of the lower "impermeable" layer, and development of new downward moving fingers of the perimeter of the layer.

Detection of DNAPLs in the unsaturated zone can be evaluated via soil gas surveys, soil borings, and to a limited extent surface geophysical techniques. However, reliable detection in the saturated zone is accomplished by means of installing monitoring wells. The evaluation of significant permeability contrasts is vitally important in conduct of subsurface assessments regarding DNAPLs.

DNAPL Apparent vs Actual Thickness

DNAPL thickness when measured in a monitoring well is not representative of the actual formation thickness. As with LNAPLs, the DNAPL thickness will be exaggerated as illustrated in Figure 9-14A. Limited laboratory studies have shown that the apparent DNAPL thickness is greater in fine sand than in coarse sand, reflecting variations in grain size. The difference in the apparent vs actual DNAPL thickness equals the DNAPL-water capillary fringe height. DNAPL thickness in hydrophobic sand, however, is greater than what is measured in a well due to the DNAPL capillary rise. Exaggerated DNAPL thickness in wells can also be due to the well partially penetrating a lower permeability layer (Figure 9-14B), or with the presence of thin low permeability laminae (Figure 9-14C).

DNAPL Recovery Considerations

Recovery of DNAPL is a very slow process which is affected by those factors encountered with LNAPL (i.e., relative permeability, viscosity, residual hydrocarbon pool distribution, and site-specific factors, etc.). Dissolution of a DNAPL pool is dependent upon the vertical dispersivity, groundwater velocity, solubility, and pool dimension. Dispersivities for charnolid solvent are estimated for a medium to coarse sand under laboratory conditions on the order of 10^{-3} to 10^{-4} m. Thus, limited dispersion at typical groundwater velocities is anticipated to be slow and may take up to decades or centuries when dealing with concentrations of a few hundred to a few thousand kilograms. To add to this dilemma, velocities are difficult to increase due to limited radial flow conditions from pumping wells, low yields,

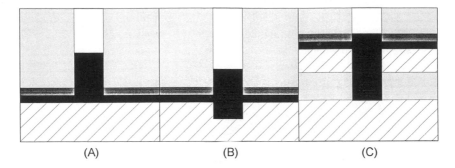

Figure 9-14   Causes for discrepancies in apparent DNAPL thickness as measured in a monitoring well in comparison with the actual thickness in the formation.

and limited coproduced water handling (notably where large water volumes yields are produced).

The residual hydrocarbons will continue to serve as a source of groundwater contamination; thus, remediation strategies for DNAPLs should be focused on long-term control (i.e., source containment and pool control and recovery) vs short-term fixes. Regardless of an increased level of effort (i.e., additional wells, increased pumping rates, etc.), the overall time for remediation is not expected to shorten by more than a factor of five.

SOURCE IDENTIFICATION

Geochemical characterization of hydrocarbons in the subsurface soil and groundwater is conducted to (1) relate the hydrocarbon or NAPL to its true source, (2) assess whether a supposed source can or cannot be the actual one, (3) differentiate between two or more NAPL releases that have commingled into one large pool, and (4) evaluate the appropriate remediation strategy to be undertaken. Some of the techniques used for characterization or "fingerprinting" are very conventional in nature, while others were specifically developed for use in the petroleum exploration industry for characterization of rock and crude oil types, and in some cases the brand, grade, and source crude. Some of these techniques have been modified and subsequently applied to environmental issues, notably in the identification and characterization of fugitive hydrocarbons, especially where a succession of ownership has occurred and the product storage history is known.

The methods employed for the characterization of fugitive hydrocarbons, from simplest to relatively more sophisticated, are:

- API gravity
- Distillation curves

- Trace metals analysis
- Gas chromatography fingerprinting
- Statistical comparisons
- Isotopic fingerprinting
  - $^{13}C/^{12}C$
  - $^{2}H/^{1}H$
  - D/H
  - $^{34}S/^{32}S$
  - $^{15}N/^{14}N$

The methods outlined above apply to different chemical classes for identification as further discussed below.

API is the simplest analysis in discriminating various NAPL product types. API gravity is similar to specific gravity but differs as follows:

$$API = \frac{141.5}{\text{specific gravity}} - 131.5 \qquad (9\text{-}12)$$

Using this conversion, water has a specific gravity of one (unit less) and an API gravity of 10°. The higher   the API gravity, the lighter the compound.

Distillation analysis of NAPLs is a common technique used to evaluate the various components of crude, but is also applicable to the characterization of product types comprising the NAPL. Distillation curves as illustrated in Figure 9-15 are plots of boiling temperature (°F) vs cumulative percent volume. Boiling points generally increase with an increase in the number of carbon atoms which comprise the compound. As a crude oil or hydrocarbon blend is incrementally heated, the hydrocarbon compounds having a boiling point at or below the current temperature volatilizes. The hydrocarbon compounds remaining in the sample will not volatilize until the temperature is raised to the boiling point of the remaining compounds. The range of boiling temperatures, from low to high, is divided into the following product types: butanes, gasoline, naphtha, kerosene, light gas-oil, heavy gas-oil, and residue.

Analysis for organic (not total) lead alkyl compounds can sometimes provide useful information on source(s) of gasoline contamination. These compounds include ethylenedibromide (EDB), tetramethyl lead, trimethylethyl lead, dimethyldiethyl lead, methyltriethyl lead, and tetraethyl lead (TEL). For example, only TEL was used as an additive in leaded gasoline in amounts ranging from one third between 1980 and the present, to an order of magnitude less prior to the early 1980s. Many of the other leaded alkyls may have also been added prior to the early 1980s. Typically, unleaded gasoline will not contain more than 0.05 g/gal (13 mg/L) organic lead, while leaded gasoline will not contain greater than 1.1 g/gal (290 mg/L), although organic lead has historically been added to gasoline fuels at concentrations up to 800 mg/L. Another example is evaluation of the distribution pattern of EDB and TEL. EDB is more soluble relative to other lead alkyl compounds. Thus, significant

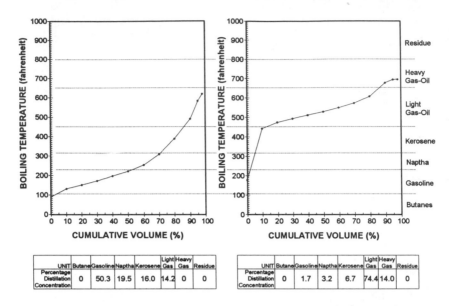

UNIT	Butane	Gasoline	Naptha	Kerosene	Light Gas	Heavy Gas	Residue
Percentage Distillation Concentration	0	50.3	19.5	16.0	14.2	0	0

UNIT	Butane	Gasoline	Naptha	Kerosene	Light Gas	Heavy Gas	Residue
Percentage Distillation Concentration	0	1.7	3.2	6.7	74.4	14.0	0

Figure 9-15   Typical distillation curves representative of two subpools that have commingled to form one LNAPL pool.

concentrations of EDB and TEL would imply that the release was very recent. Reduced concentrations of EDB or its absence would suggest that the release is 5 to 10 years old, and if TEL concentrations are also greatly reduced the release may exceed 10 years under harsh conditions. The presence of other lead alkyls in addition to TEL suggests a release prior to 1980. Other constituents of importance are manganese, nickel, sulfur, and vanadium. Nickel and vanadium are commonly associated with the petroporphrin fraction of the crude oil and increase an order of magnitude with a decrease in API gravity. Vanadium-nickel ratios are also used for oil-field correlation purposes.

Gas chromatography is conventionally used for hydrocarbon fingerprinting. The process uses a gas chromatograph equipped with a capillary column and flame ionization detector; emitted hydrocarbons are scanned in the temperature range of 10°C/min and then held at 300°C for 5 min (resulting in a total run time of 30 min). The resulting chromatograms are visually compared to standards and reported in generally no more than two of the four possible refined petroleum products in this boiling range (i.e., gasoline, kerosene, diesel, etc.). Peaks occurring in the range of C_{25} to C_{30} are above the diesel range, with those occurring during the last 5 min of the scan being the higher boiling point tars and asphaltenes and indicative of a crude oil. Ratios of certain compounds, or combinations of compounds, can be used to evaluate the degree of maturity, paraffinicity, biodegradation, and waterwashing (Figure 9-17). With refined petroleum products, similarities or differences between samples can be discerned reflecting differences in brand of gasoline. Certain product additives of

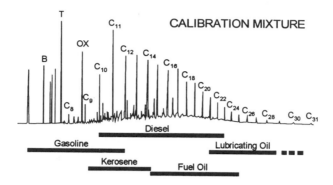

Figure 9-16   Approximate carbon number ranges for refined petroleum products showing presence of benzene (B), toluene (T), and orthoxylene (OX).

interest and of environmental significance can be qualitatively and quantitatively identified. Such additives include anti-knock compounds, blending agents, antioxidants, antirust agents, anti-icing agents, or other proprietary purposes. Common additives include 1,2-dibromoethane (EDB), 1,2-dichloroethane (EDC), isopropyl alcohol, methyl tert-butyl-ether (MTBE), p-phenylenediamine, and tert-butyl alcohol (TBA).

Ratios of stable isotopes of carbon ($^{13}C/^{12}C$) and hydrogen ($^{2}H/^{1}H$) are measured as a ratio of the heavier to the lighter (most abundant) isotope in a gas introduced into a dual-collecting mass spectrometer. Such ratios differ both among crude oils by origin, and among various fractions within the same oil. These ratios can be applied as a tracer for organic materials that have undergone partial degradation through either biological or other processes where the molecular characteristics of the original substance are obscure. Their persistence is attributable to the isotopic composition not changing to the same extent correlative to the molecular composition. Other elements such as ^{15}N and ^{34}S may also prove useful. Overall, isotope ratios have merit when extensive biodegradation has caused nearly complete alteration of the molecular composition.

Processes that may limit the application of certain methods include migration, water washing, oxidation, deasphalting, biodegradation, and the subsurface migration of hydrocarbons, result in the removal of some of the components while others are retained by the soil media. Water washing is the preferential dissolving of certain hydrocarbon constituents when in contact with nonsteady-state water conditions. Crude hydrocarbon products can undergo oxidation when exposed to dissolved oxygen, certain minerals, and bacteria, resulting in degradation of the lower molecular weight paraffinic fractions, notably the n-alkanes. Oxidation usually occurs near the surface resulting in degradation, increased viscosity, and specific gravity and conver-

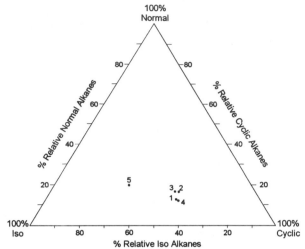

Figure 9-17   Ternary diagram showing relative percentage of normal alkanes, branched alkanes (iso-alkanes), and cyclic alkanes based on gas chromatograph results (after Testa and Winegardner, 1991).

sion of some liquid components to solids. The operation of extracting asphalt components from a petroleum product is referred to as deasphalting. For example, propane deasphalting employs propane as a solvent to separate oil and asphalt whereas the liquid propane dissolves the oil as the asphalt settles out. Biodegradation is the selective destruction of specific chemical classes via microbial consumption. Heavily biodegraded hydrocarbons are typically enriched in nitrogen, oxygen, sulfur, and polycyclic compounds, reflecting either the end product following microbial activity or removal and destruction of less stable constituents.

As with any analytical program, it is best to review the data that have already been generated, and proceed with the simplest analytical methods first. Determination of API gravity, development of distillation curves, trace metals analysis, and gas chromatography analysis should be conducted first. If the resultant data is not conclusive, more intricate and sophisticated methods can be used, such as statistical comparisons of the distribution of paraffinic or n-alkane compounds between certain C-ranges, determination of certain isotopic ratios of carbon and hydrogen for the lighter gasoline-range fractions, and nitrogen and sulfur for heavier petroleum fractions. Although in many cases the suspected source will not be available for analysis, it may be possible to characterize the hydrocarbon with enough accuracy to eliminate other sources.

BIBLIOGRAPHY

Abdul, A.S., 1988, Migration of Petroleum Products Through Sandy Hydrogeologic Systems: *Ground Water Monitoring Review,* Vol. 8, No. 4, pp. 73-81.

Abdul, A.S., Kia, S.F. and Gibson, T.L., 1989, Limitations of Monitoring Wells for the Detection and Quantification of Petroleum Products in Soils and Aquifers: *Ground Water Monitoring Review,* Vol. 9, No. 2, pp. 90-99.

Adams, T.V. and Hampton, D.R., 1992, Effects of Capillarity on DNAPL Thickness in Wells and in Adjacent Sands: In *Proceedings of the IAH Conference on Subsurface Contamination by Immiscible Fluids;* Balkema Publications, Rotterdam.

Anderson, M.R., Johnson, R.L. and Pankow, J.F., 1992, Dissolution of Dense Chlorinated Solvents into Groundwater. 3. Modeling Contaminant Plumes from Fingers and Pools of Solvent: *Environmental Science and Technology,* Vol. 26, No. 5, pp. 901-907.

Anderson, W.G., 1986a, Wettability Literature Survey—Part 1: Rock/Oil/Brine Interactions, and the Effects of Core Handling on Wettability: *Journal of Petroleum Technology,* October, pp. 1125-1149.

Anderson, W.G., 1986b, Wettability Literature Survey—Part 2: Wettability Measurement: *Journal of Petroleum Technology,* November, pp. 1246-1262.

Anderson, W.G., 1986c, Wettability Literature Survey—Part 3: The Effects of Wettability on the Electrical Properties of Porous Media: *Journal of Petroleum Technology,* December, pp. 1371-1378.

Corey, A.T., 1986, *Mechanics of Immiscible Fluids in Porous Media:* Water Resources Publications, Littleton, CO, 255 p.

Dietz, D.N., 1971, Pollution of Permeable Strata by Oil Components: In *Water Pollution by Oil* (Edited by P. Heddle), Institute of Petroleum, London, pp. 127-139.

de Pastrovich, T.L., Baradat, Y., Barthel, R., Chiarelli, A. and Fussell, D.R., 1979, Protection of Groundwater from Oil Pollution: CONCAWE Report No. 3179, The Hague, Netherlands, 61 p.

Farr, A.M., Houghtalen, R.J. and McWhorter, D.B., 1990, Volume Estimation of Light Nonaqueous Phase Liquids in Porous Media: *Ground Water,* Vol. 28, No. 1, pp. 48-56.

Faust, C.R., Guswa, J.H. and Mercer, J.W., 1989, Simulation of Three-Dimensional Flow of Immiscible Fluids Within and Below the Unsaturated Zone: *Water Resources Research,* Vol. 25, No. 12, pp. 2449-2464.

Feenstra, S., Mackay, D.M. and Cherry, J.A., 1991, A Method for Assessing Residual NAPL Based on Organic Chemical Concentrations in Soil Samples: *Ground Water Monitoring Review,* Vol. 11, No. 23, pp. 128-136.

Folkes, D.J., Bergman, M.S. and Herst, W.E., 1987, Detection and Delineation of a Fuel Oil Plume in a Layered Bedrock Deposit: In Proceedings of the National Water Well Association of Ground Water Scientists and Engineers and the American Petroleum Institute Conference on Petroleum Hydrocarbons and Organic Chemicals in Ground Water: Prevention, Detection and Restoration, pp. 279-304.

Hedgcore, H.R. and Stevens, W.S., 1991, Hydraulic Control of Vertical DNAPL Migration: In Proceedings of the Association of Ground Water Scientists and Engineers Conferences on Petroleum Hydrocarbons and Organic Chemicals in Ground Water: Prevention, Detection, and Restoration, pp. 327-338.

Huling, S.G. and Weaner, J.W., 1991, Dense Non-Aqueous Phase Liquids: U.S. Environmental Protection Agency, USEPA/540/4-91/002, 21 p.

Hunt, J.R., Sitar, N. and Udell, K., 1988, Nonaqueous Phase Liquid Transport and Cleanup, Part I, Analysis of Mechanisms: *Water Resources Research,* Vol. 24, No. 8, pp. 1247-1258.

Johnson, R.L. and Pankow, J.F., 1992, Dissolution of Dense Chlorinated Solvents in Groundwater. 2. Source Functions for Pools of Solvents: *Environmental Science and Technology,* Vol. 26, No. 5, pp. 896-900.

Kaplan, I.R., 1989, Forensic Geochemistry in Characterization of Petroleum Contaminants in Soils and Groundwater: In *Environmental Concerns in the Petroleum Industry* (Edited by S.M. Testa), Pacific Section of the American Association of Petroleum Geologists Symposium Volume, pp. 159-181.

Kaplan, I.R., 1992, Characterizing Petroleum Contaminants in Soil and Water and Determining Source of Pollutants: In Proceedings of the American Petroleum Institute and Association of Ground Water Scientists and Engineers Conference on Petroleum Hydrocarbons and Organic Chemicals in Ground Water: Prevention, Detection, and Restoration, pp. 3-18.

Karickhoff, S.W., Brown, D.S. and Scott, T.A., 1979, Sorption of Hydrophobic Pollutants on Natural Sediments: *Water Resources Research,* Vol. 13, pp. 241-248.

Kemblowski, M.W. and Chiang, C.Y., 1988, Analysis of the Measured Free Product Thickness in Dynamic Aquifers: In Proceedings of the National Water Well Association of Ground Water Scientists and Engineers and the American Petroleum Institute Conference on Petroleum Hydrocarbons and Organic Chemicals in Ground Water: Prevention, Detection and Restoration, Vol. I, pp. 183-205.

Kemblowski, M.W. and Chiang, C.Y., 1990, Hydrocarbon Thickness Fluctuations in Monitoring Wells: *Ground Water,* Vol. 28, No. 2, pp. 244-252.

Kessler, A. and Rubin, H., 1987, Relationships Between Water Infiltration and Oil Spill Migration in Sandy Soils: *Journal of Hydrology,* Vol. 91, pp. 187-204.

Kramer, W.H., 1982, Groundwater Pollution from Gasoline: *Ground Water Monitoring Review,* Vol. 2, No. 2, pp. 18-22.

Kueper, B.H., Abbott, W. and Farquhar, G., 1989, Experimental Observations of Multiphase Flow in Heterogeneous Porous Media: *Journal of Contaminant Hydrology,* Vol. 5, pp. 83-95.

Kueper, B.H. and Frind, E.O., 1988, An Overview of Immiscible Fingering in Porous Media: *Journal of Contaminant Hydrology,* Vol. 2, pp. 95-110.

Lenhard, R.J. and Parker, J.C., 1987, Measurement and Prediction of Saturation-Pressure Relationships in Three-Phase Porous Media Systems: *Journal of Contaminant Hydrology,* Vol. 1, pp. 407-424. Correction, Vol. 2, pp. 189-190.

Lenhard, R.J. and Parker, J.C., 1988, Experimental Validation of the Theory of Extending Two-Phase Saturation-Pressure Relationships to Three-Fluid Phase System for Monotonic Drainage Paths: *Water Resources Research,* Vol. 24, pp. 373-380.

Lenhard, R.J. and Parker, J.C., 1990, Estimation of Free Hydrocarbon Volume from Fluid Levels in Observation Wells: *Ground Water,* Vol. 28, No. 1, pp. 57-67.

Leverett, M.C., 1939, Flow of Oil-Water Mixtures Through Unconsolidated Sands: *Transactions of the American Mining and Metallurgical Engineers, Petroleum Development and Technology,* Vol. 132, pp. 149-171.

Leverett, M.C., 1941, Capillary Behavior in Porous Solids: *Transactions AIME, Petroleum Engineering Division,* Vol. 142, pp. 152-169.

Levy, B.S., Riordan, P.J. and Schreiber, R.P., 1990, Estimation of Leak Rates from Underground Storage Tanks: *Ground Water,* Vol. 28, No. 3, pp. 378-384.

Melrose, J.C., 1965, Wettability as Related to Capillary Action in Porous Media: *Society of Petroleum Engineers Journal,* Vol. 5, pp. 259-271.

Mercer, J.W. and Cohen, R.M., 1990, A Review of Immiscible Fluids in the Subsurface - Properties, Models, Characterization and Remediation: *Journal of Contaminant Hydrology,* Vol. 6, pp. 107-163.

Morrow, N.R., 1990, Wettability and its Effect on Oil Recovery: *Journal of Petroleum Technology,* Vol. 29, pp. 1476-1484.

Rau, B.V., Vass, T. and Stachle, W.J., 1992, DNAPL: Implications to Investigation and Remediation of Groundwater Contamination: In Proceedings of the Hazardous Materials Control Research Institute, HMC-South '92, New Orleans, pp. 84-94.

Schwille, F., 1984, Migration of Organic Fluids Immiscible with Water in the Unsaturated: *Pollutants in Porous Media* (Edited by B. Yaron, G. Dagan, and J. Goldshmid): Vol. 47, Ecological Studies, Springer-Verlag, New York, pp. 27-48.

Schwille, F., 1985, Migration of Organic Fluids Immiscible with Water in the Unsaturated and Saturated Zones: In Proceedings of the National Water Well Association Second Canadian/ American Conference on Hydrogeology, pp. 31-35.

Schwille, F., 1988, *Dense Chlorinated Solvents in Porous and Fractured Media:* Lewis Publishers, Chelsea, MI, 146 p.

Sitar, N., Hunt, J.R. and Udell, K.S., 1987, Movement of Nonaqueous Liquids in Groundwater: In Proceedings of Geotechnical Practice for Waste Disposal, ASCE, Ann Arbor, MI, pp. 205-223.

Stone, H.L., 1973, Estimation of Three-Phase Relative Permeability and Residual Oil Data: *Journal of Can. Petroleum Technology,* Vol. 12, No. 4, pp. 53-61.

Testa, S.M. and Paczkowski, M.T., 1989, Volume Determination and Recoverability of Free Hydrocarbon: *Ground Water Monitoring Review,* Vol. 9, No. 1, pp. 120-128.

Testa, S.M., Baker, D.M. and Avery, P.L., 1989, Field Studies on Occurrence, Recoverability and Mitigation Strategy for Free Phase Liquid Hydrocarbon: In *Environmental Concerns in the Petroleum Industry* (Edited by S.M. Testa); Pacific Section of the American Association of Petroleum Geologists, Palm Springs, CA, pp. 57-81.

Testa, S.M. and Winegardner, D.W., 1991, *Restoration of Petroleum-Contaminated Aquifers:* Lewis Publishers, Boca Raton, FL, 269 p.

Testa, S.M., 1991, Geochemical Characterization of Light Non-Aqueous Phase Liquid Hydrocarbon Product: In Proceedings of the Hazardous Materials Control Research Institute, Anaheim, CA, pp. 185-192.

Testa, S.M., 1992, Groundwater Remediation at Petroleum-Handling Facilities, Los Angeles Coastal Plain: In *Engineering Geology in Southern California* (Edited by B.W. Pipkin and R.J. Proctor), Association of Engineering Geologists, pp. 67-80.

U.S. Environmental Protection Agency, 1992, Dense Nonaqueous Phase Liquids - A Workshop Summary: EPS/600/R-92/030, 81 p.

Van Dam, 1967, The Migration of Hydrocarbons in a Water-Bearing Stratum: In *The Joint Problems of the Oil and Water Industries* (Edited by P. Hepple); Institute of Petroleum, London, pp. 55-96.

Weyer, K.U. (Editor), 1990, *Proceedings of the IAH Conference on Subsurface Contamination by Immiscible Fluids:* Balkema Publications, Rotterdam.

Wilson, J.L., Conrad, S.H., Hagan, E., Mason, W.R. and Peplinski, W., 1988, The Pore Level Spatial Distribution and Saturation of Organic Liquids in Porous Media: In Proceedings of the National Water Well Association of Groundwater Scientists and Engineers and the American Petroleum Institute Conference on Petroleum Hydrocarbons and Organic Chemicals in Ground Water: Prevention, Detection and Restoration, Vol. 1, pp. 107-133.

Yaniga, P.M., 1982, Alternatives in Decontamination for Hydrocarbon - Contaminated Aquifers: *Ground Water Monitoring Review,* Vol. 2, pp. 40-49.

Yaniga, P.M., 1984, Hydrocarbon Retrieval and Apparent Hydrocarbon Thickness: Interrelationships to Recharging/Discharging Aquifer Conditions: In Proceedings of the National Water Well Association of Ground Water Scientists and Engineers and the American Petroleum Institute Conference on Petroleum Hydrocarbons and Organic Chemicals in Ground Water: Prevention, Detection and Restoration, pp. 299-329.

Yaniga, P.M. and Demko, D.J., 1983, Hydrocarbon Contamination of Carbonate Aquifers: Assessment and Abatement: In Proceedings of the National Water Well Association of Ground Water Scientists and Engineers Third National Symposium on Aquifer Restoration, pp. 60-65.

10 LANDFILL DISPOSAL

*"The secret to abating pollution of the land is not to empha-
size cleaning it up but to prevent its release into the environment in the first place."*

(Stephen Testa, 1992)

INTRODUCTION

There are about 75,000 landfills in the U.S., of which about 75% have
adversely affected groundwater to varying degrees. The majority of these
landfills are sanitary landfills with little emphasis toward siting, construction,
operation, and closure of leachate generation in regards to leachate monitoring
and management. Landfills have been around for centuries. It was only in the
late 1950s, however, that landfills came under serious scrutiny. Landfills used
to be perceived as favorable since any leachate that may have migrated would
be completely attenuated by the underlying soil; thus, the potential for groundwater being adversely impacted was minimal. Over the next two decades,
however, several studies were conducted that proved otherwise. A distinction
between nonhazardous and hazardous wastes was subsequently established,
leading to separate regulations and design criteria for the two categories.

Landfills were once considered one of the least expensive alternatives for
hazardous waste disposal. Although in some cases still economically attractive,
there has been much pressure for more stringent regulation. Over the past
decade, many landfills have become hazardous waste sites. Considering the
additional costs associated with assessment, monitoring, and remediation activities, the actual cost to the public far exceeds that of other alternatives
despite Herculean efforts toward waste minimization, reuse and recycling, and
other waste reduction and treatment alternatives, in addition to severe land
disposal restriction. Of 117 landfills in Missouri, all but 5 will reach their
capacity by 1995. Despite these restrictions, landfills still remain economically
attractive, with their use anticipated to continue into the future.

1. American Ecology, Beatly, NV
2. American Ecology, Robstown, TX
3. Browning-Ferris, Livingston, LA
4. Browning-Ferris, Niagara Falls, NY
5. Browning-Ferris, Williamsburg, OH
6. Chem Waste, Adams Center, IN
7. Chem Waste, Arlington, OR
8. Chem Waste, Calumet City, IL
9. Chem Waste, Emelia, AL
10. Chem Waste, Kettleman Hills, CA
11. Chem Waste, Lake Chas., LA
12. Chem Waste, Model City, NY
13. Casmalia Disposal, Casmalia, CA
14. 4 County, Rochester, IN
15. Envirosafe Systems, Grandview, ID
16. Envirosafe, Oregon, OH
17. GSX Laidlaw, Pinewood, SC
18. International Technology, Bakersfield, CA
19. International Technology, Imperial Valley, CA
20. Peoria Disposal, Peoria, IL
21. Rollins, Baton Rouge, LA
22. Rollins, Deer Park, TX
23. USPCI, Grassy Mountain, UT
24. USPCI, Lone Mnt., OK
25. Wayne, Belleville, MI

Figure 10-1 Sites of 25 commercial hazardous waste landfills in the U.S.

Landfills today are referred to as treatment, storage, and disposal (TSD) facilities. These facilities may in turn incorporate one or more waste management units (WMU). At this writing there are only 25 commercial hazardous waste landfills throughout the U.S.; their locations are shown in Figure 10-1.

Leachate generation can be grouped into three general categories, all with the potential to adversely affect groundwater: hazardous, conventional, and nonconventional constituents. Hazardous chemicals and constituents are those which met certain regulatory criteria as discussed in Chapter 7. Not all constituents considered hazardous are noted on the EPA Priority Pollutants List. This list of 65 constituents was somewhat arbitrarily selected as part of a court order, and does not provide a representative consensus of the most important chemicals which pose a threat to groundwater resources. In addition, the list of chemicals focuses more on those chemicals that pose cancer risk concerns, not those that are known to pose a significant threat to overall groundwater quality. Conventional contaminants include constituents that are common components of a waste stream. Included in this group are total dissolved solids, hardness, alkalinity, chloride, sulfate, iron, manganese, and hydrogen sulfide. Other nondifferentiated organics belonging to this group are chemical oxygen demand (COD), biochemical oxygen demand (BOD) and total organic carbon (TOC). Parameters are important. For example, the suspected human carcinogen trichloroethylene (TCE) can convert to the highly potent, known human carcinogen vinyl chloride by the addition of oxygen-demanding material. Depletion of the dissolved oxygen in the groundwater generates anoxic (oxygen-free) conditions which allow existing bacteria to convert TCE to vinyl chloride. The last group of nonconventional contaminants is not well defined and include primarily organic chemicals routinely encountered in municipal landfill leachate. In fact, about 90 to 95% of the organic material present in leachate is derived from chemicals of unknown composition. A table of typical components of municipal landfill leachate is presented in Table 10-1.

It is generally acknowledged that all surface containment structures will eventually leak or fail. With this in mind, the overall strategy for protecting groundwater resources at TDS facilities reflects (1) leachate minimization and control through stringent design and operating requirements and (2) stringent groundwater monitoring and response requirements should leachate eventually reach the groundwater. From a geological and hydrogeological perspective, this strategy is implemented through development of a vadose (unsaturated zone) and groundwater (saturated zone) characterization and monitoring program. This chapter discusses landfill design, including liner systems and leachate collection and removal systems. Siting criteria and vadose and groundwater detection monitoring are addressed. Two case histories are then presented which illustrate the complexities involved in development of a groundwater detection monitoring strategy.

LANDFILL DESIGN

Landfills and/or WMU are classified according to their ability to contain waste. The main components of landfills are the cover or liner system, and leachate collection and removal system. All new landfills by regulation are required to have double liners, a leachate collection and removal system, and a leak detection system. The purpose of a liner system is to prevent any migration of certain wastes from the landfill into adjacent soil, surface water, or groundwater during the active and post closure life of the landfill. The liner system must also be constructed of materials that prevent wastes from passing into or through the liner. In a double-lined landfill, as illustrated in Figure 10-2, there are two liners and two leachate systems. The primary leachate system is situated above the top liner, while the second leachate system is situated between the two liners.

Liner System

Liners can consist of either flexible membrane liners (FMLs), compacted clay liners, composite liner systems (i.e., an FML overlying a compacted low permeability soil layer), or reuse/recycled cold-mix asphalt. Material specifications vary according to the type of liner. The upper FML must be a minimum 30 mm in thickness; without a soil cover layer, the specification is 45 mm. Most FMLs, however, are in the range of 60 to 100 mm in thickness. Other considerations with FMLs are chemical compatibility with the waste leachate, aging, durability, stress and strain, vapor-chemical permeation, and ease of installation. For compacted soil liners of

Table 10-1 Summary of Typical Constituents and Relative Concentration Range of Municipal Landfill Leachate[a]

Parameter	Concentration Range[b,c]	Concentration[b,c] (Average)
BOD	1,000–30,000	10,500
COD	1,000–50,000	15,000
TOC	700–10,000	3,500
Total volatile acids (as acetic acid)	70–28,000	NA[d]
Total Kjeldahl Nitrogen (as N)	10–500	500
Nitrate (as N)	0.1–10	4
Ammonia (as N)	100–400	300
Total Phosphate (PO_4)	0.5–50	30
Orthophosphate (PO_4)	1–60	22
Total alkalinity (as $CaCO_3$)	500–10,000	3,600
Total hardness (as $CaCO_3$)	500–10,000	4,200
Total solids	3,000–50,000	16,000
Total dissolved solids	1,000–20,000	11,000
Specific conductance (umhos/cm)	2,000–8,000	6,700
pH	5–7.5	63
Calcium	100–3,000	1,000
Magnesium	30–500	700
Sodium	200–1,500	700
Chloride	100–2,000	980
Sulfate	10–1,000	380
Chromium (total)	0.05–1	0.9
Cadmium	0.001–0.1	0.05
Copper	0.02–1	0.5
Lead	0.1–1	0.5
Nickel	0.1–1	1.2
Iron	10–1,000	430
Zinc	0.5–30	21
Methane gas	60%	NA
Carbon dioxide	40%	NA

[a] Modified after Lee and Jones (1991).
[b] Information based on data compiled for 83 landfills.
[c] All units in milligrams per kilogram unless otherwise noted.
[d] NA = not available

low permeability, considerations are similar to that for FMLs but also include certain characteristics such as plasticity index, Atterburg limits, grain size, clay mineralogy, and attenuation properties. Leachate migration through a liner is controlled by the liner's hydraulic conductivity, the head of the leachate on top of the liner, and the liner's total area. Overall,

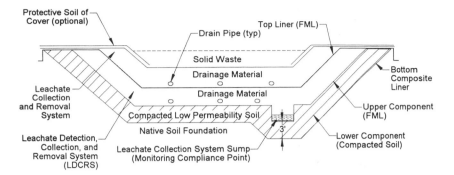

Figure 10-2 Typical double-lined landfill design (modified after USEPA, 1989).

composite liner should outperform either FMLs or clay liners, reflecting reduced spreading of the leachate through a hole or defect between the FML and clay liner since the area of flow is much smaller.

The utilization of contaminated soil as an ingredient in cold-mix asphalt for use as a liner material also has merit. Although not currently used as a liner material for landfills, cold-mix asphalt exhibits many favorable characteristics such as excellent durability, minimal adverse effects due to aging, very low permeability, and minimal leachability. Thus, what would be considered a waste is now a product. The best indication of long-term durability is the existence of surviving asphaltic structures from antiquity (about 2000 B.C.). Asphalt liners are extremely stable both chemically and physically, outperforming clay liners. Permeability tests have been conducted on a variety of liners after being subjected to aging tests. The permeabilities obtained for each liner are presented in Table 10-2. Accelerated aging tests of an asphalt liner at 20°C under oxygen partial pressures of 0.21, 1, and 1.7 atm, with continuous exposure to an acidic leachate of the composition $CaSO_4$ (500 g/L), $CaCO_3$ (18.1 g/L), $MgSO_4$ (8.6 g/L), Na_2SO_4 (7.4 g/L), NaCl (7.4 g/L), Fe_2O_3 (2.4 g/L), $NaCO_3$ (2.3 g/L), and Al_2O_3 (1.2 g/L) have been performed. Solution pHs of 2.5, 2.0, and 1.5 were designated as normal, intermediate, and highly accelerated conditions. Acidity levels were shown to have an unmeasurable effect on asphalt aging. Permeability was also used as a means to measure the immediate effectiveness of the asphalt liner. Liner permeabilities under these exposure conditions are shown in Figure 10-3, with the permeability appearing to be relatively unaffected.

It is anticipated that future landfills will have liner and cover systems with multiple hydraulic barriers. Several triple-lined units are, in fact, already in existence (Figure 10-4A). To better assure that leachate control will extend through and beyond the 30-year post-closure period, future liner systems may evolve to include gravel top to minimize erosion, a capillary barrier system, biotic barrier system and double composite hydraulic barrier system, as illustrated in Figure 10-4B.

Table 10-2 Anticipated Field Liner Permeabilities

Liner	Average Final Permeabilities (K, cm/s)	Assumed Field Thickness (L, cm)	Effectiveness Factor (K/L, S⁻¹)
Asphalt concrete	7×10^{-8}	10	7×10^{-9}
Hypalon	2×10^{-10}	0.12	2×10^{-9}
Asphalt rubber membrane	4×10^{-6}	0.8	5×10^{-6}
Catalytic airblown membrane	7×10^{-9}	0.9	8×10^{-9}
Sodium bentonite	1×10^{-7}	10	8×10^{-8}
Saline seal 100	8×10^{-6}	10	8×10^{-7}
GSR-60	6×10^{-6}	10	6×10^{-7}
Soil (as a liner)	1×10^{-5}	10	1×10^{-6}

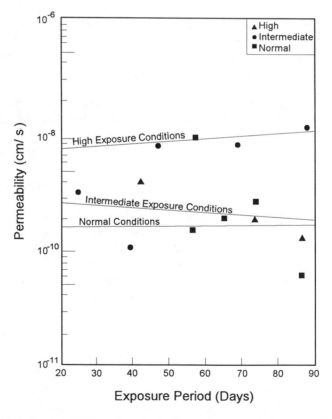

Figure 10-3 Permeability of asphalt liner at normal, intermediate, and highly accelerated conditions (after Buelt, 1985).

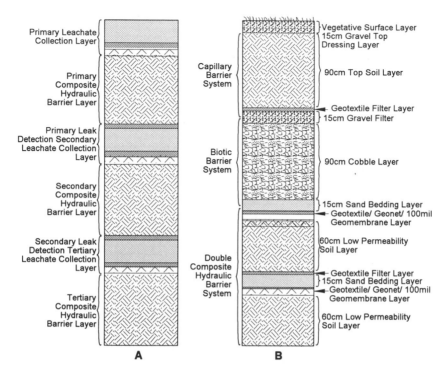

Figure 10-4 Triple-composite multiple-liner system (A) and potential future landfill cover system incorporating capillary, biotic, and hydraulic barriers (B) (modified after Anderson, 1991).

Leachate Collection and Removal System

Double-lined landfills contain both primary and secondary leachate collection and removal systems which includes most importantly, the drainage layer, filters, cushions, sumps, pipes, and other appurtenances. The drainage layer may consist of either a granular (i.e., clean sand, gravel with little fines, etc.) or synthetic material designed to facilitate rapid collection and removal of liquids, thus reducing the hydraulic head on both liner systems. The minimum design standards for the granular drainage layer requires a minimum thickness of 1 ft, minimum (high) hydraulic conductivity of 1 cm/sec, and overall low capillary tension. For synthetic drainage materials, design criteria is expressed in hydraulic transmissivity which refers to the thickness of the drainage layer multiplied by the hydraulic conductivity. This provides assurance that the design performance for a synthetic drainage layer (or geocomposite or geonet) is comparable to that for a 1-ft-thick granular drainage layer. The minimum hydraulic transmissivity criteria for synthetic drainage layer is 5×10^{-4} m²/sec, which is within an order of magnitude of that for granular drainage layers.

SITING CRITERIA

The siting of landfills prior to the 1970s was based upon business and economic considerations with minimal technical and regulatory restrictions, and minimal public involvement. Regulatory aspects were limited to land-use zoning issues and certain air and water discharge permitting requirements. Public involvement was limited, but often favorable since certain economic benefits to the community at large could be anticipated. This is certainly not the case today where the siting of a landfill includes more rigorous oversight and restrictions imposed by the regulatory community, and interaction and public acceptance by the host community.

Certain areas are considered inappropriate for the siting of a TSD facility. For example, sites excluded from consideration are those situated in flood-hazard areas, beneficial-use groundwater watershed areas, moderate to high seismic risk zones, and sites requiring natural preservation or having historical significance. On the other hand, potential TSD sites would include existing facilities, areas zoned as industrial, areas near major transportation routes, and areas close to where large quantities of waste are generated. However, rural (low population density) areas typically have been favored.

Providing the considerations previously discussed are addressed, landfill siting criteria should incorporate, but not be limited to, certain hydrologic and geologic environmental characteristics. Favorable environmental characteristics include semiarid or arid climate, low precipitation rates, high evapotranspiration rates, low rates of infiltration, deep water table or upper zone of saturation (or thick vadose zone), abundant low permeability and moisture deficient soils, lack of nearby surface water bodies, absence of beneficial-use groundwater, and water wells used for drinking water or industrial purposes. Not all TSD facilities attain all of these characteristics, especially those that were in existence prior to the mid 1980s. Landfills excavated and extending below the water table or zones of saturation has been conducted as a matter of practice in some areas, e.g., Wisconsin. This design concept reflects a preponderance of low-permeability soils (i.e., glacial tills), and the hydraulic situation wherein the excavation below the water table serves as groundwater discharge areas, such that the excavation forms a depression in the water table with groundwater flow toward the excavation in response to the lower hydraulic head. Thus, potential for leachate migration from the landfill is controlled as long as the hydraulic gradients at the perimeter and base of the landfill continue toward the landfill.

In any case, these same favorable environmental characteristics that make a site suitable for hazardous and toxic waste disposal often are very difficult to demonstrate in an unequivocal manner due to the tenuous reliability of subsurface testing methods available under such conditions. In addition, these characteristics often conflict with the "demonstration" of the site to be "properly characterized" and the ability to perform immediate leak detection monitoring. Such situations are actually common in cases where soils are moisture deficient such that vadose monitoring is not technically feasible, heterogeneous aniso-tropic conditions prevail, and/or the depth to groundwater is on the order of hundreds of feet below ground surface.

VADOSE DETECTION MONITORING

The vadose zone is that portion of the subsurface geologic profile between the earth's surface and the water table or principle water-bearing strata. Within the vadose zone, varying degrees of partial saturation occur under negative atmospheric pressure, with intervening periods of preferential saturated condi-tions in response to hydrologic events or the release of liquids or leachate. The vadose zone thus serves as a buffer zone between a WMU and underlying groundwater resources, and provides a means for early detection of a leak in a WMU liner prior to groundwater being adversely affected. Moisture mea-surements can be made using several techniques, notably, utilizing either lysimeters or neutron moisture probes.

Lysimeters

Lysimeters are used for the collection of in situ soil water; they include suction and pan-type lysimeters. A typical suction lysimeter consists of a porous cup, a PVC sample accumulation chamber, and sampling and air tubes (Figure 10-5). The cups are usually comprised of a hydrophilic material such as ceramic or an aluminum oxide (alundum), or polytetrafluorethylene, which is hydrophobic. During installation, the porous cup is enveloped by silica flour to prevent plugging and maintain sufficient vacuum (i.e., 10 centibars). The flexible sampling tube terminates at the bottom of the lysim-eter. The air tube terminates immediately below the top of the sample accumulation chamber.

Soil water samples can be obtained to depths of about 55 ft, and to a depth of 100 ft with modification. The sampling radius of a suction lysimeter is on the order of centimeters; thus, numerous lysimeters are required to adequately serve the function of an early-detection warning system.

Pan-type lysimeters or free-drainage samplers are used to obtain water samples which collect by gravity drainage. The pan lysimeter typically consists of a shallow sand-filled cone with a drain in the center. A lysimeter is placed on top of the drainage layer, but below the geofabric and liner material. Pan-type lysimeters have also been installed in buried culverts and trenches.

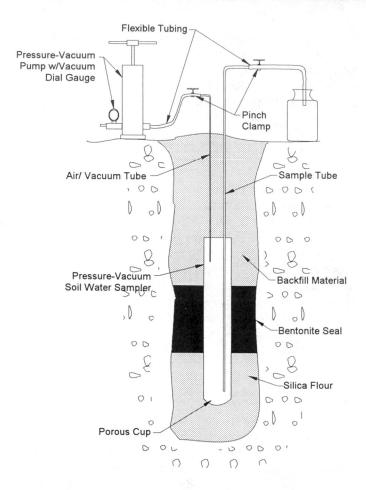

Figure 10-5 Schematic of an installed suction lysimeter.

Neutron Moisture Probes

Neutron moisture probes have only recently been increasingly used for vadose zone monitoring, although they have been used in agricultural studies since the early 1960s. Neutron moisture probes are used to measure in situ soil moisture by lowering the neutron probe down the access tubes, then slowly raising the probe and measuring the count rate at discrete intervals (i.e., 2.5 ft). Calibration curves are based on plotting of the ratio of counts recorded vs the laboratory moisture contents on samples retrieved during drilling, followed by calculating a linear regression for a unique solution. Once baseline soil moisture and soil chemistry conditions are established, monitoring of soil moisture content is then performed at a set frequency (i.e., quarterly), the results of which are compared with earlier measurements and the calibration curve. A typical vadose zone access tube construction is shown in Figure 10-6.

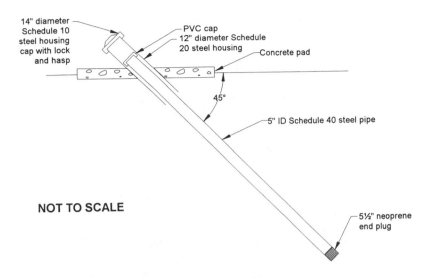

Figure 10-6 Typical vadose zone access tube construction for neutron probe measurements.

GROUNDWATER DETECTION MONITORING

Prior to design and development of a vadose zone and groundwater monitoring network, subsurface geologic and hydrogeologic conditions are required under RCRA to be characterized. The level of effort put forth for site characterization will undoubtedly vary from site to site; however, this site characterization process needs to be adequate to the extent necessary for design and development of a monitoring network. Subsurface geologic and hydrogeologic characterization typically incorporates several phases of investigation. The regional geology within 5 mi of the facility is compiled and reviewed. This regional review is typically combined with site visits to the important outcrops. The field investigation is an iterative process, whereas the initial drilling phases provide geologic information on the stratigraphy, soil and bedrock characteristics, significant structural elements, and general hydrogeologic conditions. The latter phases focus on demonstrating lateral and vertical consistency, and addressing areas of uncertainty.

How many borings or groundwater monitoring wells will be required for this purpose is always a difficult issue to address and will depend upon a variety of factors including the complexity of subsurface conditions. Uncomplicated geology (i.e., thick horizontal bedding, homogeneous and isotropic, and lack of structural features such as faults, fractures, joints, etc.), supported by surface and downhole geophysical data, and relatively constant hydraulic conductivity values and flow direction with minimal offsite influences (i.e., tidal effects, water pumpage, etc.) would support a wider spacing of borings and wells. Factors that would support a closer spacing include:

- Heterogeneous, anisotropic conditions
- Presence of structural elements such as fractures, joints, faults, etc.
- Significant lateral and vertical discontinuities
- Variable hydraulic conductivity values
- Inconsistencies in groundwater flow directions
- Within or in close proximity to a recharge zone
- Indication of preferred migration pathways not defined by wider spacing
- Indication of off-site influences.

Other site-specific factors may also influence the horizontal spacing of monitoring wells. A closer spacing of wells may be deemed necessary if the facility manages or has managed liquid waste, contains waste which is incompatible with the liner material, is old and design features and past waste disposal practices are less certain, or contains underground utilities where a point-source leak may occur (i.e., buried pipes, utility trenches, etc.).

Groundwater monitoring as required under RCRA (40 CFR Part 264, Subpart F) is designed to establish background water quality within the uppermost aquifer beneath the site, and to detect groundwater contamination, should it occur in the future. To meet these objectives, a network of groundwater monitoring wells is designed such that if a release of leachate did occur from waste management unit, it would be detected by the presence of certain contaminants in representative groundwater samples retrieved from a well or wells which comprise the network. It is thus very important that the screened interval is both horizontally and vertically strategically located within the uppermost zone of saturation. A number of sophisticated analytical methodologies have been proposed for the demonstration of an adequate groundwater sampling network design as summarized in Table 10-3. Less sophisticated and more practical approaches are further discussed below.

Point of Compliance

From a regulatory perspective, detection monitoring is conducted at the point of compliance. The point of compliance as defined under RCRA is the loci of points located within a narrow vertical zone extending from the hydraulically downgradient limits of the waste management area down to and into the uppermost aquifer underlying the regulated unit or units. The point of compliance also serves as the location where any contaminants, in the event that they would reach the water table in detectable concentrations, could be intercepted and prevented from migrating off site. The point of compliance, although meeting its regulatory objective as defined in the regulations, may not serve its intended purpose when significant permeability contrasts exist within the vadose zone. Such conditions can result in significant lateral spreading of a

Table 10-3 Summary of Groundwater Sampling Network Design Strategies

Method	Reference
Mixed-integer programming	Hsu and Yeh (1989)
Kriging and cokriging	Carrera et al. (1984), McLaughlin and Graham (1986), Wolf and Testa (1985)
Variance reduction analysis	Rouhani (1985)
Nearest neighbor approach	Olea (1984)
Optimization	Hsueh and Rajagopal (1988), Loaiciga (1989), Andricevic (1990a), ASCE (1990b)
Simulation	Meyer and Brill (1988), Massmann and Freeze (1987), Van Geer (1987)
Transfer function approach	Andricevic and Georgiou (1991)

contaminant plume within the vadose zone such that by the time the plume reaches the water table of the uppermost zone of saturation, the plume has migrated laterally beyond the hydraulically downgradient perimeter of the waste management unit.

Well Placement

The number and spacing of detection monitoring wells along the point of compliance necessary to provide a high probability that groundwater quality degradation would be detected, in the unlikely event that this occurs, are dependent on several factors. Most important is consideration of the homogeneity or heterogeneity of the aquifer and its hydraulic transmissivity and storage characteristics: the rate, concentration and nature of the contaminants ultimately reaching the water table, and the potential risk to human health and the environment.

Karst terrains present a unique set of concerns and have always been regarded as problematic due to their overall heterogeneity and anisotrophy, and high secondary permeability and porosity (primarily via fissures, fractures, and solution cavities), resulting in the potential for rapid unattenuated migration of contaminants. There is thus a high potential for groundwater contamination in such areas as the southeastern U.S., England, and Wales. Due to the relatively low probability of having a monitoring well intercepting the actual conduit through which contaminants would migrate, detection monitoring in karstified areas include a combination of both monitoring wells and the sampling of springs.

Parametric Studies

The spacing of detection monitoring wells is also dependent on the concentration level at which the potential contaminants can be detected in groundwater. For example, many organics can be detected at concentrations of a few parts per billion, whereas other constituents may only be detectable in the parts per million range. From this perspective, dispersion and dilution phenomena, both within the unsaturated zone and below the water table, can result in a significant decrease in concentration levels as can be illustrated by parametric studies.

Parametric model analysis can be used to estimate the probable lateral extent of a potential contaminant plume emanating from a discrete source within the vadose zone during downward seepage to and within the uppermost aquifer. This information in turn is used to define the horizontal placement of detection groundwater monitoring wells. Parametric model analysis is usually conducted in phases. Phase I focuses upon the analysis of downward contaminant migration and lateral dispersion within the vadose zone. Phase II focuses upon saturated flow and contaminant migration within the uppermost aquifer. For simplicity, parametric model analysis typically does not include more than one source, nor take into account chemical, physical, and biological interactions (i.e., ion exchange capacity) with the native soils. Because these interactions are likely, the parametric approach is even more conservative; thus, applied use of the results presented herein must be considered in this perspective.

Within the vadose zone, a two-dimensional, vertical-plane, variably saturated base case model is selected. A base case is defined as the case using the recharge and flow parameters judged to be most typical. A discretized calculation domain (i.e., finite difference grid) is established as shown in Figure 10-7. No-flow boundaries are established, as in the example presented in Figure 10-7, which include the upgradient boundary above the water table (i.e., assume flow in the vadose zone at the upstream boundary to be vertically downward) and the base of the uppermost aquifer. Hydraulic head data is typically provided by field measurements. Under an assumed recharge rate in this example of 0.5 in./year, the following type of data is supplied to the model:

- Horizontal hydraulic conductivity   0.032 ft/day (1.0×10^{-5} cm/s)
- Vertical hydraulic conductivity   0.032 ft/day (1.0×10^{-5} cm/s)
- Porosity   0.40
- Specific storativity 1.0×10^{-4} 1/ft
- Longitudinal dispersivity   30 ft
- Transverse dispersivity   3 ft

The selected dispersivity values and their probable range are based on a literature survey on field and laboratory dispersivity values. Field dispersivity

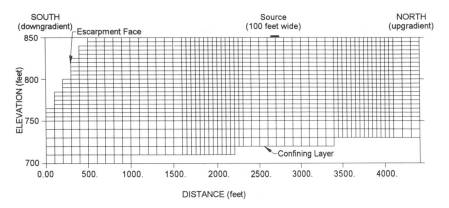

Figure 10-7 Discretized calculation domain showing hydraulic head distribution.

values typically range over 2 orders of magnitude from about 1 to 100 ft. The dispersivity is dependent on the particular chemical constituents involved and their relative concentrations, as well as the site-specific soil characteristics. Because of this variability, representative values which are believed to be conservative are assumed. Further, for slow rates of groundwater flow and the width of a contaminant plume is relatively insensitive to the assumed hydraulic conductivity values that are based on field measurements performed as part of the site characterization.

The results of the modeling using the base parameters are shown in Figures 10-8A and B. The normalized concentrations of a contaminant release from a source (i.e., lagoon) 100 ft wide at a rate equal to the assumed recharge rate are shown (Figure 10-8A). These concentrations can also be conservatively viewed by adjusting the source rate, for example, at five times that value (Figure 10-8B). The downgradient contaminant plume width is seen to be much greater than the upgradient width. In the upgradient direction, the plume width is primarily the result of lateral diffusion. However, the downgradient plume clearly shows the influence of an escarpment face to the south; therefore, the distance of the contaminant release from the escarpment face will determine both the concentrations reaching the face and the lateral displacement of the isocons in that direction.

Sensitivity analysis can also be carried out to evaluate the effect of variation of key parameters on the predicted base case results. Specifically, the ratio of horizontal to vertical permeability (i.e., ratios of 1:1, 5:1, and 10:1), the rate of downward infiltration to the water table (i.e., 0.02, 0.063, and 0.2 in./year), and the dispersivity (i.e., 10, 30, and 100 ft longitudinal and 1, 3, and 10 ft transverse) are varied to assess their relative influence on the shape of a potential contaminant plume emanating from a discrete source. The width of the contaminant source at the ground surface for the purpose of this sensitivity study was assumed to be 100 ft.

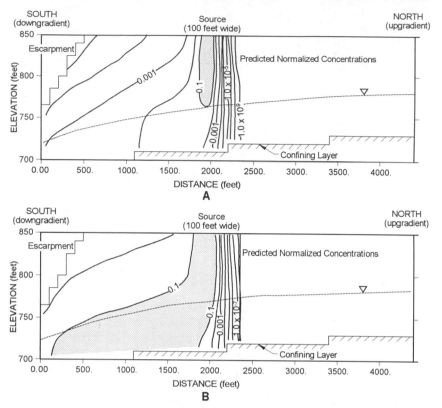

Figure 10-8 Normalized concentrations of a modeled contaminant release (A) at assumed recharge rate and (B) at five times that rate.

The two rates of contaminant release used in the base case areas were considered: (1) equal to the natural rate of groundwater recharge and (2) five times the natural groundwater recharge rate. The analysis as presented herein, therefore, is conservative with respect to the width of the contaminant plume during downward migration through the unsaturated zone.

The results of the sensitivity analysis are shown in Figure 10-9. This figure shows the total width of the 10% plume at the water table, as a function of the recharge rate, ratio of horizontal to vertical conductivity, and dispersivity. The results of the sensitivity analysis and the upgradient and downgradient 10% plume widths are presented in Table 10-4. In this case, the 10% plume is defined as the plume which encompasses contaminant concentrations greater than or equal to 10% of the normalized source concentration.

At both discharge rates, the ratio of the horizontal to vertical hydraulic conductivity has very little influence on the 10% plume width at the water table (Figure 10-9). The very small lateral hydraulic gradient is responsible for this result. The rate of infiltration has a greater effect upon the 10% plume width, as also evident in Figure 10-9. In the case of the discharge rate equal to the recharge rate, the plume width increases as the infiltration rate increases. The apparent reduction in the plume width at the higher infiltration rates for the five times discharge rate example is an effect of the higher water table which allows less distance (time) for dispersion. The relatively small effect of dispersivity upon the plume width as shown in Figure 10-9 is also due to the relatively small lateral hydraulic gradients.

The results of the vertical-plane studies show that significant lateral dispersion occurs during downward migration of a contaminant plume within the vadose zone. Furthermore, the extent of lateral dispersion increases as the rate of contaminant release exceeds the natural rate of groundwater recharge. The sensitivity studies indicated that only the rate of infiltration has a significant effect upon the plume behavior. As can be seen in Table 10-5, the upgradient 10% plume width is less than 200 ft in all cases while the downgradient 10% plume width is typically on the order of 150 ft for the discharge rate equal to the recharge rate and 1500 ft for the discharge rate five times the natural recharge rate.

The analysis of groundwater flow and contaminant transport within the uppermost aquifer analysis were obtained from steady-state (vertical-plane) flow rates and concentrations predicted at the water table by the unsaturated zone analysis previously discussed. Since the unsaturated zone analysis incorporates conservative assumptions, the results of the saturated model calculations are also conservative. Principal inputs to the horizontal plane model include the measured top of the lower confining layer (i.e., base of saturated zone), upper surface elevations, and the water levels in the overlying saturated zone. As in the vertical calculations, the confining layer was assumed to be impermeable.

Using the measured water levels to drive the saturated aquifer hydrodynamics, the steady-state flow patterns are calculated. The result of the horizontal plane transport calculations for contaminant sources equal to the natural recharge rate are shown in Figures 10-10A and B. These figures show the predicted normalized concentrations of contaminants in the uppermost aquifer at the base of the saturated zone of 10, 20, and 30 years after the contaminants reach the water table. The contours indicate the predicted concentration values relative to the concentration of contaminated water at the top of the uppermost aquifer. Details of the plume dimensions shown in Figure 10-10 are listed in Table 10-5. Figure 10-10 indicates that the predicted plumes dilute and disperse very slowly due to the low permeability of the uppermost aquifer. With time, the plume would continue to enlarge if the source were to continue unabated.

In summary, for rates of contaminant release equal to or at five times the natural groundwater recharge rate, the width of the resultant contaminant plume was defined by the modeling study to be 10% of the initial source strength. If the source strength is diluted by a factor of 1000, the width of the resultant plume is on the order of 170 and 250 ft for rates of contaminant

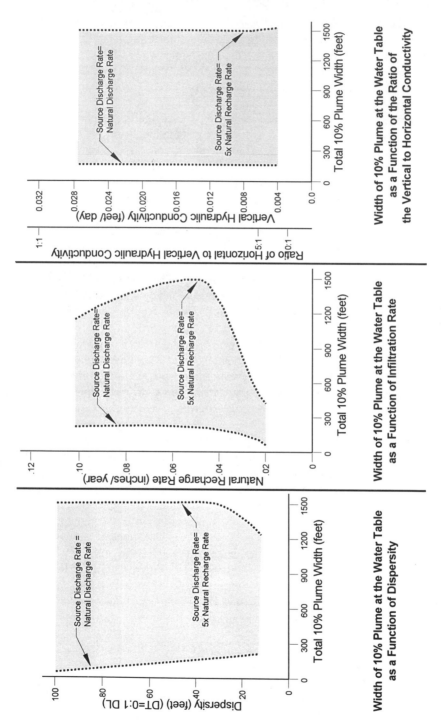

Figure 10-9 Results of sensitivity analysis showing effects of variation of dispersivity at natural recharge rates and five times the natural recharge rates.

Table 10-4 Summary of Vertical-Plane Results

Case	Source Recharge Rate[a]	Horizontal/ Vertical Conductivity	Lateral/ Transverse Dispersivity	Recharge Rate (in./yr)	10% Plume Width (ft)[b]		
					Down	Up	Total
1a	1	0.032/0.032	30:3	0.05	119	52	171
1b	5	0.032/0.032	30:3	0.05	1400	105	1505
2a	1	0.032/0.0064	30:3	0.05	119	52	171
2b	5	0.032/0.0064	30:3	0.05	1398	124	1522
3a	1	0.032/0.0032	30:3	0.05	119	52	171
3b	5	0.032/0.0032	30:3	0.05	1398	162	1560
4a	1	0.032/0.032	10:1	0.05	171	50	221
4b	5	0.032/0.032	10:1	0.05	1098	106	1204
5a	1	0.032/0.032	100:10	0.05	19	14	33
5b	5	0.032/0.032	100:10	0.05	1397	119	1516
6a	1	0.032/0.032	30:3	0.10	171	53	224
6b	5	0.032/0.032	30:3	0.10	998	107	1105
7a	1	0.032/0.032	30:3	0.02	8	4	12
7b	5	0.032/0.032	30:3	0.02	323	64	387
8a	1[c]	0.032/0.032	30:3	0.05	116	65	170

[a] Recharge rates are given as times natural rate.
[b] Up and down refer to the upgradient and downgradient directions.
[c] Upgradient source location.

Table 10-5 Horizontal-Plane Plume Dimensions

Time (years)	Normalized Concentration	Plume Width (ft)	Plume Length (ft)
10	10^{-1}	—	—
	10^{-2}	—	—
	10^{-3}	—	—
	10^{-4}	156	406
	10^{-5}	281	681
	10^{-6}	296	781
20	10^{-1}	—	—
	10^{-2}	—	—
	10^{-3}	31	78
	10^{-4}	218	609
	10^{-5}	281	703
	10^{-6}	328	832
30	10^{-1}	—	—
	10^{-2}	—	—
	10^{-3}	63	125
	10^{-4}	250	625
	10^{-5}	296	781
	10^{-6}	344	844

release equal to and 5 times the natural groundwater recharge rate, respectively. The width of the contaminant plume within both the unsaturated zone above the water table and below the water table is substantially greater for dilution factors greater than 1000 times. For concentration levels equal to 1×10^{-6} that of the source concentrations, the width of the contaminant plume is on the order of 300 to 500 ft or more, depending on the rate at which the contaminants are released (Figure 10-10).

CASE HISTORIES

Subsurface hydrogeologic conditions and subsequent development of a groundwater detection monitoring strategy for several major TSD facilities is discussed in the following subsections. A system design under heterogeneous, anisotropic conditions is exemplified by the Arlington facility located in north-central Oregon. A system design under fractured flow conditions is exemplified by the Casmalia facility located in Southern California.

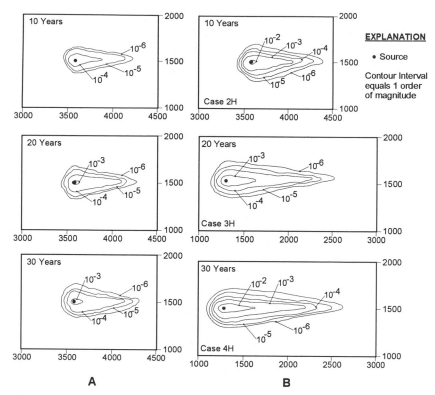

Figure 10-10 Predicted plume dimensions at the natural recharge rates (A) and at five times the natural recharge rate (B).

Arlington Facility, North-Central Oregon

Site Description

The Arlington facility is located in north-central Oregon, about 6.5 mi south of the Columbia River, and southwest of the town of Arlington, OR (number 7 in Figure 10-1). Situated on a 640-acre parcel, the eastern 320 acres are currently used for waste management operations (Figure 10-11). The property is bounded on the south by the east-west trending Alkali Canyon at an elevation of approximately 700 ft (site datum). The upland plateau is at an elevation of approximately 850 to 995 ft. Waste management activities are limited to that area above 920 ft on the eastern tract property. In the southern portion of the property bounded by Alkali Canyon, the relief between the valley floor and the upland plateau is approximately 280 ft.

The area is characterized by upland areas of sandy deserts separated by relatively wide, deep-to-moderate ephemeral stream drainages such as the Alkali Canyon which borders the south side of the property. In addition to the semiarid climate, several factors account for the physiography of the area including the presence of extensive flood basalts, subsurface geologic structure, and catastrophic floods of glacial meltwater.

Facility Layout and Operation

The Arlington facility was opened in 1976 and provides hazardous waste treatment, storage, and disposal services primarily to the Pacific Northwest, Alaska, and Hawaii, although it also receives hazardous wastes from other western states and Superfund-related activities. The facility does not accept explosive, radioactive, or infectious wastes. Wastes that cannot be treated or disposed of at the facility, or that can be reused or recycled, are temporarily stored at the facility and then shipped elsewhere for treatment, recycling, disposal, or beneficial use.

The existing WMU that require groundwater monitoring under RCRA at the Arlington facility include surface impoundment, reactive solids hydrolysis, and landfills. Sixteen waste impoundments (of which a minimum of eight are undergoing closure), nine landfills (of which four are complete), seven container storage areas, and four storage tanks comprise the WMU at the site; however, additional facilities are planned for the future. A liquid-waste solidification system and a truck-wash operation are also in use. The WMU of most importance at the facility, notably surface impoundments and landfills, are shown in Figure 10-11.

Geologic Setting

An extensive field investigative program to characterize existing site geologic and hydrogeologic conditions was conducted during the period from December 1983 through November 1986 in support of a Part B Application under RCRA and included:

- Detailed geologic mapping including trenching to evaluate and characterize faults
- Drilling of over 200 boreholes and installation of over 120 wells/ piezometers to depths ranging to 363 ft below ground surface to evaluate subsurface hydrogeologic conditions and the groundwater flow regime
- Collection of undisturbed samples for detailed laboratory evaluation and testing
- Performance of both pumping/packer and slug tests to assess in-situ hydraulic characteristics and possible intercommunication between the uppermost aquifer and the underlying basalts
- Performance of both surface and borehole geophysics

Figure 10-11 Aerial view of the Arlington facility, east up Alkali Canyon in Oregon (dated 1985).

- Performance of analytical testing of groundwater samples to evaluate general water chemistry and water quality, including the presence of tritium to assess recent recharge
- Development of a groundwater detection monitoring network.

The subsurface geology of the facility and surrounding areas consists of a thick, accordantly layered sequence of basalt flows and sedimentary interbeds, collectively known as the Columbia River Basalt Group. The basalt flows are part of the Columbia Plateau geological flood-basalt province of Miocene to lower Pliocene age (8 to 17 million years old). This sequence is unconformably overlain by younger intercalated and suprabasalt sedimentary units of Miocene to Holocene age.

Within the site area the formations which comprise the Columbia River Basalt and the Ellensburg Formation include several members of regional extent. These are the Frenchman Springs and Priest Rapids Members of the Wanapum Basalt, the Pomona Member of the Saddle Mountains Basalt, and the Selah and Rattlesnake Ridge Members of the Ellensburg Formation. Several informal units useful to comprehension of site structure and stratigraphy are also defined. These include several unnamed interbeds within the Priest Rapids Member, and an areally extensive vitric tuff which occurs at the top of the Selah Member. In addition to the informal units are several facies of local extent. These include three facies of different lithology within both the Selah Member of the Ellensburg Formation and the Dalles Formation. An on-site geologic and geophysical log showing the representative units which comprise the Columbia River Basalt Group in the site area is shown in Figure 6-10. A generalized geologic cross section developed north-south across the site is shown in Figure 10-12.

Of most importance and relevant to this site is the Selah Member of the Ellensburg Formation since the uppermost aquifer or zone of saturation beneath the facility lies at the lower portion of this member, 100 to 200 ft beneath the existing ground surface. The Selah Member of the Ellensburg Formation occurs as an interbed between the Priest Rapids Member and the Saddle Mountains Member (Figure 10-13). Deposited directly on the underlying Priest Rapids Basalt, it is the most prominent interbed in the area and is comprised of weathered tuffs and fluviolacustrine tuffaceous sediments that accumulated during the volcanic hiatus between extrusion of the Priest Rapids and Pomona basalts. The thickness of the Selah varies from approximately 115 to 160 ft in boreholes at the site where the top has not been eroded. Where it has been eroded, channels of Dalles or glaciofluvial gravel exist. The uneroded Selah averages approximately 138 ft in thickness.

The Selah contains three facies which are readily distinguishable on geophysical logs (Figure 6-12). Due to severe weathering, and zeolitic and clayey alteration of the member, they are not as discernible in outcrop or borehole samples. The facies include (1) a lower facies comprised primarily of floodplain deposits derived from the Columbia Plateau and adjacent areas, but also containing three to four airfall tuff units; (2) a middle facies similar to the lower, but containing large amounts of silicic-volcanoclastic material derived from volcanic areas; and (3) an upper or channel facies which results, in part, from reworking of the lower two facies. Natural gamma and neutron-neutron geophysical cross sections aligned north-south across the site are presented in Figure 6-12. The lower facies of the Selah is distinguished from the overlying

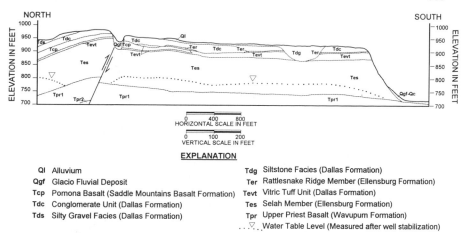

EXPLANATION

Ql	Alluvium	Tdg	Siltstone Facies (Dallas Formation)
Qgf	Glacio Fluvial Deposit	Ter	Rattlesnake Ridge Member (Ellensburg Formation)
Tcp	Pomona Basalt (Saddle Mountains Basalt Formation)	Tevt	Vitric Tuff Unit (Dallas Formation)
Tdc	Conglomerate Unit (Dallas Formation)	Tes	Selah Member (Ellensburg Formation)
Tds	Silty Gravel Facies (Dallas Formation)	Tpr	Upper Priest Basalt (Wavupum Formation)
		..∇..	Water Table Level (Measured after well stabilization)

Figure 10-12 Hydrogeologic cross section oriented north-south.

middle and upper channel facies by generally lower density and higher porosity, and from the middle facies by lower potassium content as indicated by the gamma-gamma, neutron-neutron, and natural gamma logs, respectively (Figure 6-12). Where the middle facies overlies the lower facies a marked decrease in natural gamma occurs at the boundary. Samples of the lower lithologies have textures which range from yellow-green massive clayey siltstone to laminated sandy siltstone, and yellow-brown to blue-green silty sandstone. Bedding is generally preserved in the sandy layers but obscured by weathering and possibly by bioturbation in silty claystone and clayey siltstone intervals. Tuff layers in the lower member are altered to massive silty claystone; their origin is indicated only by their high natural gamma and the presence of accretionary lapilli. In many of the boreholes, a blue-green, fine-grained sandstone occurs at the base of the Selah. The thickness and distribution of the sandstone is variable. The middle facies of the Selah consists of soft to medium-hard yellow-green to brown, tuffaceous, silty claystone and clayey siltstone. It is characterized by higher natural gamma, higher density, and lower porosity than the lower facies. Samples of the middle facies from boreholes frequently contain finely laminated, weathered tuffs. Locally, the middle or highly tuffaceous facies of the Selah Member appears to contain broad shallow channels filled with flood plain deposits ranging from clayey siltstone to fine sandy siltstone. These channel-fill sediments have low natural gamma similar to the lower facies. However, unlike the lower facies, they are dominantly weathered tuff with the consistency of clayey siltstone.

A gray, dacitic, vitric tuff bed attaining a maximum thickness of about 30 ft occurs at the top of the Selah Member of the Ellensburg Formation. Texturally, the vitric tuff is soft to medium hard, gray, medium-grained sandstone (tuff) which is well sorted and massive in the central 20 to 24 ft of the deposit. The lower 1 to 5 ft consist of soft to medium hard, gray to dark gray, laminated, silty sandstone (tuff) and sandy siltstone (tuff). The finer laminae are weathered

Figure 10-13 Core of tuffaceous siltstone of the Selah overlying the upper Priest Rapids basalt. Note lithologic and textural variations ranging from finely laminated to pebbly within the Selah.

to soft, clayey siltstone or silty claystone. The upper 3 to 8 ft consist of silty, fine to medium, vitric sandstone that occurs as cross-laminated and/or thinly graded beds. These fine-grained layers consist of weathered clayey silt/silty clay similar to the bottom laminated portion.

Overlying the vitric tuff bed from oldest to youngest is the vitric tuff which comprises the Rattlesnake Ridge Member of the Ellensburg Formation; basalt of the Pomona Member; basal cemented sandstone and conglomerate, and overlying tuffaceous siltstone facies of the Dalles Formation; and surficial materials including Pleistocene glaciofluvial stackwater, torrential flood, colluvial, alluvial, and Holocene eolian deposits.

Hydrogeologic Setting

The hydrogeologic conditions at the Arlington facility were evaluated by installation of over 120 perimeter and interior wells and/or piezometers. The conditions encountered are complex, consisting of multiple zones of saturation with varying degrees of interconnection. The uppermost zone of saturation beneath the Arlington facility exists under water-table conditions at the base of the Selah, 100 to 200 ft beneath the existing ground surface (Figure 10-12). This zone of saturation at the base of the Selah is the first detectable zone encountered during drilling capable of yielding even small quantities of water to an open borehole and it is, therefore, the uppermost zone capable of being monitored. Because of stratification and marked permeability contrasts that exist within and between the overlying Dalles Formation, Pomona Basalt, Vitric Tuff, and the Selah, it is reasonable to expect that isolated, perched zones of saturation could exist above the base of the Selah; however, no such zones have been detected. Furthermore, no perched zones of saturation were identified from the downhole geophysical logs although variation in soil moisture content with depth is evident (Figure 6-10).

Groundwater occurs under water-table conditions at the base of the Selah. It also occurs under both water-table and partially confined conditions within the upper Priest Rapids flow above the interbed between upper and lower flows, and within the interflow zone between the two Priest Rapids basalt flows. Barometric efficiency tests were performed to assess whether there were measurable fluctuations in water levels as a result of changes in atmospheric pressure. The test indicated that water-level fluctuations in the uppermost aquifer within the lower portion of the Selah are on the order of 0.25 ft, and that these fluctuations are in part a function of variations in atmospheric pressure. Thus, the uppermost aquifer is partially confined and therefore responds to changes in atmospheric pressure, although these fluctuations are minimal (0.1 to 0.5 ft).

The uppermost zone of saturation is located physically within the Selah and above the top of the Priest Rapids Basalt (Figure 10-14). This saturated zone

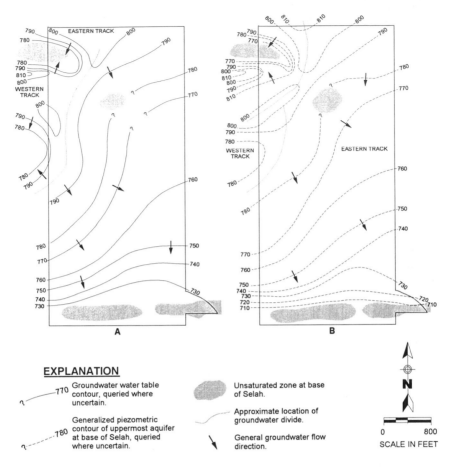

EXPLANATION

—770— Groundwater water table contour, queried where uncertain.

--780-- Generalized piezometric contour of uppermost aquifer at base of Selah, queried where uncertain.

(shaded) Unsaturated zone at base of Selah.

--- Approximate location of groundwater divide.

↓ General groundwater flow direction.

N

0   800
SCALE IN FEET

Figure 10-14 Generalized groundwater table (A) and piezometric (B) contour map of uppermost aquifer at base of Selah.

is continuous across the southern two thirds of the site (north of Alkali Canyon) where waste management activities occur. It is either thin or absent beneath the northwestern corner and in the north-central part of the site. Groundwater that would otherwise be present in these areas is believed to flow downward from the Selah into the Priest Rapids, in contrast to other areas of the site where groundwater is perched on top of the Priest Rapids and forms a continuous saturated zone. An isopach map of the uppermost aquifer within the lower portion of the Selah is shown in Figure 10-15.

Beneath the northern two thirds of the site and along its southern margin, an unsaturated zone exists within the upper part of the upper Priest Rapids basalt flow. Thickness of this unsaturated zone ranges from a few feet near the southern boundary of the property to greater than 80 ft in the northern portion of the site. The lower portion of the upper Priest Rapids basalt flow is saturated.

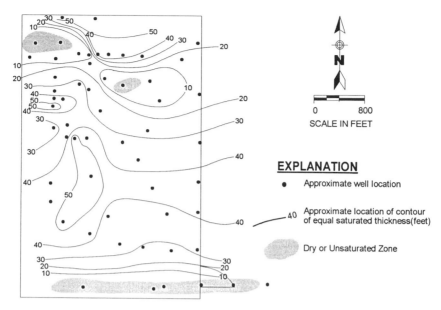

Figure 10-15 Isopach map of the uppermost aquifer at base of Selah.

In the southeastern portion of the site, continuous saturation appears to exist from the base of the Selah downward to the top of the interbed. The existence of the saturated zone within the Selah is thus a manifestation of the anisotrophic nature of the Selah as well as the existence of a low permeability zone(s) at the Selah/Priest Rapids interface.

Groundwater also occurs under both confined and water-table conditions within the interflow zone at the top of the lower Priest Rapids basalt flow. Groundwater within the interflow zone beneath the southcentral portion of the site, in general, is confined or partially confined. In the northern portion of the site where the interbed rises toward the anticline, groundwater within it exists under water-table or unconfined conditions.

The existence of a zone of saturation at the base of the Selah directly overlying an unsaturated zone within the upper Priest Rapids Basalt suggests either a permeability contrast between the Selah and the Priest Rapids or the presence of a lower hydraulic conductivity layer or zone. The latter hypothesis is consistent with the observed pressure head distribution within the Selah.

In situ falling head permeability (slug) tests in open boreholes terminated at the base of the Selah/top of the Priest Rapids indicate that the horizontal hydraulic conductivities at the base of the Selah range over two orders of magnitude from about 1×10^{-6} to 1×10^{-4} cm/s. Laboratory tests on undisturbed cores from the saturated zone at the base of the Selah indicate that the vertical hydraulic conductivity, which ranges from 1×10^{-8} to 1×10^{-5} cm/s, is one to several orders of magnitude less than the horizontal hydraulic conductivity as determined from slug tests. The apparent difference between field horizontal

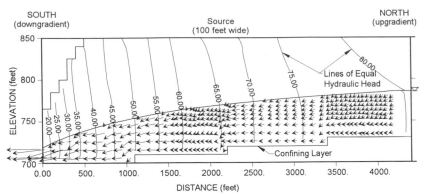

NOTE: Length of flow arrow represents relative magnitude of velocity.

Figure 10-16 Hydraulic head distribution in oriented north-south cross section.

hydraulic conductivity and laboratory vertical hydraulic conductivity may be attributed in part to the difference in measurement techniques.

Groundwater flow is affected by the thickness and anisotrophy of each hydrostratigraphic unit, and the position and continuity of each layer within the units. In a layered media, such as the Selah, the hydrostratigraphic unit with the lowest vertical hydraulic conductivity (K_v) is the controlling factor for groundwater flow in the vertical direction (normal to bedding). Groundwater flow in the horizontal direction (parallel to bedding) is controlled by the hydrostrati–graphic unit with the highest horizontal hydraulic conductivity.

This condition exists within the transmissive zones of the saturated portion of the Selah. The dominant mode of groundwater flow is down-dip parallel to bedding (1 to 2° from horizontal) as shown in Figure 10-16. This is due to the confining beds that are a naturally occurring portion of the bedding and the anisotrophy. The Selah is anisotropic in that the horizontal hydraulic conductivity is greater than the vertical hydraulic conductivity. However, due to the discontinuous and lenticular nature of much of the saturated portion of the Selah, combined with hydraulic head conditions and inferred hydraulic conductivity differences, a component of cross-bed flow can occur as illustrated in the conceptual groundwater flow net (Figure 10-17).

In situ and laboratory permeability tests indicated that horizontal and vertical hydraulic conductivities of the saturated portion of the Selah range over at least four orders of magnitude (1×10^{-8} to 1×10^{-4} cm/s). This range of variability is characteristic of stratified sedimentary interbeds in the Columbia Plateau region. The ratio of horizontal to vertical hydraulic conductivity is at least 175:1 and, locally, could be greater than 500:1. Storativity values obtained from long-term pumping tests show a wide range that spans those typical of both unconfined and confined aquifer conditions. In summary, the Selah was characterized as follows:

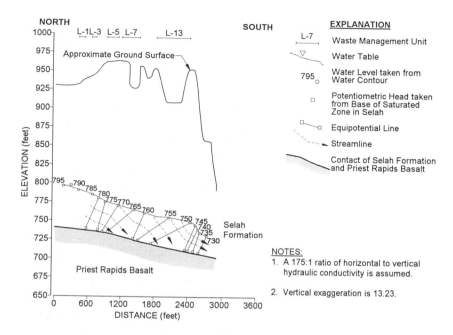

Figure 10-17 Flow net oriented north-south across the site.

- The Selah stratigraphy consists of alternating beds of sands and silts which are indicative of layered heterogeneity. Such heterogeneities result in variations in horizontal groundwater flow velocities due to variations in horizontal hydraulic conductivity.
- The horizontal hydraulic conductivity in the Selah ranges from 1×10^{-8} to 1×10^{-4} cm/s with a geometric mean of 6×10^{-5} cm/s.
- The layered nature of the Selah results in high horizontal to vertical anisotrophy ratios with overall horizontal hydraulic conductivity estimates to be on the order of approximately 200 to over 500 times the vertical hydraulic conductivity.
- The anisotrophy ratio indicates that potential contaminants entering the Selah from the vadose zone will flow horizontally for significant distances in the uppermost saturated zone before any vertical gradients could cause them to migrate downward into deeper aquifers.
- The storativity values obtained from the long-term pumping test show a wide range that spans those typical of both unconfined and confined aquifer conditions.

Groundwater Monitoring System Design

In developing a groundwater detection monitoring system, the results of the site characterization showed that the geology and existing groundwater conditions are relatively uniform across the site and beneath the waste management areas in particular. The exception to this is in the extreme northwest corner of the site where anomalous conditions were encountered. In this area, the groundwater table is depressed below that encountered elsewhere beneath the site as an apparent result of the higher hydraulic conductivity of the under-lying Priest Rapids in this area. The groundwater regime in the extreme northwest corner of the site is hydraulically separated from the flow regime beneath the remainder of the site by a groundwater divide. The uppermost aquifer or continuous zone of saturation of the Arlington facility is at a depth of 110 to 225 ft beneath the existing ground surface within the lower portion of the Selah Member of the Ellensburg Formation and above the top of the Priest Rapids basalt. South of this divide, the groundwater table (or potentio-metric surface) reflects the configuration of the top of the Priest Rapids and slopes to the south and southeast at 1.7 to 2.6°. The rates of groundwater flow across the site to the south and southeast are estimated to range from less than 1 to about 36 ft/year.

The strata above the existing groundwater table at the base of the Selah are generally unsaturated. Although thin, perched isolated zones of saturation may exist within the unsaturated zone due to stratification and permeability con-trasts, which are known to exist, none capable of producing even small quan-tities of water was encountered during drilling. The uppermost aquifer, within the southern two thirds of the site, is both physically and hydraulically isolated from the regional aquifers which are associated with interflow zones within the Priest Rapids and Frenchman Springs basalt flows. The physical separation between the uppermost aquifer and the regional aquifers within the Priest Rapids is at least 70 ft, and that within the Frenchman Springs basalts is on the order of 300 to 400 ft.

Groundwater recharge to the uppermost aquifer is by direct infiltration of incident precipitation and accumulated surface runoff. Based on preliminary water balance calculations, the rates of natural groundwater recharge are esti-mated to be in the range of several hundredths to a few tenths of an inch per year. The closed depressions present within the northern portion of the site, because of the accumulation of surface runoff, are believed to be preferential areas of groundwater recharge. Groundwater discharge from the uppermost aquifer is to the unconsolidated alluvial and colluvial materials infilling Alkali Canyon. Groundwater flow within the alluvium/colluvium within Alkali Canyon is gen-erally believed to be down valley and ultimately to the Columbia River.

For the purpose of designing a detection monitoring network, each regu-lated WMU was considered to be a discrete entity with its own point of compliance. A primary monitoring network within the saturated zone at the base of the Selah was thus prepared for each unit.

The detection monitoring network developed the Arlington facility consists of a system of three wells located upgradient of the waste management area,

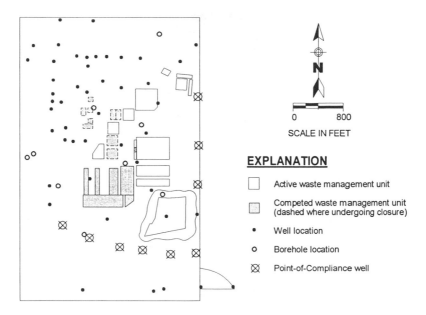

Figure 10-18 Location of groundwater detection monitoring wells.

and ten wells along the point of compliance on the south and east boundaries of the waste management area (Figure 10-18). The horizontal well spacing along the south and eastern boundary is about 500 ft. The detection monitoring wells were completed at the top of the Priest Rapids basalt and screened over a 20-ft saturated thickness at the base of the Selah. Since some degree of uncertainty exists with respect to location of the groundwater divide in the northeastern corner of the site. Given the density of boreholes in this area, the degree of uncertainty with regard to location of the groundwater divide is estimated to be on the order of 50 ft. To account for the uncertainty with respect to location of the groundwater divide and seasonality effects, if any, a setback of 250 ft has been proposed. This setback has been incorporated into the location of the waste management area boundary and the design of future landfill units.

In the location and design of future landfill unit, the proposed setback has been applied to the bottom of the landfill cell. To prevent potential contaminant release from the side slopes which would encroach into the setback zone, a double synthetic liner system has been incorporated into the engineering design of the landfill. The double-liner system would assure that any leak from the inner liner would be routed to the leachate collection system located south of the 250-ft minimum setback. In so doing, the requisite buffer zone from the groundwater divide will be maintained.

Casmalia Resources Facility, Southern California

Site Description

The Casmalia Resources Facility is a hazardous waste management facility located in the northwest portion of Santa Barbara County, CA (number 13 in Figure 10-1). The facility provides services for the treatment and disposal of hazardous wastes that are generated off-site. The facility started receiving wastes for disposal in 1972. A wide spectrum of chemical and industrial wastes has been received, a significant portion by volume of which was generated by oil and gas exploration and development activities and are considered nonhazardous waste under the RCRA. The remainder consists of solvents, pesticides, alkalines, acids, cyanides, resins, phenols, laboratory wastes, heavy metals, and PCB-contaminated equipment.

Extensive investigations have been conducted at and in the vicinity of the facility since the early 1980s to characterize geologic and hydrogeologic conditions, and develop a groundwater monitoring program including the installation of a groundwater monitoring system designed to detect the immediate release of constituents from the WMU.

The Casmalia Resources Facility is situated in the Casmalia Hills, a region of low, rolling hills which lie between the Santa Maria Valley and the Pacific Coast. The facility is located about 1.5 mi north-northwest of the town of Casmalia, and about 5 mi southwest of the city of Santa Maria. Generally situated with south-facing exposures, ground surface elevations generally decrease from north to south ranging from about 835 to 405 ft above mean sea level, respectively. The northern portion of the facility is situated between northwest- and northeast-trending ridges; the southern portion is situated between two low, adjacent knolls. Casmalia Creek and Casmalia Canyon, and an unnamed drainage are located west and northeast of the facility, respectively. Both drainages are relatively broad and eventually merge with Shuman Creek and Shuman Canyon about 2 mi south of the facility, and about 1 mi west of the town of Casmalia. Toqnuzini Creek and Toqnuzini Canyon also merge with Shuman Creek south of the facility.

The overall climate is generally semiarid to arid, resulting in sparse vegetation and intermittent streams. The vast majority of the surrounding area is open grassland with shrubs, live oak, and scattered willow and eucalyptus trees found along some of the intermittent drainages along Casmalia and Shuman Creek. Some land to the southwest of the facility is used for dry farming, notably, garbanzo beans and wheat. Overall land use within 1 mi of the facility is developed as large parcel agricultural consisting mostly of livestock grazing and associated ranching operations. Oil and gas developments and a few residences are located southeast of the facility along Black Road.

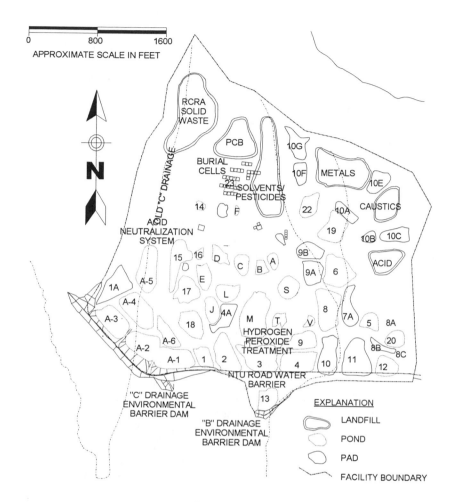

Figure 10-19 **Facility layout map showing the approximate location of major ponds, evaporation pads, landfills, and associated structures.**

Facility Layout and Operation

At present, the facility has four operating landfill units, identified by the types of wastes they receive (i.e., acids, caustics, metals, and solvents/pesticides), two operating process treatment systems, and approximately 43 ponds and 15 evaporation pads that are currently or will be undergoing closure in the near future (Figure 10-19). The ponds and pads, which comprise the liquid waste management facilities, are presently being closed. Substantial modification of the facility is anticipated to occur within the near future to maximize the use for solid waste management.

Clay barriers (or containment structures) have been installed in six locations at the facility. These structures were designed and installed to control the migration of potential leachate both within and from the facility. The clay material used in constructing these dams was obtained on site or adjacent to the facility.

Four of the six clay barriers have been installed with drainage collection galleries that have pump-out facilities. The largest of these three is called "C" Environmental Barrier Dam and is located south of Ponds A-1, A-2, and A-3 along the southern perimeter of the facility. The "B" Environmental Barrier Dam is located at the southern edge of Pond 13, along the southern perimeter of the facility. The remaining two clay barriers with drainage collection galleries are located at the foot of the Solvents/Pesticides Landfill Unit and at the southern edge of the Catch Basin in West Canyon. Two other clay barriers (without drainage collection galleries) are located at the southwestern edge of the PCB Landfill Unit and around the southern and eastern perimeter of Pond 20.

Geologic Setting

The facility is situated within the Casmalia Hills which is characterized as a structurally uplifted region situated between the Santa Maria Valley and the Pacific Coast. Regionally, late Tertiary marine sedimentary rocks overlie Mesozoic Franciscan basement rocks. The late Tertiary sedimentary formation within a several-mile radius of the facility generally appears to become progressively younger from southwest to northeast. In order of decreasing age, the sedimentary formations include the Miocene Series, Miocene-Pliocene Series, Pliocene Series, Pliocene and Pleistocene Series, and the Holocene Series. Geologic conditions at and within a 1-mi radius of the site are shown in Figure 10-20. An east-west and north-south trending hydrogeologic crosssection is presented in Figure 10-21.

The stratigraphic units characterized at the facility include the Miocene-Pliocene Todos Santos Claystone Member of the Sisquoc Formation, Pleistocene terrace deposits, and Quaternary alluvium and colluvium. Artificial fill was also characterized to a limited extent. Pertinent to this discussion is the Todos Santos Claystone Member.

The entire facility resides on thin surficial deposits and weathered portions of the Todos Santos Claystone Member of the Sisquoc Formation (hereinafter Todos Santos). Based on site-specific studies, the Todos Santos is divided into two informal units: the lower, unweathered claystone and the upper, weathered claystone. This division does not reflect two distinct faces with stratigraphic and structural relations within the Todos Santos, but instead the two units reflect a pervasive weathering profile.

The Todos Santos forms low, smoothly rounded hills which are covered by thin, predominantly clayish soil that locally supports a variety of grasses,

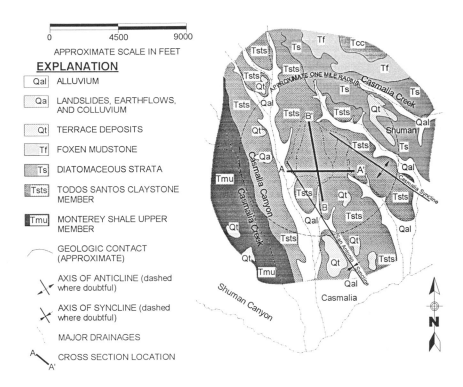

EXPLANATION

Qal	ALLUVIUM
Qa	LANDSLIDES, EARTHFLOWS, AND COLLUVIUM
Qt	TERRACE DEPOSITS
Tf	FOXEN MUDSTONE
Ts	DIATOMACEOUS STRATA
Tsts	TODOS SANTOS CLAYSTONE MEMBER
Tmu	MONTEREY SHALE UPPER MEMBER

GEOLOGIC CONTACT (APPROXIMATE)

AXIS OF ANTICLINE (dashed where doubtful)

AXIS OF SYNCLINE (dashed where doubtful)

MAJOR DRAINAGES

CROSS SECTION LOCATION

Figure 10-20 Geologic map of area within a 1-mi radius of the Casmalia facility.

isolated shrubs, and small groves of live oak, eucalyptus, and willow trees. The Todos Santos is exposed along fairly extensive cut-slopes and cut-areas within the facility boundary.

Along the southeastern facility boundary, bedding attitudes strike generally east-west and dip 5 to 15° southward. Toward the northeast corner of the facility, the dip angles decrease to 3 to 5° but maintain their southerly direction before gradually dipping to the north. In the central portion of the facility, near pond 22 and along the western facility boundary, bedding generally strikes within 20° east or west of north, and dips from 3 to 20° eastward.

Because unweathered claystone comprises less than 10% of the exposures on site, most of the lithologic descriptions are from core samples. Based on visual examination of core samples and surface exposures, the unweathered unit typically consists of olive-black to gray olive-green (wet), and medium bluish-gray to greenish-gray (dry) claystone to silty claystone. It is generally massive with some thin laminations and fissile zones.

Estimates of strength, based on gross field tests, show the unweathered claystone as firm to strong. This unit is less fractured than the weathered claystone and contains a small percentage of diatoms.

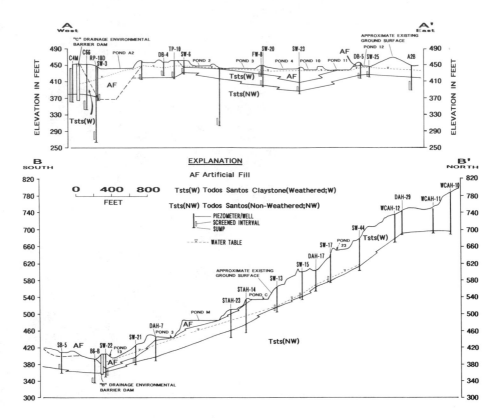

Figure 10-21 Hydrogeologic cross sections oriented east-west (A-A') and north-south (B-B').

Within the facility, the weathered claystone accounts for most of the exposed, mappable bedrock. The weathered claystone is present over most of the facility with the exception of areas beneath the environmental barrier dams in the B- and C-drainages, portions of West Canyon, and possibly beneath many of the solid waste disposal units.

Because the division between the weathered and unweathered claystone is based on the presence of weathering rather than depositional relationships, the petrography of the weathered claystone is essentially the same as that of the unweathered claystone. The most obvious differences between the weathered and unweathered units is a distinctly different color, the presence of prevalent discontinuities (fractures and joints), and secondary mineralization within the weathered unit.

Hydrogeologic Setting

Groundwater occurs under water-table conditions at depths of about 150 ft below ground surface in the northern portion of the facility, to about 40 ft below most impoundments, to surface springs within the Todos Santos Member of the Sisquoc Formation. The Todos Santos is characterized as, vertically and laterally, the most extensive hydrostratigraphic unit (HSU) underlying and in the vicinity of the facility. For the purposes of hydrogeologic characterization, an HSU is defined as a body of rock having considerable lateral extent and composing a geologic framework for a reasonably distinct hydrologic system. The presence of a pervasive weathering profile within the upper portion of this unit allows the Todos Santos to be subdivided into distinct units. Hereinafter, the upper portion of the Todos Santos hydrostratigraphic unit is referred to as the upper HSU, while the lower portion is referred to as the lower HSU. A water table contour map is presented in Figure 10-22.

The upper HSU is composed of the weathered claystone portion of the Todos Santos. The transition zone of the weathering profile between the weathered and unweathered claystone observed in piezometers and core borings is relatively thin, ranging from about 2 to 5 ft in thickness. Since this zone exhibits some weathering characteristics similar to those of the weathered claystone, it is included within the upper HSU. Near-surface geologic deposits (i.e., surficial clay-rich soil and colluvium) are thought to be of minor importance to groundwater flow at the facility, as compared to the weathered claystone, due to their relatively limited extent beneath and in the immediate vicinity of the facility. In addition, the waste material itself is not considered to be part of the upper HSU due primarily to its discontinuous subsurface extent between WMU within the facility. The lower HSU consists entirely of the unweathered portion of the Todos Santos. The classification of the Todos Santos HSU into two units reflects the following:

- Stratigraphic data allowing identification and verification of the presence of a laterally extensive weathering profile within the claystone
- Structural data including seismic refraction surface geophysical surveys, borehole geophysical logs, and the substantial decrease in fracture density below the weathering profile observed during detailed on-site geologic mapping and in the recovered cores
- Contrasting hydraulic conductivity values measured between the upper and lower HSU from short- and long-term hydraulic tests. Although permeabilities vary widely, the geometric means of the hydraulic conductivity values based on all available borehole hydraulic tests of the lower HSU (1.5×10^{-6} cm/s) is approximately 45 times lower than that of the upper HSU (6.8×10^{-5} cm/s). This contrast in mean hydraulic conductivity is expected to be sufficient to retard groundwater flow on a large scale between the upper and lower HSU as compared to the flow rates within the upper HSU
- Hydrochemical data from adjacent clustered piezometers completed to different depths showing distinct differences in water quality between the upper and lower HSU. Both the natural geochemistry of the groundwater as well as the levels of site-related constituents detected are dissimilar for each unit.

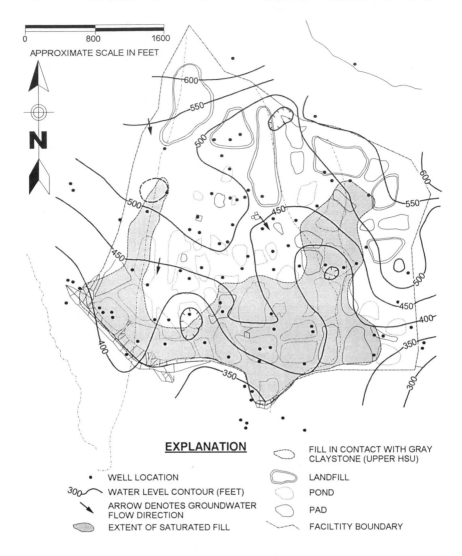

Figure 10-22 Contour map showing groundwater occurrence under water-table and perched conditions.

Perched water, defined as unconfined groundwater that is separated from an underlying main zone of groundwater by an unsaturated zone, occurs when vertically migrating water from the ground surface encounters zones of relatively lower permeability which retard its migration. The water then accumulates above the low-permeability material, causing the matrix above to become saturated. Such perched conditions have been observed at some locations within the northern portions of the facility at depth intervals near the contact between the upper and lower HSU.

Minor zones of seepage have been observed and described in the east-side excavation of the Caustics Landfill Unit. These seeps, and similar ones in West Canyon, occur near the weathered/unweathered claystone contact and are laterally discontinuous over distances of several feet to several tens of feet. These seeps are also vertically discontinuous over distances of less than an inch to several feet. Zones of moisture and minor seepage that occur along bedding planes tend to be associated with intersections of geologic features which create contrasting hydraulic conductivities. Bedding in the claystone is typically not well developed and bedding plane partings, where they are discernible, are often laterally discontinuous and obscured by subhorizontal fractures related to erosional unloading. However, in some localized zones, bedding may influence flow pathways over short distances.

An excavation in the claystone on the east side of the Caustics Landfill Unit exposed a subhorizontal alignment of minor seeps. These seeps occur along a very thin discontinuous zone (in places approximately 1/2 in. thick) that occurs near the base of the excavation. Although bedding is not clearly discernible in this excavation, these seeps may coincide with a subhorizontally oriented, bedding-related zone or discontinuity. Some of the seeps appear to occur where steeply dipping fractures intersect this subhorizontal discontinuity. In the northern portion of the excavation, the apparent line of seeps is approximately 10 to 12 ft below the contact between the weathered and unweathered claystone. However, in the southern portion of the excavation, the seepage line appears to coincide with the weathered/unweathered claystone contact. This alignment of seeps represents a minor zone of perched water that is formed by groundwater that has been restricted in its downward movement by a subhorizontal zone of relatively low permeability. These seeps are probably related to the position of the contact between the weathered and unweathered claystone, the possible presence of a bedding plane discontinuity, and intersecting fractures.

Other than the weathered/unweathered contact zone, it is unclear which stratigraphic zones, beds, or horizons within the weathered or unweathered claystone could predictably enable water to be perched. While it also may be possible that a certain geometry or set of fractures could enable some groundwater to be isolated or essentially perched at any depth interval above the water table, perched groundwater, if it were to occur in the claystone, is mostly like to occur near the contact between the upper and lower HSU.

Groundwater Monitoring System Design

Although geologic materials underlying the facility do not meet prescriptive permeability criteria for the disposal of hazardous materials, provisions for alternative standards consistent with the performance goal afforded equivalent protection of water quality. The groundwater monitoring system was thus designed to monitor off-site migration of groundwater from the entire site rather than from specific WMU; therefore, individual monitoring wells have not been used to identify vertical or lateral migration of constituents from specific waste management units within the facility boundary.

The detection monitoring program incorporates surface water, groundwater, and vadose zone soil pore liquids monitoring. The system itself is comprised of approximately 24 surface water sampling locations, most of which are situated offsite along Casmalia Creek and its tributaries. These locations, situated both upstream and downstream, constitute the points of compliance for surface water. The groundwater detection monitoring program includes approximately 18 background monitoring wells and approximately 59 downgradient monitoring wells which constitute the points of compliance with respect to groundwater. These wells are screened at various intervals, including opposite the water table within weathered claystone (upper HSU), fill, and alluvium; at the contact between the weathered and unweathered claystone (upper and lower HSU, respectively), and within the unweathered claystone (lower HSU) at depths of about 30 to 40 ft and 100 ft below ground surface. The vadose zone monitoring program being prepared is also planned for the facility. Approximate location of surface water and groundwater monitoring wells that comprise the system are shown in Figure 10-23.

Relative to the overall operation of the facility, the detection monitoring program design is consistent with the established zero discharge site boundary. The environmental barriers and collection galleries were installed to improve upon the favorable hydrogeologic characteristics of the site. The design of the detection groundwater monitoring system with regard to selection of points of compliance is not significantly dependent upon horizontal to vertical hydraulic conductivity values within the uppermost aquifer. Instead, most of the wells comprising the groundwater system are downgradient of the facility and located in respect to base flow toward and within the four drainages that have been identified as being the primary potential groundwater flow paths. These monitoring wells thus intersect the paths of potential fluid migration and allow detection of off-site migration of site-specific chemicals.

Overall, the facility maintains numerous favorable environmental characteristics which make it suitable for the siting and operation of a hazardous waste disposal facility. These characteristics are:

- Abundant low permeability soil which underlie impoundments
- Relatively slow groundwater flow rates on the order of 5 to 30 ft per year
- Natural groundwater quality beneath the facility exceeds EPA drinking water standards for several constituents, thus limiting potential groundwater uses
- Average evapotranspiration rates exceed average rainfall rates by approximately 40 in. per year. This condition in conjunction with the water needed for on-site reuse will minimize the amount of water

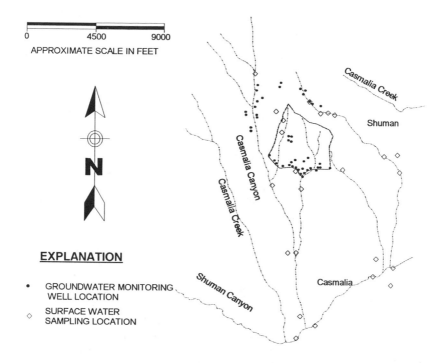

Figure 10-23 Detection monitoring wells and surface water sampling locations for the Casmalia facility.

stored in the impoundments, thus lowering the possibility of the impoundments impacting states' water
- Liner systems will exceed prescribed standards and be retrofitted with new containment systems every 10 years, if required
- Liquids in impoundments will be limited to less than 1% halogenated organic compounds, pH between 5.0 and 9.0, and total dissolved solids on the order of 10,000 ppm
- Existing cut-off trenches and extraction systems which capture and retain groundwater underlying the facility
- Underdrain systems installed under each new impoundment will maintain a 5-ft separation between waste and highest anticipated groundwater, and serve as an additional collection system in the unlikely event of a failure of the primary, secondary and tertiary containment system
- Establishment of a closure fund.

Although beneficial use or potential beneficial use groundwater occurs within 1/2 mi downgradient of the facility, the potential for adverse impact on this groundwater resource is considered very low.

BIBLIOGRAPHY

Anderson, D.C., 1991, Future of Hazardous Waste Landfills: In Proceedings of the Seventh Annual Hazardous Materials and Environmental Management Conference West, Long Beach, CA, pp. 165-175.

Andricevic, R., 1990a, Cost-effective Network Design for Groundwater Flow Monitoring: *Stochastic Hydrology Hydraulics,* Vol. 4, No. 1, pp. 27-41.

Andricevic, R., 1990b, A Real-time Approach to Management and Monitoring of Groundwater Hydraulics: *Water Resources Research,* Vol. 26, pp. 2747-2755.

Andricevic, R. and Foufoula-Georgiou, E., 1991, A Transfer Function Approach to Sampling Network Design for Groundwater Contamination: *Water Resources Research,* Vol. 27, No. 10, pp. 2759-2769.

ASCE Task Committee on Geostatistical Techniques in Geohydrology, Review of Geostatistics in Geohydrology, I, Basic Concepts: *Journal of Hydraulic Engineering,* Vol. 116, No. 5, pp. 612-632.

ASCE Task Committee on Geostatistical Techniques in Geohydrology, Review of Geostatistics in Geohydrology, II, Applications: *Journal of Hydraulic Engineering,* Vol. 116, No. 5, pp. 633-659.

Bagchi, A., 1990, *Design, Construction, and Monitoring of Sanitary Landfill:* John Wiley & Sons, New York, NY, 284 p.

Ball, J. and Coley, D.M., 1986, A Comparison of Vadose Monitoring Procedures: In Proceedings of the National Water Well Association Sixth National Symposium and Exposition on Aquifer Restoration and Ground Water Monitoring, pp. 52-61.

Buelt, J.L., 1983, Liner Evaluation for Uranium Mill Tailings: Final Report PHL-4842, Pacific Northwest Laboratory, Richland, WA.

Carrera, J., Usunoff, E. and Szidarovszky, F., 1984, A Method for Optimal Observation Network Design for Groundwater Management: *Journal of Hydrology,* Vol. 73, pp. 147-163.

Delhomme, J.P., 1978, Kriging in the Hydrosciences: In *Advances in Water Resources,* Vol. 1, pp. 251-266.

Devary, J.L. and Doctor, P.G., 1982, Pore Velocity Estimation Uncertainties: *Water Resources Research,* Vol. 18, pp. 1157-1164.

Dorrus, A.M., Conrad, R.C. and Nonro, L.M., 1992, RCRA Facility Investigation for the Townsite of Los Alamos, New Mexico: In Proceedings of the Hazardous Materials Research Control Institute HMC South '92 Conference, pp. 371-376.

Edwards, A.J. and Smart, P.L., 1989, Waste Disposal on Karstified Carboniferous Limestone Aquifers of England and Wales: In proceedings of the Third Multidisciplinary Conference on Sinkholes and the Engineering and Environmental Impacts of Sinkholes and Karst (Edited by Beck, B.F.), pp. 165-182.

Everett, L.G., 1980, *Groundwater Monitoring:* Genium Publishing, Schenectady, NY.

Everett, L.G., Wilson, L.G. and Hoylman, E. W., 1984, *Vadose Zone Monitoring for Hazardous Waste Sites:* Noyes Data Corp., Parkridge, NJ.

Everett, L.G., Hoylman, E.W., Wilson, L.G. and McMillion, L.G., 1984, Constraints and Categories of Vadose Zone Monitoring Devices: *Ground Water Monitoring Review,* Vol. 4, No. 1, pp. 26-32.

Everett, L.G., McMillion, L.G. and Eccles, L.A., 1988, Suction Lysimeter Operations at Hazardous Waste Sites: In *Ground-Water Contamination: Field Methods, American Society for Testing and Materials STP 963* (Edited by Collins, A.G. and Johnson, A.I.), Philadelphia, pp. 304-327.

Germann, P.F., 1988, Approaches to Rapid and Far-Reaching Hydrologic Processes in the Vadose Zone: *Journal of Contaminant Hydrology,* Vol. 3, pp. 115-127.

Glass, R.J., Steenhuis, T.S. and Parlange, J.Y., 1988, Wetting Front Instability as a Rapid and Far-Reaching Hydrologic Process in the Vadose Zone: *Journal of Contaminant Hydrology,* Vol. 3, pp. 207-226.

Green, W.H. and Ampt, G.A., 1911, Studies on Soil Physics: I. Flow of Air and Water Through Soils: *Journal of Agricultural Sciences,* Vol. 4, pp. 1-24.

Hannah, E.D., Pride, T.E., Ogden, A.E. and Paylor, R., 1989, Assessing Ground Water Flow Paths from Pollution Sources in the Karst of Putnam County, Tennessee: In Proceedings of the Third Multi-disciplinary Conference on Sinkholes and the Engineering and Environmental Impacts of Sinkholes and Karst (Edited by Beck, B.F.), pp. 183-188.

Hsu, N.-S. and Yeh, W.W.-G., 1989, Optimum Experimental Design for Parameter Identification in Groundwater Hydrology: *Water Resources Research,* Vol. 25, pp. 1025-1040.

Hsueh, Y.W. and Rajagopal, R., 1988, Modeling Groundwater Quality Sampling Decisions: *Ground Water Monitoring Review,* Vol. 8, pp. 121-134.

Hudak, P.F. and Loaiciga, H.A., 1992, A Location Modeling Approach for Groundwater Monitoring Network Augmentation: *Water Resources Research,* Vol. 28, No. 3, pp. 643-649.

Johnson, K.S., 1991, Regional Hydrogeologic Screening Characteristics Used for Siting Near-Surface Waste-Disposal Facilities in Oklahoma, U.S.A.: *Environmental Geology and Water Sciences,* Vol. 17, No. 1, pp. 3-7.

Kramer, J.H., Everett, L.G. and Cullen, S.J., 1991, Innovative Vadose Zone Monitoring at a Landfill Using the Neutron Probe: In Proceedings of the Fifth National Outdoor Action Conference on Aquifer Restoration, Ground Water Monitoring and Geophysical Methods, NWWA, Dublin, OH.

Kramer, J.H., Cullen, S.J. and Everett, L.G., 1992, Vadose Zone Monitoring with the Neutron Moisture Probe: *Ground Water Monitoring Review,* Vol. 12, No. 3.

Lee, G.F. and Jones, P.A., 1991, Groundwater Pollution by Municipal Landfills - Leachate Composition, Detection and its Water Quality Significance: In Proceedings of the Association of Ground Water Scientists and Engineers, Las Vegas, NV, pp. 257-271.

Lifer, D.W. and Erchul, R.A., 1989, Sinkhole Dumps and the Risk to Groundwater in Virginia's Karst Areas: In Proceedings of the Third Multi-disciplinary Conference on Sinkholes and the Engineering and Environmental Impacts of Sinkholes and Karst (Edited by Beck, B.F.), pp. 207-212.

Litaor, M.I., 1988, Review of Soil Solution Samplers: *Water Resources Research,* Vol. 24, No. 5, pp. 727-733.

Loaiciga, H., 1989, An Optimization Approach for Groundwater Quality Monitoring Network Design: *Water Resources Research,* Vol. 25, pp. 1771-1782.

Massmann, J. and Freeze, R.A., 1987, Groundwater Contamination from Waste Management Sites: The Interaction Between Risk-Based Engineering Design and Regulatory Policy: *Water Resources Research,* Vol. 23, pp. 351-367.

Matheron, G., 1971, *The Theory of Regionalized Variables and Its Applications:* Ecole Des Mines, Fontainebleau, France.

McKinney, D.C. and Loucks, D.P., 1992, Network Design for Predicting Groundwater Contamination: *Water Resources Research,* Vol. 28, No. 1, pp. 133-147.

McLaughlin, D. and Graham, W., 1986, *Design of Cost-Effective Programs for Monitoring Groundwater Contamination:* IAHS Publishers, Vol. 158, pp.

Meier, J.A., 1990, Characterization of Preferred Groundwater Migration Pathways - Weldon Spring Site: In Proceedings of the Association of Groundwater Scientists and Engineers Fourth National Outdoor Action Conference on Aquifer Restoration, Groundwater Monitoring and Geophysical Methods, Las Vegas, NV, pp. 1277-1289.

Meyer, P.D. and Brill, E.D., 1988, A Method for Locating Wells in a Groundwater Monitoring Network Under Uncertainty: *Water Resources Research,* Vol. 24, pp. 1277-1282.

Morrison, R.D. and Lowery, B., 1990, Sampling Radius of a Porous Cup Sampler - Experimental Results: *Ground Water,* Vol. 28, No. 2, pp. 262-267.

Nielsen, D.M. (ed.), 1991, *Practical Handbook of Ground-Water Monitoring:* Lewis Publishers, Boca Raton, FL, 717 p.

Olea, R.A., 1984, Sampling Design Optimization for Spatial Functions: *Mathematical Geology,* Vol. 16, No. 4, pp. 365-391.

O'Leary, P.R., Walsh, P.W., and Ham, R.K., 1988, Managing Solid Waste: *Scientific American,* Vol. 259, No. 6, pp. 36-42.

Pfannkuch, H.O., 1981, Problems of Monitoring Network Design to Detect Unanticipated Contamination: *Groundwater Monitoring Review,* pp. 11-20.

Pfannkuch, H.O. and Labro, B.A., 1976, Design and Optimization of Groundwater Monitoring Networks for Pollution Studies: *Ground Water,* Vol. 14, No. 6.

Price, D.J., 1989, Anatomy of a Hazardous Waste Site in a Paleosink Basin: In Proceedings of the Third Multi-disciplinary Conference on Sinkholes and the Engineering and Environmental Impacts of Sinkholes and Karst (Edited by Beck, B.F.), pp. 189-196.

Reaber, P.W. and Stein, T.L., 1990, Design and Installation of a Detection Monitoring Network at a Class I Facility in an Arid Environment: In Proceedings of the Association of Groundwater Scientists and Engineers, Fourth National Outdoor Conference on Aquifer Restoration, Groundwater Monitoring, and Geophysical Methods, pp. 299-311.

Reeder, P.P. and Crawford, N.C., 1989, Potential Groundwater Contamination of an Urban Karst Aquifer - Bowling Green, Kentucky: In Proceedings of the Third Multi-disciplinary Conference on Sinkholes and the Engineering and Environmental Impacts of Sinkholes and Karst (Edited by Beck, B.F.), October 2-4, 1989, pp. 197-206.

Silkwork, D.R. and Grigal, D.R., 1981, Field Comparison of Soil Solution Samplers: *Soil Science Society of America Journal,* Vol. 45, No. 2, pp. 440-442.

Smith, E.D., 1984, Hydrogeologic Assessment of Zone-of-Saturation Landfill Design: Oak Ridge National Laboratory Report CONF8405144-3, DE84 012356.

Testa, S.M., 1988, Hazardous Waste Disposal Site Hydrogeologic Characterization: In Proceedings of the Second International Conference on Case Histories in Geotechnical Engineering, St. Louis, Paper No. 1.26, pp. 97-106.

Testa, S.M., 1991, Site Characterization and Monitoring Well Network Design, Columbia Plateau Physiographic Province, Arlington, Oregon: *Geological Society of America Abstract* No. 12357, Vol. 23, No. 5, p. A325.

Testa, S.M., Patton, D.L. and Conca, J.L., 1991, Fixation of Petroleum-Contaminated Soils via Cold-Mix Asphalt for Use as a Liner: In Proceedings of the Hazardous Material Control Research Institute South, '91 Conference, New Orleans, pp. 30-33.

Testa, S.M., 1992, Hydrogeologic Characterization and Groundwater Monitoring System Design, Casmalia Resources Facility: In *Association of Engineering Geologists Special Volume on Engineering Geology in Southern California* (Edited by Pipkin, B. and Proctor, R.), in press.

Thackston, J., Meeks, Y., Strandberg, J. and Tuchfeld, H., 1989, Characterization of a Fracture Flow Groundwater System at a Waste Management Facility: In Proceedings of the Association of Groundwater Scientists and Engineers, Third National Outdoor Conference on Aquifer Restoration, Groundwater Monitoring, and Geophysical Methods, pp. 1079-1093.

U.S. Environmental Protection Agency, 1989, Requirements for Hazardous Waste Landfill Design, Construction, and Closure: Seminar Publication, EPA/625/4-89/022, 127 p.

VanGeer, F.C., 1987, Applications of Kalman Filtering in the Analysis and Design of Groundwater Monitoring Network, Tech. Rep. PPN 87-05, TNO-DGV Institute of Applied Geosciences, Delft, Netherlands.

Wilson, L.G., 1980, Monitoring in the Vadose Zone: A Review of Technical Elements and Methods: EPA-600/70-80-134, U.S. Environmental Protection Agency, Las Vegas, NV.

Wilson, L.G., 1981, Monitoring in the Vadose Zone, Part 1 - Storage Changes: *Ground Water Monitoring Review,* Vol. 1, No. 3, pp. 32-41.

Wilson, L.G., 1990, Methods for Sampling Fluids in the Vadose Zone: In *Ground Water and Vadose Zone Monitoring* (Edited by Nielsen, D.M. and Johnson, A.I.), American Society for Testing and Materials STP 1053, Philadelphia, pp. 7-24.

Wolf, F.G. and Testa, S.M., 1985, Kriging - A Geostatistical Technique for Evaluating Groundwater Monitoring Network: In Proceedings of the Hazardous Materials Control Research Institute Hazardous Wastes and Environmental Emergencies Conference, Cincinnati, pp. 168-172.

Wood, W.W., 1973, A Technique Using Porous Cups for Water Sampling at Any Depth in the Unsaturated Zone: *Water Resources Research,* Vol. 9, No. 2, pp. 486-488.

11   UNDERGROUND INJECTION

*". . . rocks generally are but filters of various degrees of
porosity, and only capable of permitting water to pass through
them in a given quantity and time."*

(De La Beche, 1851)

INTRODUCTION

Underground injection for the disposal of hazardous substances is based on simple hydrogeologic principles. Sedimentary basins are characterized by thousands of feet of relatively undisturbed, layered, water-bearing sedimentary rocks. Total dissolved solids (TDS) concentration in subsurface waters generally increases with depth, with the direction of groundwater flow from higher to lower TDS concentrations. Significant differences in water quality exist between deep (saline waters) and shallow fresh groundwater, and layers of relatively impermeable rocks act as a barrier to the upward movement of the saline waters. These saline water-bearing deposits are considered of nonbeneficial use and thus under certain geologic and hydrogeologic conditions may serve as receptors of hazardous and toxic waste. As such, subsurface injection is defined as the subsurface emplacement of fluid through a well or dug hole whose depth is greater than its width. Injection wells essentially serve as liquid waste disposal facilities where liquid or liquefiable wastes are pumped into confined geologic formations via gravity flow or under pressure, thus providing an alternative to surface water discharge.

The use of wells for the subsurface disposal of industrial wastes, notably brine disposal, has been known since the early 1930s as an alternative to surface discharge of saline solutions (brine) associated with oil and gas production. Only four injection wells have been reported to be constructed prior to 1950. The injection of hazardous waste in deep underground aquifers was more prevalent since about 1950, reflecting improved injection technology and the

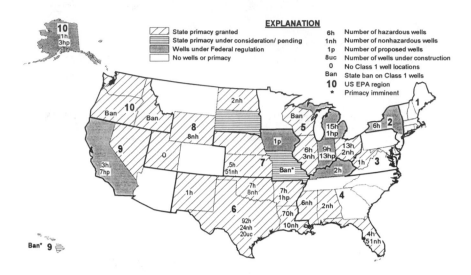

Figure 11-1 Location and regulatory status of Class I injection wells in the U.S.

inability to safely discharge wastes to surface water even after dilution and treatment. From 1950 to 1965, an average of only two wells per year were constructed. By 1960, approximately 30 industrial hazardous waste injection wells were in operation. Between 1965 and 1980, more stringent regulation of industrial waste disposal into surface water bodies resulted in the construction of more than 130 deep-injection Class I wells for the disposal of hazardous waste. As of 1975, 322 industrial and municipal injection wells had been constructed, of which 209 were considered operational. In 1985, EPA identified 112 facilities, of which 90 were active and injecting hazardous waste through 195 wells; the remaining 57 wells were inactive.

Hazardous waste injection is currently performed in 15 states (Figure 11-1). The majority of these wells are located in industrialized areas along the Gulf Coast and in the vicinity of the Great Lakes in the north-central U.S. Louisiana and Texas account for about 66% of the wells, with subordinate but sizable numbers in Michigan, Indiana, Ohio, Illinois, and Oklahoma. Deep-well injection disposal is used by the chemical, petrochemical, and pharmaceutical industries, among others. Depth generally varies between 1000 and 10,000 ft, most often between 2000 and 6000 ft. Many of these wells are located in states with a long history of oil and gas exploration.

In 1985, the EPA reported that the chemical industry was the largest user of injection wells, with 64% of all wells. The petroleum-refining industry accounted for about 25% of the wells. Although over half of the fluids injected are considered nonhazardous, volumes injected are, overall, considered

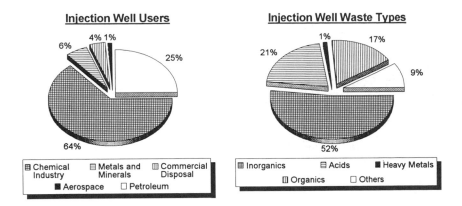

Figure 11-2 Injection well users and waste types.

hazardous. The major users of injection wells for disposal purposes and major waste types are shown in Figure 11-2.

Presented in this chapter is discussion of the various classes of injection wells, siting criteria, site-specific hydrogeologic considerations, design criteria, and reporting and monitoring requirements. The geochemical fate of injected hazardous waste is also discussed.

CLASSES OF INJECTION WELLS

Wells under the Underground Injection Control (UIC) program are categorized by class of well. There are five classes as summarized in Table 11-1. Current UIC technical classification scheme and requirements per class are discussed below.

Class I Wells

Class I wells are used by generators of hazardous waste, or by owners or operators of hazardous waste management facilities, to inject hazardous waste beneath the lowermost formation containing, within 1/4 mi of the well bore, an underground source of drinking water (USDW). Also included are other industrial, municipal, and agricultural waste disposal wells. Class I wells are further divided into two categories: nonhazardous and hazardous. Active hazardous waste injection systems with single or multiple injection wells currently operate in 15 states, (Figure 11-1), with the largest user being the chemical industry (Figure 11-2). An inventory of all Class I wells was completed by EPA in 1985. This inventory identified 112 facilities which inject hazardous waste through

Table 11-1 Injection Well Classification

Class	Application
I	Injection of hazardous, industrial, and municipal wastes
II	Injection of fluids brought to surface in connection with oil and gas production, for enhanced oil recovery, or for storage of hydrocarbons
III	Injection of fluids for solution mining of minerals
IV	Injection of hazardous and radioactive wastes
V	Injection of fluids and wastes not addressed under Class I, II, III, or IV such as drainage wells and air conditioning return flow wells

252 Class I wells; 90 of these facilities were active, comprising 195 wells as of 1984. The majority of the wells are situated in either heavily industrialized states such as Illinois, New York, Michigan, Indiana, and Ohio, or in oil- and gas-producing states such as Texas, Louisiana, Oklahoma, and California (Figure 11-1). It is estimated that a total volume of 11.5 billion gal of waste was injected in 1983.

Class I wells that inject nonhazardous waste are basically subject to the same technical standards as for all Class I wells. These standards include:

- The well must not be located where another well penetrates that part of the injection zone expected to be influenced by the Class I well, if the other well could act as a conduit for wastes to escape from the injection zone
- The well must inject below the lowest USDW, must be cased and cemented, and have a packer or approved fluid seal set between the injection tubing and the casing, immediately above the injection zone
- Mechanical integrity must be determined from both construction logs and mechanical integrity testing
- Mechanical integrity demonstration must be performed once every five years
- Periodic sampling and chemical testing of injected wastes must be conducted
- Annual monitoring of pressure build-up in the injection zone must also be conducted.

Monitoring wells are also required in aquifers that serve as a source of drinking water, to detect fluid movement or pressure from the injection. Monitoring may also be required in the first aquifer overlying the confining zone, as well as quarterly sampling for groundwater quality. To prevent fractures of the injection zone and the overlying confining zone, a maximum operating pressure for the injection tubing is established; controlled fracturing is also allowed.

Class I wells used for the injection of hazardous waste are considered hazardous waste disposal facilities and, thus, are subject to certain RCRA requirements. It must be shown that the area is "geologically suitable" for injection with requirements for both the injection and confining zones. In addition, the upper layer of the confining zone must be separated from the lowermost USDW by a buffer zone of additional strata, at least one relatively permeable stratum and one relatively impermeable stratum. A USDW is defined as an aquifer or its portion with a TDS value of less than 10,000 mg/L, and which is capable of supplying a public water supply system.

The area of review for Class I wells is defined by a 2-mi radius from the well, or a 12 1/2-mi area. If the cone of influence for the injection well is greater than 2 mi, then the area of review can be increased. Thus, corrective action must be proposed for potentially many more wells than other injection wells. Corrective action means assurance that other wells penetrating the confining or injection zones have been adequately plugged and sealed, preventing the potential for contamination of USDWs. Limiting injection pressure is an alternative means of addressing corrective action-related issues. With new Class I hazardous waste injection wells, assurance that the well materials are compatible with the injection wastes, and that the geological information presented in the permit application is correct and an accurate baseline is established.

Class II Wells

Class II wells include those wells which inject fluids that are brought to the surface in connection with conventional oil and natural gas production (unless classified as a hazardous waste at the time of injection), for enhanced oil or natural gas recovery and for storage of hydrocarbons which are liquids at standard temperature and pressure. In 1986, there were approximately 879,000 active oil and gas wells which generated about 60 million bbl of oilfield fluids that were subsequently injected through 166,000 injection wells in the continental U.S. (Table 11-2). The average daily production was on the order of 7.6 million bbl of oil, 40 billion ft^3 of natural gas, and 61 million bbl of produced water. Wastewater that may be disposed of in a Class II injection well also includes produced water, scrubber blowdown waters, and water softener regeneration brine water from steam generators and cogeneration facilities, and drilling waters. These disposal volumes are likely to increase significantly in the future as producing fields eventually become depleted.

Table 11-2 1986 Production and Injection Statistics by State for Continental U.S.[a,b]

State	Number Active Wells				Average Production[c]			Average Injection[c]	
	Oil	Gas	Water Injection	Swd[d]	Oil	Gas[e]	Water Injection	Water Injection	Swd
Alabama	861	1,029	147	94	57,947	401	242,934	39,934	203,000
Alaska	1,191	104	327	9	1,866,602	3,788	473,577	1,666,373	53,295
Arizona	23	0	0	7	439	1	862	0	645
Arkansas	9,700	2,550	239	979	43,340	429	859,964	171,798	688,166
California	54,629	1,546	14,825	641	1,037,089	1,141	7,244,747	4,816,438	1,507,342
Colorado	5,598	4,580	825	158	81,262	480	971,398	937,437	33,961
Florida	113	0	56	18	25,707	27	188,504	149,057	118,736
Illinois	31,100	190	12,631	1,917	74,644	4	3,135,048	2,978,296	156,752
Indiana	7,164	1,220	2,919	386	13,038	1	547,596	520,216	27,380
Kansas	49,906	12,057	9,366	5,536	183,655	1,274	2,357,615	943,046	1,414,569
Kentucky	21,844	9,515	5,311	106	17,740	200	212,880	134,114	78,766
Louisiana	26,418	14,436	1,275	3,149	4,984,000	5,066	3,984,000	400,000	2,160,000
Michigan	5,125	755	1,028	627	70,378	361	197,058	128,088	68,970
Mississippi	3,732	717	304	677	82,182	570	910,054	273,016	637,038
Missouri	300	0	450	10	510	0	2,041	1,904	137
Montana	4,680	2,023	1,196	256	74,424	132	445,613	434,000	40,000
Nebraska	1,814	16	523	98	19,446	4	205,765	80,000	123,000
Nevada	27	0	0	8	7,961	0	1,000	0	1,000
New Mexico	17,557	16,762	3,855	307	206,864	1,898	936,858	777,592	159,266

New York	4,400	5,038	3,248	6	2,337	94	24,134	22,841	1,293
North Dakota	3,838	98	155	252	125,075	199	271,506	77,362	182,636
Ohio	29,659	32,721	127	3,829	36,828	499	40,000	4,579	28,574
Oklahoma	103,000	24,500	14,895	7,897	408,506	5,282	7,753,000	7,054,000	699,000
Pennsylvania	24,000	24,000	4,315	1,868	10,364	412	155,460	93,276	62,184
South Dakota	146	42	30	11	4,346	7	8,045	1,600	800
Tennessee	873	921	8	3	1,766	9	3,000	2,000	1,000
Texas	200,055	46,080	36,117	15,532	2,229,096	15,516	25,198,863	19,225,828	5,973,035
Utah	3,773	160	456	46	107,319	653	314,395	225,311	62,507
Virginia	28	581	0	0	50	41	50	0	0
West Virginia	15,895	32,500	687	74	8,616	397	25,848	18,094	7,754
Wyoming	15,122	2,161	5,257	679	334,900	1,636	4,520,016	1,306,110	2,578,730
Total U.S.	642,571	236,302	120,572	45,180	7,630,430	40,521	61,177,831	42,482,310	17,069,537

a After Michie (1988).
b Totals do not include federal offshore production and injection statistics.
c Barrels per day unless otherwise noted.
d Swd = saltwater disposal.
e Million cubic feet per day.

Class III Wells

Class III wells are restricted to particular types of operations such as solution mining for salt and sulfur, and mining of minerals. The number of Class III wells is uncertain, although considered not to be very large and concentrated in certain states. Class III wells include those wells which inject for extraction of minerals. Included are fluids injected from mining of sulfur by the Frasch process and in situ production of uranium or other metals although restricted only to in situ production from ore bodies which have not been conventionally mined. Solution mining of conventional mines such as stopes leaching is included under Class V wells. Wells used for fluids injected for solution mining of salts or potash are also categorized as Class III wells.

Class IV Wells

Class IV wells are those wells which inject hazardous wastes or radioactive wastes into or above a formation which, within 1/4 mi of the well, contains USDW, or for disposal of hazardous waste not classified as previously defined. These wells are now banned under RCRA.

Class V Wells

Injection wells not included in Classes I, II, III, or IV that inject nonhazardous fluids into or above formations that contain USDW are categorized as Class V wells. Injection wells which fall into this category comprise at least 25 distinct types, including:

- Air conditioning return flow wells used to return to the supply aquifer the water used for heating or cooling in a heat pump
- Cesspools, including multiple dwelling, community, or regional cesspools, or other devices that receive wastes, which have an open bottom and sometimes have perforated sides. The UIC requirements do not apply to single-family residential cesspools or to nonresidential cesspools which receive solely sanitary wastes and have the capacity to serve fewer than 20 persons a day
- Cooling water return flow wells used to inject water previously used for cooling
- Drainage wells used to drain surface fluid, primarily storm runoff, into a subsurface formation
- Dry wells used for the injection of wastes into a subsurface formation
- Recharge wells used to replenish the water in an aquifer
- Saltwater intrusion barrier wells used to inject water into a freshwater aquifer to prevent the intrusion of saltwater

- Sand backfill and other backfill wells used to inject a mixture of water and sand, mill tailings, or other solids into mined-out portions of subsurface mines whether what is injected is a radioactive waste or not
- Septic system wells used to inject the waste or effluent from a multiple dwelling, business establishment, community or regional business establishment septic tank. The UIC requirements do not apply to single-family residential septic system wells, or to nonresidential septic system wells which are used solely for the disposal of sanitary waste and have the capacity to serve fewer than 20 persons a day
- Subsidence control wells (not used for the purpose of oil or natural gas production) used to inject fluids into a nonoil- or gas-producing zone to reduce or eliminate subsidence associated with the overdraft of fresh water
- Radioactive waste-disposal wells other than Class IV
- Injection wells associated with the recovery of geothermal energy for heating, aquaculture or production of electric power
- Wells used for solution mining of conventional mines such as stopes leaching
- Wells used to inject spent brine into the same formation from which it was withdrawn after extraction of halogens or their salts
- Injection wells used in experimental technologies.

These wells have not been extensively regulated in the past, so that little inventory data exists for this class as a whole. In addition, unlike other classes of injection wells, many owners and operators are unaware of the UIC regulation and furthermore do not consider their facility (i.e., cesspools, heat pump system wells, etc.) to be an injection well.

Some of the states and EPA regions have conducted initial inventories. While some of the inventories are fairly accurate, other inventories are limited, for a variety of reasons. For example, the Maricopa County (Arizona) Health Department estimates that 30,000 septic system wells and 75,000 to 100,000 drainage wells exist in the county. Although the Health Department holds records on septic systems, it does not have funding available to excerpt the necessary inventory information from its files. The drainage wells are regulated by a variety of local agencies and the County Flood Control District, so the inventory data is scattered among agency files.

About 173,000 Class V shallow injection wells have been identified and divided into 32 subcategories. This, however, is not a conservative value and the actual number of Class V wells may be as high as a million. The impact of Class V wells and the potential for this source of potential groundwater contamination may rival that of underground storage tanks (UST), of which an estimated 2.5 million exist — an issue which must be addressed as part of the due diligence investigations conducted, for example, during property transfers.

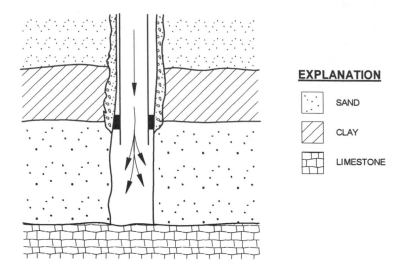

EXPLANATION

SAND

CLAY

LIMESTONE

Figure 11-3 Cross section illustrating idealized hydrogeologic conditions and terms.

Because of their potential widespread impact, new and more elaborate regulations are being proposed by EPA.

SITING CRITERIA

The siting of an underground injection well initiates at the regional level, is followed by localized studies, and then focuses upon the immediate well location. Siting must include the fulfillment of certain general geologic and hydrogeologic requirements. These general requirements are illustrated in Figure 11-3 and include:

- The presence of a water-bearing zone of nonbeneficial use (i.e., saline, economically of little or no value, etc.) that is sufficiently thick and laterally extensive, and of sufficient porosity and permeability to accept injected fluids at the required rate
- The presence of enveloping confining beds both above and below the potential zone of injection that is sufficiently thick of laterally extensive and of sufficient impermeability to confine the injected fluids to the zone of injection
- The absence of zones of preferred migration pathways such as faults, fractures, and joint systems, solution collapse features, zones of relatively higher permeability (i.e., depositional features), and unplugged or poorly abandoned wells which could result in the migration of injection fluids into adjacent aquifers.

The fundamental regional hydrogeologic consideration in underground injection involves the designation of USDWs, including both currently used and potential drinking water sources. The identification of such aquifers or portions of aquifers is a key element to the UIC program. In 1984, the EPA released its Ground-Water Protection Strategy (GWPS) with the aim of achieving consistency in overall groundwater efforts and developed a groundwater classification system to provide guidelines for EPA programs.

Designation of Underground Sources of Drinking Water

Classes are based on the use and value of the resource, with drinking water being the highest beneficial use. For the most part, the concept of the USDW fits well into the classification system. However, there are apparent inconsistencies with respect to exemption criteria that may need to be clarified. Most notably, exemptions for mining and production of mineral, hydrocarbon, or geothermal energy may not fit well in any class.

A USDW correlates with Class I and Class II of the EPA Ground Water Classification System. Class I, special groundwaters, have the highest value and, therefore, should receive extraordinary levels of protection. The GWPS indicates that special permit conditions (i.e., special grouting requirement) will be considered for wells penetrating Class I waters. Class II waters encompass those groundwaters currently or potentially serving as a source of drinking water and having TDS concentration less than 10,000 mg/L. Class II ground water thus embraces the USDW concept to the fullest extent. Furthermore, a potential source of drinking water, Class IIB, would be less valued compared to a current source, Class IIA. The GWPS indicates that where the potential is low for ground water to be used as drinking water (for example, when TDS levels are between 3000 and 10,000 mg/L, mineral production is competing use, or the aquifer is inaccessibly deep), EPA will apply existing UIC requirements. Class III groundwaters are those with greater than 10,000 mg/L TDS (and therefore not a USDW) or those otherwise likely to meet the exemption criteria. The Class III criteria for low TDS waters is nearly identical to the exemption criterion of economic or technological impracticability to treat for human use.

Aquifer Exemption

Injection wells typically may not inject into an aquifer classified as an USDW unless the aquifer has been exempted. The rationale adopted by EPA to support the concept of "aquifer exemption" is that a competing use of the aquifer (it is being mined for minerals, for example) may be more important than, or otherwise take precedence over, the use of the aquifer as a source of water supply for human consumption. Regulations provide that a USDW may be exempted from coverage if (1) the aquifer does not currently serve as a source of drinking water and (2) the aquifer cannot now and will not in the future serve as a source of drinking water based on certain criterion. An aquifer can be exempted if it does not currently serve as a source of drinking water, or if the TDS content for ground water is greater than 3000 and less than 10,000 mg/L, then the aquifer may be exempted since it is not reasonably expected to supply a public water system.

In addition, an aquifer may be exempted if it cannot now and will not in the future serve as a source of drinking water because (1) it is mineral, hydrocarbon, or geothermal energy producing, or contains minerals or hydrocarbon that, because of the quality and location, are expected to be commercially producible; (2) it is so contaminated that it would be economically or technologically impractical to render the water fit for human consumption; or (3) it is located over a Class III well mining area subject to subsidence or catastrophic collapse.

Aquifer exemptions are intended to apply to cases where an aquifer or portion of an aquifer is so inaccessible or disrupted by mining, energy production, or contamination that it would make little sense to develop the aquifer as a source of drinking water. General guidance has been provided to address exemption demonstrations. The advent of the GWPS and classification system may result in the need to update this guidance. For example, the Class III test for waters with less than 10,000 mg/L TDS could rely on the concept of an "economic burden", a somewhat different test than the cost/benefit test in the exemption guidance.

A wealth of information exists on U.S. ground waters. In 1923 the late O.E. Meinzer, considered the father of groundwater hydrology in the U.S., divided the country into 21 groundwater provinces. Later, in 1952, H.E. Thomas consolidated and rearranged Meinzer's 21 provinces into 10 groundwater regions. The most up-to-date delineations of groundwater regions was assembled by Heath (1984). The distribution of the 15 regions across the U.S., Puerto Rico, and the Virgin Islands is shown in Figure 11-4. The physical and hydrologic characteristics of the groundwater regions are presented in Figure 11-5.

The approach usually taken to arrive at an aquifer designation is a refinement of the techniques used by Thomas and Health as applied solely at the state level. The total dissolved solids status in the regional aquifers is evaluated. Areas which are not capable of yielding groundwater or which are contaminated beyond use are identified to the extent possible and excluded. This exclusion is also applied to those aquifer areas set aside because they are used to produce minerals, oil or gas, or geothermal energy, and are exempted as stated by the UIC regulations. While current regulations do not explicitly call for a determination of groundwater class according to the GWCS, the information needs are similar and such a class determination can usually be made.

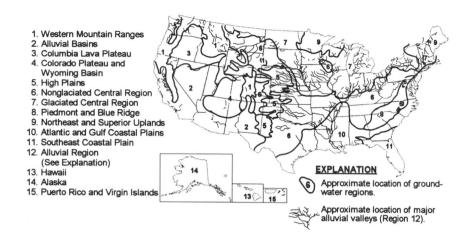

1. Western Mountain Ranges
2. Alluvial Basins
3. Columbia Lava Plateau
4. Colorado Plateau and
   Wyoming Basin
5. High Plains
6. Nonglaciated Central Region
7. Glaciated Central Region
8. Piedmont and Blue Ridge
9. Northeast and Superior Uplands
10. Atlantic and Gulf Coastal Plains
11. Southeast Coastal Plain
12. Alluvial Region
    (See Explanation)
13. Hawaii
14. Alaska
15. Puerto Rico and Virgin Islands

EXPLANATION

Approximate location of groundwater regions.

Approximate location of major
alluvial valleys (Region 12).

Figure 11-4 Groundwater regions of the U.S. (modified after Heath, 1984).

HYDROGEOLOGIC CONSIDERATIONS

Understanding of subsurface geologic and hydrogeologic conditions and characteristics is essential to the assessment of site suitability, and to the successful design, construction, and operation of underground injection wells. Geologic and hydrogeologic factors to be considered are summarized in Table 11-3.

Structural contour and isopach maps of each injection and confining zone in the project area are typically developed. At least one geologic cross section through at least one injection well in the project area is also required. The cross section should identify all formations, aquifer salinity, freshwater aquifers, and hydrocarbon-producing zones, if applicable. Representative geophysical logs to a depth below the deepest producing zone identifying all formations, fresh-water aquifers, hydrocarbon zones, injection zone(s), and confining zone(s) are also developed.

Site-specific geologic and hydrogeologic considerations include characterization of the injection zone(s), confining zone(s), and the area of review. These considerations are further discussed below.

Injection Zone Characteristics

The injection zone is the stratum that receives the injected fluid. Geologic and hydrogeologic (reservoir) characteristics of each injection zone must be assessed with regard to:

Characteristics of the principal physical and hydrologic characteristics of the groundwater regions of the U.S.

Components of the system

- Unconfined aquifer
  - Hydrogeologically insignificant
  - Minor aquifer or not very productive
  - Dominant aquifer
- Confining beds
  - Hydrogeologically insignificant
  - Thin, discontinuous, or very leaky
  - Interlayered with aquifers
- Confined aquifers
  - Hydrogeologically insignificant
  - Not highly productive
  - Multiple productive aquifers
  - The dominant productive aquifer
- Presence and arrangement
  - Single unconfined aquifer
  - Two interconnected aquifers
  - Unconfined aquifer, confining bed, confined aquifer
  - Complex interbedded sequence

Characteristics of the dominant aquifers

- Water-bearing openings
  - Primary
    - Pores in unconsolidated deposits
    - Pores in semi-consolidated rocks
    - Tubes and cooling cracks in lava
  - Secondary
    - Fractures and faults
    - Solution-enlarged openings
- Composition
  - Degree of solubility
    - Insoluble
    - Mixed soluble and insoluble
    - Soluble
- Storage and transmission properties
  - Porosity
    - Large (>0.2)
    - Moderate (0.01-0.2)
    - Small (<0.01)
  - Transmissivity
    - Large (>2,500 m^2-day)
    - Moderate (250-2,500 m^2-day)
    - Small (25-250 m^2-day)
    - Very small (<25 m^2-day)
- Recharge and discharge conditions
  - Recharge
    - Uplands between streams
    - Losing streams
    - Leakage through confining beds
  - Discharge
    - Springs and surface seepage
    - Evaporation and basin sinks
    - Into other aquifers

Regions:

Region No.	Name
1	Western Mountain Ranges
2	Alluvial Basins
3	Columbia Lava Plateau
4	Colorado Plateau and Wyoming Basin
5	High Plains
6	Nonglaciated Central Region
7	Glaciated Central Region
8	Piedmont and Blue Ridge
9	Northeast and Superior Uplands
10	Atlantic and Gulf Coastal Plain
11	Southeast Coastal Plain
12	Alluvial Valleys
13	Hawaii
14	Alaska
15	Puerto Rico and Virgin Islands

Figure 11-5 Summary of the principal physical and hydrologic characteristics of the groundwater regions of the U.S. (after Heath, 1984)

Table 11-3 Summary of Geologic and Hydrogeologic Factors for Siting an Underground Injection Well[a]

Geology	Hydrogeologic
Regional factors	
Geologic history	Beneficial use groundwater
Lithology and characteristics	
Mineral resources	Occurrence (depth, thickness)
Physiography	Quality
Seismicity	Usage (current and future)
Stratigraphy	
Structure	Groundwater occurrence
	Hydraulic intercommunication
	Hydrodynamics
	Recharge/discharge area
Local factors	
Geochemical characteristics	Fluids
Geology	Fracture pressure
Lithology and characteristics	Geochemical characteristics of reservoir
Mineral resources (oil, gas,	Groundwater occurrence
coal, brines, etc.)	Hydrodynamics
Structure	Reservoir pressure
Stratigraphy	Reservoir temperature
Confining beds	
Injection zones	
Lateral distribution	
Lithology	
Permeability	
Porosity	
Thickness	

[a] Modified after Warner and Lehr (1977).

- Hydraulic conductivity and permeability
- Total porosity and effective porosity
- Stratigraphy including thickness, vertical and lateral extent, and continuity
- Formation fracture gradient
- Reservoir pressure and temperature (original and current)
- Residual oil, gas, and water saturations.

On a site-specific basis, this information is used to demonstrate that the injection zone is capable of receiving the proposed volume of injected fluid over the life of the project without excessive compression of the fluids, which may fracture, or otherwise compromise, the injection and confining zones, thus possibly contaminating overlying beneficial-use ground water. The geologic characterization demonstrates the geometry of the injection zone, thus indicating which model of wastewater plume expansion may be appropriate for consideration in the area of review calculations as discussed later.

Geologic maps need to illustrate the characteristics of the injection zone. Cogeneration wastewaters are slightly denser than most formation waters due to their TDS content. Since the injection plume may not be radial due to the tendency of the injected fluid to migrate, e.g., in a down dip direction, a structural contour map will help to determine this preferred migration direction over time. The migration effects over time, however, will also be decreased by diffusion of the TDS from the injection fluid into the formation fluid and dispersion due to groundwater flow.

A net unit isopach (or isochore) map is used to demonstrate either the lateral continuity of the injection zone, or that lateral stratigraphic boundaries exist which will help to contain the injection fluid. A net sand isopach (or isochore) map of the injection zone also aids in the determination and calculation of the injection zone hydrologic characteristics.

Confining Zone Characteristics

The confining zone, above the injection zone, essentially confines the injected wastes within the injection zone and isolates the wastes from overlying groundwater and/or hydrocarbon-producing zones. One of the primary concerns of regulatory agencies is whether or not the confining zone is aerially extensive, competent as a confining layer, and will not react with the injected wastewaters. The operator thus needs to consider the geologic and hydrogeologic characteristics of the confining zone when designing an injection project. Geologic characteristics of concern include lithology, hydraulic conductivity or permeability, porosity, thickness, stratigraphy, description of vertical and lateral extent and continuity, and confining zone and wastewater compatibility.

The presence of a fault (or fractures) through the confining zone does not necessarily invalidate the potential permitting of an injection well. If a fault is present, demonstration that the fault is sealed and will not provide a conduit or preferred migration pathway for hazardous substances and injected wastewater through the confining zone. It must also be demonstrated that the injection zone is not offset across the fault by sands which are stratigraphically above the confining zone. Appropriate engineering tests (i.e., injection pressure buildup tests, injectivity/falloff tests, pulse tests, etc.) can also be used to demonstrate that a fault is unlikely to serve as a conduit or migration pathway for hazardous substances. Areas of potential problems are illustrated in Figure 11-6.

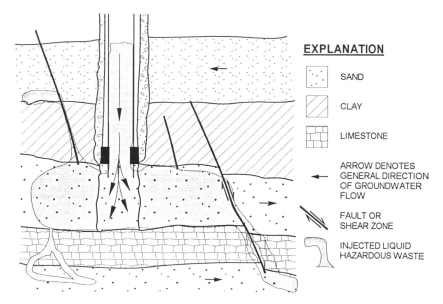

EXPLANATION

SAND

CLAY

LIMESTONE

ARROW DENOTES
GENERAL DIRECTION
OF GROUNDWATER
FLOW

FAULT OR
SHEAR ZONE

INJECTED LIQUID
HAZARDOUS WASTE

Figure 11-6 Schematic showing areas of concern during injection well installation and operation.

It is important to demonstrate that all well bores which penetrate the confining zone within the area of review have been properly plugged across the confining zone to preclude vertical waste migration through the confining zone. If an abandoned well exists within the area of review which is not plugged across the confining zone, it will need to be plugged prior to commencing injection.

Area of Review

The area of influence of the waste front and the pressure front constitutes the area of review around the injection well (Figure 11-6). During and after the life of the injection well, the injected wastewaters migrate outward away from the well bore, producing a waste front. Preceding the waste front, as the formation fluids are displaced away from the well bore, is a zone of increased fluid pressure referred to as the pressure front.

The area of review is normally considered as being within a 1/4-mi radius of the well bore, or the area of influence of the waste and pressure fronts, whichever is greater. The objective of this review is to ensure that all wells of public record (i.e., producing, injection, abandoned, dry, water, etc.) are identified in the application for a UIC permit. Geologic and hydrogeologic characteristics of the injection zone may cause the injected waste to migrate further than 1/4 mi from the well bore. If the 1/4-mi area of review is deemed

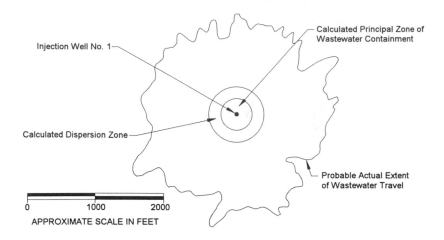

Figure 11-7 Estimated and probable actual extent of wastewater migration for a well completed in a carbonate injection zone (modified after Warner and Lehr, 1981).

insufficient, a larger area of review may be imposed. The anticipated migration of injected fluid and the displaced formation fluid in the injection zone and the effects on other injection and production wells within the area of review are addressed. Standard calculations have been devised to estimate the extent of wastewater travel and dispersion. The actual extent of wastewater migration is probably much greater than that estimated over time, as illustrated in Figure 11-7.

Estimates of the area of influence for both the waste front and the pressure front expected over the life of the project are typically conducted. Respective calculations include the volumetric method and the pressure build-up method (i.e., modified Theis equation).

The volumetric method is used to estimate the minimum distance of injection-fluid migration radially outward from the well bore. The estimate is calculated by assuming that the injection fluid will uniformly occupy an expanding cylinder with the well at the center. The equation used for the volumetric calculation is

$$r = (V / \pi b \phi)^{1/2} \tag{11-1}$$

where r = radial distance of injection fluid front from the well bore (ft), V = cumulative volume of injected fluid (bbl), b = effective injection zone thickness (ft), and ϕ = average effective porosity (decimal).

The minimum distance of injection-fluid migration is anticipated to be exceeded due to dispersion, density segregation, and channeling through high-permeability zones. The estimated influence of dispersion is thus calculated using the following equation.

$$r' = r + 2.3 [Dr]^{1/2}$$ (11-2)

where r' = radial distance of injection fluid travel with dispersion (ft), r = 100% injection fluid invasion radius from previous calculation, and D = dispersion coefficient (3 ft for sandstone aquifers).

Calculations for both the nondispersed and the dispersed radius over the project life and/or a period of 25 years are typically required.

Wastewater injected into an aquifer does not move into empty voids, but rather displaces existing fluids. The displacement process requires exertion of some pressure in excess of ambient formation pressure. The pressure increase is greatest at the injection well and decreases away from the well in an approximately logarithmic manner. The amount of additional pressure required, and the distance to which it extends, depend on the characteristics of the formation and the fluids, the amount of fluid being injected, and the length of time that injection has continued.

The rate of pressure change throughout the project life is estimated by using a modified Theis equation as follows:

$$\Delta P = \frac{162.6Q\mu}{Kb} \left(\log \frac{K_t}{\phi\mu cr^2} - 3.23 \right) [psi]$$ (11-3)

where ΔP = reservoir pressure change at radius r and time t (psi), Q = injection rate (bbl/d), μ = viscosity (centipoise), K = average reservoir permeability (millidarcys), b = reservoir thickness (ft), t = time since injection began (h), c = reservoir compressibility (psi^{-1}), r = radial distance from well bore to point of interest (ft), and ϕ = average reservoir porosity (decimal).

The zone of influence of an injection well can also be calculated using the Bernard pressure build-up equation.

$$P(r,t) = Pi + \frac{5575Q\mu}{KH} \left[\log t + \log\left(\frac{K}{\phi\mu Cr^2} \right) - 3.23 + 0.87S \right]$$ (11-4)

where P(r,t) = pressure (psi) at radius (r) at a given time (t), Pi = initial zone pressure (psi), Q = injection rate (gpm), μ = injection fluid viscosity (centipoise), K = zone permeability (millidarcys), H = net zone thickness (ft), t = time (h), ϕ = zone porosity (decimal), C = compressibility of injected fluids (psi^{-1}), S = skin factor (ratio, no dimension), and r = radius (ft).

Equation 11-4 is used to calculate the amount of time required after initial injection to influence a possible problem well (unplugged abandoned well, producing well, etc.) or cross a lease line. If the purpose is to calculate the radius (r) where there is no pressure effect, then P(r,t) should be equal to Pi in the calculation.

The calculations indicate the anticipated pressure increase over a given time period assuming a homogeneous formation, constant injection rate, radial flow (no influence from other injection wells), and no gas saturation. It is not sufficiently dependable to be relied upon to predict the exact lifetime reservoir pressure changes resulting from an injection well, but does provide the best available indication of what should occur. If there is any gas in the injection zone, it will act as a pressure sink and tend to cause the calculated time to be shorter than it actually is.

If more than one injection well is planned, the area of influence of all wells and their overlapping zones of influence is estimated. Two important characteristics of the pressure change equation above are that hydrologic boundaries (such as faults or facies changes) can be simulated by a properly located imaginary well and that individual solutions can be superimposed. The superimposing of solutions allows the effects of multiple wells to be easily analyzed by estimating the combined pressure effects of multiple wells. The separate effects of two or more wells at any point of interest is estimated by adding their individual pressures to obtain their combined effect. Since the effect of a boundary is analogous to a properly located pumping or injection well, the existence of boundaries can be detected by observing reservoir response to injection or pumping. Conversely, the effects of known or suspected boundaries can be estimated.

Part of the area of review process requires that the records of existing wells penetrating the injection zone be examined to ensure that wells are properly constructed or abandoned; however, records for many of these wells are unavailable. The number of abandoned wells between 1859 and 1974 is estimated at 1,647,661 to 1,930,000 wells. Of these, records are available on only about 1,200,000 abandoned wells. It is also estimated that most wells abandoned before 1930 were probably improperly abandoned in comparison with current standards. Since thousands of these wells penetrate both fresh and saline water-bearing formations, the leakage of contaminated or highly mineralized water via poorly abandoned wells or unplugged exploration boreholes have led to numerous instances of degradation of beneficial-use ground water and soil, and the subsequent failure of many deep-well injection projects (Figures 11-8 and 11-9). Various methods for locating abandoned wells are summarized in Table 11-4.

DESIGN CRITERIA

Injection wells must be designed in a manner to protect all formations containing useable waters penetrated by the well. Several methods for well completion have

Figure 11-8 Brine-induced corrosion of production casing at the Martha Oil Field, June 1987, showing "moth-eaten" appearance (photo courtesy of Eger and Vargo, 1989).

been devised based upon formation and waste type, among other factors (Figure 11-10). An injection well typically consists of concentric pipes. The outermost piping, referred to as surface casing, usually extends below the deepest usable water aquifer, and is cemented from its terminus to the ground surface. Two strings of piping extend to the injection zone: the outer long string and inner production (injection) tubing. The long string is cemented back to the surface casing. The production tubing is used to conduct the fluids to be injected under pressure or gravity flow. Injected fluids may also be injected through perforations at the bottom of the long string. A packer is usually set at the bottom of the well between the production tubing and long string casing (annular space) to prevent waste from backing up into the annulus. The annulus is then filled with an inert fluid, and maintained under pressure slightly higher than the waste injection pressure to prevent leaks into the annular space. The wellhead is capped and equipped with an automatic shut-off valve, and gauges to monitor injection pressure, injection rate, and annulus pressures. A typical hazardous waste injection well is shown in Figure 11-11.

REPORTING AND MONITORING

Reporting requirements for underground injection usually include preparation of a permit application, and one or several technical reports which address

Figure 11-9 Petroleum and brine outflows induced by water flooding activities emanating from an abandoned oil well (photo courtesy of Eger and Vargo, 1989).

Table 11-4 Summary of Methods for Locating Abandoned Wells

Method	Limitation
Records search	Records poor or missing for many pre-1930 wells
Personal interviews	Difficult to identify and locate
Historical aerial photographs review	Not widely employed until the late 1920s
Geophysical methods	
Magnetic, electrical resistivity, and electromagnetic surveys	Limited to cased wells with metallic casings; whether a well is properly plugged or subject to leakage cannot be ascertained
Ground penetrating radar	Can be used for both cased and uncased wells; whether a well is properly plugged or subject to leakage cannot be ascertained
Hydrologic methods	
Water level monitoring[a]	Feasible only if several water wells are already present in vicinity of unknown abandoned well
Injection of fluid into injection zone[b]	If conduit is not open to surface, or induced pressure is insufficient, no observable leakage to the surface may be observed
Analytical solution[c]	Not applicable where the confining layer leaks or the abandoned well is filled with very low permeability materials (i.e., 0.1 darcy or less)

[a] Water levels are monitored in wells penetrating the freshwater aquifer overlying the injection formations in the vicinity of the abandoned well. Any major leakage from the abandoned well should produce a water level anomaly in the freshwater aquifer.

[b] The presence of a leaky abandoned well is indicated if pressure resulting from fluid injection causes fluid follow up through the abandoned well to the ground surface.

[c] Conduct a pumping test in the injection and/or observation wells in conjunction with set of select type curves should reveal distance of a leaky abandoned well to an injection well.

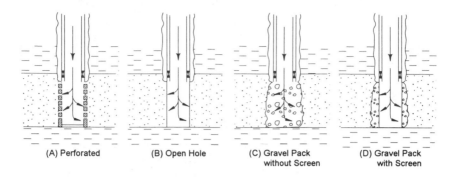

Figure 11-10 Typical injection well completion methods (modified after Donaldson, 1974).

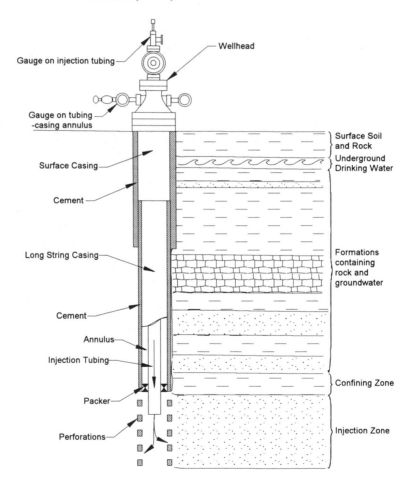

Figure 11-11 Typical injection well construction detail.

the engineering and mechanical aspects of the project. Application for injection wells requires documentation that a usable USDW will not be degraded by a wastewater injection well. It must be shown that the injection interval will contain the injected fluids and that the injection and formation fluids are compatible. The confining layer must be shown to contain the injected fluids and not react with the injection fluids. Furthermore, the injection well must also initially have mechanical integrity.

The application also requires that an engineering study be prepared to describe the primary purpose of the project, injection, and confining zone characteristics, and the drilling and abandonment program of the injection well project. An injection plan is designed to describe the mechanical aspects of the injection project. Included in this report, at minimum, is an injection facility layout plan and discussion of the maximum anticipated surface injection pressure and daily injection rate. Description of the proposed monitoring system is presented to ensure that no damage to the environment or facilities occurs and that the injection fluid is confined to the injection zone(s). Method of injection (i.e., down tubing set in packer, etc.) and list of proposed cathodic protection measures, if warranted, for the plant, pipelines, and wells is provided. A description of wastewater treatment prior to injection is also addressed.

Injection wells are monitored for mechanical integrity and injected wastewater quality. The mechanical integrity tests are conducted annually and include:

- Radioactive tracer survey
- Spinner survey
- Temperature survey
- Annular space pressure test.

If the mechanical integrity tests indicate a problem (e.g., leak in the casing, tubing, packer, etc.) then the injection well will need to be shut-in until the problem is corrected. These tests are normally required annually.

The wastewater is generally monitored by analyzing a grab sample, taken at the well head, for various characteristics and chemical constituents. Normally, samples are collected and results reported at initiation of injection to establish the wastewater quality. An operator may be required by the lead agency to collect and analyze samples at any time. A new waste stream analysis every 3 years or when a change is typically required made in the waste stream composition. The injection well operator will also monitor and report the daily injection pressure, injection fluid volume, injection rate, and annular space pressure. These are a daily check on the mechanical integrity of the injection well and compliance on permit limits of injection pressure and volume.

GEOCHEMICAL FATE OF INJECTED HAZARDOUS WASTE

Commercial underground injection disposal facilities inject about 300 different varieties of waste. Common waste types include oil-field brine, which is generated during the production of oil and gas; inorganic and organic industrial liquid wastes, and radioactive wastes. Brines are considered waste products due to their high concentration of dissolved solids and salts. Most brines are comprised primarily of sodium and calcium chlorides, magnesium, bicarbonate, and sulfates. Dissolved solids range from less than 30,000 to more than 100,000 ppm. Problems encountered with the injection of brine include precipitation of salts and hydroxides, corrosion, and adequate filtration of suspended solids. Inorganic wastes are categorized as either neutral, basic, or acidic. While neutral wastes are similar to oil-field brine in chemical behavior, filtration prior to injection is usually all that is required. Basic wastes are usually incompatible with brine, which results in the precipitation of hydroxides and salts upon mixing, causing swelling of certain clay minerals and polymerization of resin-like materials, all of which reduces permeability. Acidic inorganic wastes are highly corrosive to the well components and include aluminum hydroxide, ammonia liquor, boiler water, brine, cooling tower water, chlorides, process liquors, pickle liquor, and sulfates, among numerous others. Organic waste is divided into seven categories: aldehydes, alcohols, glycols, phenols, acids, nitrites, and disulfides. These types of waste are derived from primary, secondary, and tertiary sewage treatment facilities. Radioactive wastes are divided into low, intermediate and high-level wastes. Both low- and intermediate-level wastes have been disposed of via injection. Low-level radioactive wastes, derived from nuclear reactor process water, contain a few curies per cubic meter. Intermediate-level radioactive waste contains thousands of curies per cubic meter and is derived from various stages of the nuclear fuel cycle. High-level radioactive waste contains millions of curies per cubic meter, may contain uranium and plutonium, and is derived from the fuel-processing process which contains most of the fission products from spent fuel. Problems associated with the injection of radioactive waste are confinement, dissipating the heat that is generated, system corrosion, and radiation protection.

Geochemical concerns regarding the injection of hazardous and toxic waste evolves around the mobility of the waste once injected, and the interactions that take place between the waste and the formation material or water within the injection zone. Mobility of the waste is governed by its diffusion rate, degree of sorption, or interaction with the formation material itself, and the distance the injected waste travels in the formation water. Diffusion, as discussed in Chapter 8, is perceived as the rate each constituent is carried by the formation water. During injection, diffusion is considered negligible since the groundwater velocity during injection (i.e., radial groundwater flow) is greater than that during noninjection periods (i.e., a few to 10 ft per day vs fractions of an inch to a few inches per day, respectively).

Sorption results in the retardation of waste constituents relative to the advancing water front. Sorption can occur via several means. Heavy metals can be significantly retarded (i.e., up to 10 times slower than the advancing water front), primarily by ion exchange with clay minerals, whereas organic waste constituents are retarded by sorption to the formation material. The latter case is not desirable since the end result is a decrease in permeability; thus, the need for higher injection pressures to achieve the desired injection rate.

Geochemical interactions may be beneficial, neutral, or of detrimental consequences. Potential problems associated with waste and formation interactions include changes in pH and ionic strength, bond destruction of clay particles, chemical solution of various clays, dissolution, precipitation, and adsorption or exchange. Such reactions could result in pore blockage, changes in waste character, and reduced rates of movement of certain waste components, or formation of aluminum and ferric oxide gels which can reduce the permeability of the injection zone formation.

Very small changes in pH can generate profound changes in the subsurface environment. Certain chemical processes are strongly influenced by pH changes. Partition processes such as adsorption and precipitation-dissolution reactions are strongly influenced, affecting heavy metal and organic mobility. Transformation processes such as complexation, hydrolysis, and oxidation-reduction are also strongly influenced by changes in pH (i.e., position of equilibria involving complexions, rate of aliphatic and alkylic halides being optimum at neutral to basic conditions, faster hydrolysis reaction rates at either high or low pH, and more reduced conditions with increasing pH).

Reactions such as dissolution can result in beneficial effects by increasing the effective porosity, and thus the injection potential of the formation, while concurrently reducing the acidity of the waste stream. Dissolution reactions can also result in negative effects. For example, when acidic wastes are injected into dolomite and limestone formations, dissolution is the prevalent reaction that takes place. Dissolution of calcium carbonate or calcium-magnesium carbonate forms water, carbon dioxide, calcium or magnesium chlorides, and sulfates or fluorides. Ultimately, the acidic solution is neutralized. However, if large quantities of carbon dioxide gas are produced such that the solubility limit and/or critical temperature and pressure of the system is exceeded, a blowout can occur where geyser-like ejection of gas and liquid is vented at the surface. Dissolution can also result in enlarged flow paths or caverns (i.e., secondary porosity type features) with the potential to adversely affect the competence of overlying units by structurally weakening the confining layer. Other chemical incompatibilities between the injected waste and interstitial water can occur when chemical reactions between the two liquids result in gas evolution, precipitation, or some type of deposition. Since wastewater injection has the potential for adverse chemical reactions with the formation water or confining

Table 11-5 Injection Fluid Characteristics

Property	Characteristics
Volume	NA[a]
Physical	Density, gas content, suspended solids, temperature, viscosity
Chemical	Chemical stability, dissolved constituents, pH, reactivity, toxicity
Biological	NA

[a] NA = not applicable.

zone, resulting in the possible degradation of usable or potentially usable ground water, the operator must demonstrate in a technical report that the injection fluids are compatible with the formation fluids and confining zone, and that a usable, or potentially usable, ground water will not be degraded. This evaluation includes characterization of the formation water from the injection zone, the proposed injection wastewater, and, if the injection zone qualifies as a USDW, a demonstration that the injection zone qualifies as an exempt aquifer or that it will not be degraded by the injectate.

A representative sample of the formation water from the injection zone is analyzed for water quality (a standard geochemical water analysis). The wastewater is also characterized using the same chemical analyses that were conducted on the formation water. This in turn provides an indication of the amount of degradation, if any, to be expected of the formation water by the injected fluid, and by which constituents. Evaluation of the suitability of certain wastes for injection include estimated volume of fluids to be injected, and assessment of certain physical and chemical characteristics. The management of an underground injection program is constrained by the volume of fluids that can be safely injected over a given period of time. The injection rate is limited by the physical characteristics of the zone of injection and the pressure capacity of the injection system (i.e., pumps, tubing, and casing). In addition, the injection well is conservatively designed to avoid hydraulic fracturing and potential adverse effects to the confining beds.

Compatibility of the injection fluid with the receiving formation and confining beds, natural formation water and injection-well system requires evaluation by an accepted testing method. Injected fluid characteristics to be considered are summarized in Table 11-5. Compatibility tests are used to demonstrate that the wastewater will not react with the minerals in the confining layer (i.e., dissolution of carbonate minerals), thus compromising its effectiveness, and that excess swelling of interstitial clays in the injection zone, for example, will not cause injection pressures to exceed the permitted limit. Compatibility tests on formation waters assures that precipitates (i.e., alkaline earth such as Ca, Ba, and Mg, or heavy metals such as Al, Cd, Fe, or Zn) or polymerization of injected chemicals which can cause plugging of the injection interval are avoided. Depending upon grain size of the injection interval, TDS may also be of concern. Compatibility tests are also performed to assure that the injected fluids will be compatible and not adversely affect the mechanical components of the injection well system. The most undesirable reaction is corrosion.

BIBLIOGRAPHY

Aller, L., Methods for Determining the Location of Abandoned Wells, 1984, EPA-600/2-83-123: Robert S. Kerr Environmental Research Laboratory, U.S. Environmental Protection Agency, Ada, OK, 130 p.

Anzzolin, A.R. and Graham, L.L. , 1984, Abandoned Wells - A Regulatory Perspective: In *Proceedings of the First National Conference on Abandoned Wells: Problems and Solutions* (Edited by Fairchild, D.M.), Environmental and Ground Water Institute, University of Oklahoma, Norman, OK, pp. 17-36.

Barlow, A.C., 1972, Basic Disposal Well Design: In *Underground Waste Management and Environmental Implications* (Edited by Cook, T.D.), American Association of Petroleum Geologists, Memoir No. 18, pp. 72-76.

Bear, J., 1972, *Dynamics of Fluids in Porous Media:* Elsevier Publishing Co., New York, 764 p.

Belitz, K. and Bredehoeft, J., 1986, Comment and Reply on "Safe Disposal of Toxic Radioactive Liquid Wastes": *Geology*, Vol. 14, pp. 266-267.

Bradley, J.S., 1985, Safe Disposal of Toxic and Radioactive Liquid Wastes: *Geology*, Vol. 13, pp. 328-329.

Brower, R.D. and Visocky, A.P., et al., 1989, Evaluation of Underground Injection of Industrial Waste in Illinois: Illinois State Geological Survey, Hazardous Waste Research Information Center.

Chen, C.S., 1989, Solutions Approximating Solute Transport in a Leaky Aquifer Receiving Wastewater Injection: *Water Resources Research,* Vol. 25, No. 1, pp. 61-72.

Donaldson, E.C., 1964, Subsurface Disposal of Industrial Waste in the U.S.: U.S. Bureau of Mines Information Circular, No. 8212, 34 pp.

Donaldson, E.C., 1974, Subsurface Waste Injection in the U.S.: U.S. Bureau of Mines Information Circular, No. 8636, 72 p.

Eger, C.K. and Vargo, J.S., 1989, Prevention-Ground Water Contamination at the Martha Oil Field, Lawrence and Johnson Counties, Kentucky: In *Environmental Concerns in the Petroleum Industry* (Edited by Testa, S.M.), Pacific Section American Association of Petroleum Geologists, Palm Springs, CA, pp. 83-105.

Fryberger, J.S. and Tinlin, R.M., 1984, Pollution Potential from Injection Wells via Abandoned Wells: In *Proceedings of the First National Conference on Abandoned Wells: Problems and Solutions* (Edited by Fairchild, D.M.), Environmental and Groundwater Institute, University of Oklahoma, Norman, OK, pp. 84-117.

Harlow, I.F., 1939, Waste Problem of a Chemical Company: *Industrial Engineering and Chemistry,* Vol. 31, pp. 345-349.

Heath, R.C., 1982, Classification of Ground-Water Systems in the U.S.: *Ground Water,* Vol. 20, No. 4, pp. 393-401.

Heath, R.C., 1984, Ground-Water Regions of the U.S.: U.S. Geological Survey Water-Supply Paper 2242, 78 p.

Javandel, I., Tsang, C.F., Witherspoon, P.A., Morganwalp, D., 1988, Hydrologic Detection of Abandoned Wells for Hazardous Waste Disposal: *Water Resources Research,* Vol. 24, No. 2, pp. 261-270.

Johnson, D.S., 1989, Disposal of Cogeneration Wastewaters from Enhanced Recovery Operations: In *Environmental Concerns in the Petroleum Industry* (Edited by Testa, S.M.), Pacific Section American Association of Petroleum Geologists, Palm Springs, CA, pp. 183-197.

Matthews, C.S. and Russell, D.G., 1967, Pressure Buildup and Flow Tests in Wells: American Institute of Mining, Metallurgy and Engineering Monograph, Vol. 1.

Meinzer, O.E., 1923, The Occurrence of Ground-Water in the U.S., With a Discussion of Principles: U.S. Geological Survey Water-Supply Paper 489, 321 p.

Michie, T., 1988, Oil and Gas Industry Water Injection Well Corrosion Study: In Proceedings of the Underground Injection Protection Conference, Portland, OR, pp. 47-67.

Mogharabi, I. 1989, Multiple Criteria Optimization of Hazardous Waste Disposal by Deep-Well Injection: Ph.D. dissertation, University of Oklahoma, Norman, OK, 225 p.

Office of Technology Assessment, 1983, Technologies and Management Strategies for Hazardous Waste Control: U.S. Government Printing Office, Washington, D.C., 407 p.

Reeder, L.R., et al., 1977, Review and Assessment of Deep-Well Injection of Hazardous Waste, Volume IV, Appendices E, F, G, H, I & J: U.S. Environmental Protection Agency, EPA-60012-77-029d, 1446 p.

Roy, W.R., Mravik, S.C., Krapac, I.G., Dickerson, P.R. and Griffin, R.A., 1988, Geochemical Interactions of Hazardous Wastes with Geological Formations in Deep-Well Systems: Illinois State Geological Survey, Water Survey Division, Hazardous Waste Research and Information Center, HWRIC RR-032, 61 p.

Roy, W.R., Seyler, B., Steele, J.D., Moore, D.M. and Krapac, I.G., 1991, Geochemical Interactions of Two Deep-Well Injected Wastes with Geological Formations, Long-Term Laboratory Studies: Illinois State Geological Survey, Hazardous Waste Research and Information Center, Environmental Geology 137, HWRIC RR 054, 17 p.

Syed, T., 1989, An Overview of the Underground Injection Control Regulations for Class II (Oil and Gas Associated) Injection Wells - Past, Present and Future: In *Environmental Concerns in the Petroleum Industry* (Edited by Testa, S.M.), Pacific Section American Association of Petroleum Geologists, Palm Springs, CA, pp. 199-207.

Thomas, H.E., 1952, Ground-Water Regions of the U.S. - Their Storage Facilities: In *The Physical and Economic Foundation of National Resources, Vol. 3,* U.S. 83d Congress, House Committee on Interior and Insular Affairs, pp. 3-78.

Thornhill, J.T., Short, T.E. and L. Silka, 1982, Application of the Area of Review Concept: *Ground Water,* Vol. 20 No. 1, pp. 32-38.

U.S. Environmental Protection Agency, 1985, Report to Congress on Injection of Hazardous Waste, EPA 570/9-85-003, Office of Drinking Water, Washington, D.C.

U.S. Environmental Protection Agency, 1987, Injection Well Mechanical Integrity: EPA/625/9-87/007, pp. 52-64.

U.S. Environmental Protection Agency, 1990, Assessing the Geochemical Fate of Deep-Well-Injected Hazardous Waste: A Reference Guide: EPA/625/6-89/025a, 183 p.

U.S. Environmental Protection Agency, 1990, Assessing the Geochemical Fate of Deep-Well-Injected Hazardous Waste: Summaries of Recent Research: EPA/625/6-89/025b, July, 1990, 97 p.

Warner, D.L. and Orcutt, D.H., 1973, Industrial Wastewater-Injection Wells in the U.S. - Status of Use and Regulations: In *Underground Waste Management and Artificial Recharge* (Edited by Braunstein, J.), Vol. 2, George Banton, Menasha, WI, pp. 687-694.

Warner, D.L. and Lehr, J.H., 1981, *Subsurface Wastewater Injection, The Technology of Injecting Wastewater into Deep Wells for Disposal:* Premier Press, Berkeley, CA, 334 p.

Wolff, R.G., Bredehoeft, J.D., Keys, W.S. and Shater, E., 1975, Stress Determination of Hydraulic Fracturing in Subsurface Waste Injection: *Journal of the American Water Works Association,* pp. 519-523.

12 UNDERGROUND GEOLOGIC REPOSITORIES

"Now, the curtains are drawn, and the fire's bright, and here's your armchair — and you're to tell us all about what you promised."

(Ruskin, 1866)

INTRODUCTION

Underground geologic repositories are currently considered one of the better alternatives to the subsurface disposal of hazardous and toxic waste. Disposal of hazardous waste and toxic materials in underground geologic repositories is relatively new, although disposal of radioactive waste in such repositories has received attention in the U.S. since the early 1950s. Much of this attention has been on the engineering and geotechnical aspects pertaining to the ability of the host rock type to contain radionuclides. For about two decades, salt was investigated almost to the exclusion of other rock types because of its physical, chemical, and mechanical properties. It was only in the early 1970s that an increasing number of investigators questioned the emphasis on solely one host rock type and by the mid 1970s, other rock types were being investigated. Since the intense radioactivity of fission products can sustain itself over several hundreds of years and the persistence of actinides over many thousands of years, the need for the isolation of wastes from the biosphere for a very long time was needed. Thus, emplacement of such waste in underground, conventionally mined geologic repositories provided the most direct, readily available means for isolation. In the 1980s, in fact, the U.S. Department of Energy (DOE) established geologic repositories as the most viable option in the management of commercially generated radioactive waste, including wastes stored as liquids and sludges, reprocessing effluents, and spent fuel-rod assemblies.

Such waste would essentially be emplaced in drill holes in the floors of excavated caverns at depths of 300 to 900 m below the ground surface.

Since the early 1980s, in an effort to offer other alternatives to the handling and disposal of hazardous and toxic waste, other alternatives including the use of underground geologic repositories were considered. With increasing environmental pressure against disposal of such waste materials in landfills, injection wells or via ocean disposal, geologic repositories exhibit much merit, notably, for those wastes which technically or economically can be destroyed, altered to less toxic forms, incinerated, eliminated during the manufacturing process, or further treated, reused, or recycled. Underground repositories are also attractive as land disposal and underground injection prohibitions become effective. There is little question, however, that the social and political aspects, as with other disposal alternatives, must be given adequate attention. In the case of certain wastes, concerns about their fate in the distant future is dependent upon predictions with some acceptable degree of uncertainty. In lieu of selection of new repository sites and subsequent excavation, the utilization of existing mined space is currently being considered. In the case of Missouri, about 20 million ft^2 of mined space has already been converted for commercial uses, leading the world in the secondary use of underground mined waste. The advantages of this utilization is many. Excavation costs would be minimal. Leachate production in a properly sited and designed mine would also be minimal. Waste types could easily be segregated and readily accessible for periodic inspection or for the purpose of future reuse and recycling. In the case of shallow mines (less than 100 ft below ground surface), accessibility would not present a problem. Surface land would also be available for more productive uses.

Presented in this chapter is discussion of siting criteria, various host rock types used and for consideration, and design considerations.

SITING CRITERIA

The chief geologic barrier for the safe, long-term underground disposal of waste materials is the surrounding rock into which the repository is excavated. At minimum, certain geologic criteria must be met prior to implementation of any underground disposal strategy. A favorable host rock must preferably occur as a stratified unit or nonbedded rock mass and maintain a high degree of homogeneity, both laterally and vertically. The host rock must also be laterally continuous over an area of several square kilometers in order to accommodate the actual repository and provide an adequate buffer zone. A high degree of predictive geologic stability should exist such that the repository will not be affected by erosional effects, uplift, and frequent seismic and volcanic activity over the anticipated period of concern. The host rock should have adequate rock strength such that openings can be maintained for years and even decades. In addition to adequate rock strength, the host rock should be characterized by high conductivity in order to keep the overall temperatures low, maintain low permeability, and contain few structural elements (i.e., fractures, joints, brecciated zones, etc.) which could serve as potential preferential pathways for groundwater flow.

In siting underground geologic repositories, detailed information demonstrating the suitability of the site for the storage and/or disposal of hazardous and toxic material over the active life of the repository is needed. As with other subsurface disposal alternatives, the stability of the repository structure and degree of waste isolation is directly affected by the site geology. The regional and local geologic setting must thus be characterized to the degree necessary to predict the long-term performance of the repository in regards to stratigraphy, structure and tectonics, mineralogy, geochemistry, rock characteristics, and natural resources.

From a hydrogeologic perspective, a dry climate with low precipitation is preferred. Groundwater should be absent or scant, and noncorrosive. Any potential groundwater pathways should be engineered such that the potential for groundwater to come in contact with the waste materials is minimal to nonexistent. The repository should also be located in areas of minimal geologic exploration and production areas — in other words, situated far from ore deposits and oil and gas fields, and other potential economic resources. The location should be situated in an area of low relief to allow for easy access, and of low population density. Permitting for geologic repositories is complicated. UIC regulations apply to fluid waste, whereas solid waste comes under RCRA. Land disposal restrictions under RCRA require that any hazardous waste placed in a repository must be treated using best demonstrated available technology (BDAT) unless a no-migration variance is petitioned. Required components of a no-migration variance include waste description, facility descriptions and operating plans, repository and waste monitoring plans, site hydrogeologic characterizations, waste mobility modeling, environmental risk assessment, infrequent and uncertain events analysis, and quality assurance and control plans. To a reasonable degree of certainty, the no-migration variance must be demonstrated for as long as the waste remains hazardous based on (1) repository setting relative to groundwater, (2) degree of hydrologic isolation achieved reflecting location and engineered features of the repository, and (3) immobility of wastes once placed in the repository. Such variances are attractive due to the confidence in predicting long-term performance in properly sited, well-designed repositories. A summary of hydrogeologic information required is presented in Table 12-1.

HOST ROCK TYPES

Several host rock types have been considered for suitability as geologic repositories. These have included rock salt, rocks of igneous origin (i.e., granite, basalt, and volcanic tuff), sedimentary rocks (i.e., argillaceous or

Table 12-1 Summary of Hydrogeologic Information Required for Siting and Performance Analysis for Underground Repositories

Property	Information Needs	Characterization Methods	Limitation to the Characterization	Presentation of Data
Natural resources	Number, location, depth, and present state of artificial penetrations Geologic/hydrogeologic resources Type of resource Accessibility, quality, and present and future demands Mineral rights ownership in the area	Literature searches Visual inspection of site Geophysical methods	Quality of records for penetration Seismic data may not resolve small-scale features	Geologic maps/cross sections for seismic data Reports on penetrations/backfilling activities should be provided Raw data from geophysical surveys Listing of all penetrations through repository formations
Mineralogy/ geochemistry	Mineral content and composition Heterogeneity of rock composition Heterogeneity of rock composition Presence of fluid inclusions, fracture filling, and alteration zones Rock textural features	Laboratory studies Petrographic analysis X-ray diffraction Electron microprobe analysis Geophysical sampling	Solution-mined cavern Limited number of borehole sampling points near repository site Existing mine Poor definition	Graphics and figures for geophysical and rock coring data Graphs for mineral component percentages Photos of petrographic thin sections Figures for microprobe analysis

	Geomechanical parameters / data	Methods	Limitations	Presentation
Rock characteristics	Geomechanical parameters Density, porosity, and water content Plasticity Undrained shear strength Compressibility Swelling potential Angle of internal friction Maximum low strain shear modules Soil corrosiveness	Laboratory tests Triaxial/uniaxial compressive load test Creep test Stress-strain test Strain rate test Direct shear test In situ stress tests In situ stress test In situ creep test	of the formation away from the mine itself Inability to accurately recreate in situ conditions Poor understanding of in situ stress mechanisms Uncertainties in laboratory values for various parameters	Geologic cross sections for location of fluid or gas pockets Standard stress-strain diagrams should be used Creep curves plotted Creep measurements plotted vs time Detailed text
Stratigraphy	Lateral and vertical extent of unit Nature of the contacts between units Other stratigraphic discontinuities	Review existing geologic maps/crosssections Field mapping Surface geophysical survey	Structural complexity of the region Poor structural/chemical contrast between units Limited number of geophysical or borehole sampling points	Geological maps with regional strike/dip, surface elevations, major structural features Site geological map at scale of about 1:5,000–1:12,000 Cross-sectional maps, depth interval of about 50–100 ft to sufficient depth: Sedimentary facies Structural features Petrological facies (ligneous/meta rocks)

Table 12-1 Summary of Hydrogeologic Information Required for Siting and Performance Analysis for Underground Repositories (continued)

Property	Information Needs	Characterization Methods	Limitation to the Characterization	Presentation of Data
Structure and tectonics	Structural discontinuities Fault folding of regional and local area Past tectonic activity Regional plate tectonics Present temperature, confining pressure conditions, pore fluid pressures In situ state of stress	Field mapping Drill core and sampling Geophysical sampling Remote image interpretation Hydraulic fracturing (for in situ stress measurement)	Small scale variability of subsurface structures Limitations on the number of geophysical or borehole sampling points	Isopach maps with about 10-ft contour intervals and each stratigraphic member Geologic maps at scale of about 1:5,000–1:12,000 Structural and tectonic features Rock outcrop patterns Structural fence diagrams Rose diagrams or stereonets for surface and drill-core fracture orientations Core analysis and fracture histograms for each drill hole Detailed test and tables describing all observed structural features, their mode of formation, distribution, and relation to the site's

| Groundwater | Groundwater location
Aquifer characteristics
General flow direction with respect to the repository
Flor type
Horizontal and vertical flow velocities
Hydraulic interconnectivity points between aquifers
Drawdown around wells in the area
Groundwater chemistry
Salinity in total dissolved solids
Mineralization
Dissolved gases
Age dating
pH, Eh
Drinking water quality parameters
Microbiological activity
Temperature
Indicator chemicals | Geophysical surveys
Field groundwater testing and sampling
Piezometric head readings
General hydraulic testing methods
Laboratory analyses | Geophysical resolution of subsurface structures and groundwater
Location of well boreholes and rock coring activities
Inaccuracy of groundwater flow models | containment qualities, and the methods by which they were mapped and analyzed
Groundwater
Hydrologic setting information
Piezometric surface maps
Flow nets
Geophysical data should be provided
Stratigraphic cross section
Tabulated chemical data
Hydraulic conductivities and transmissivities specified |

Table 12-1 Summary of Hydrogeologic Information Required for Siting and Performance Analysis for Underground Repositories (continued)

Property	Information Needs	Characterization Methods	Limitation to the Characterization	Presentation of Data
Surface water	Regional precipitation and E.T. rates Location and size of all surface water bodies Water volume, flow rate, and direction Drainage network Hydraulic connection between surface water and groundwater Soil hydraulic properties General water chemistry and use	Field survey at site to characterize surface water and drainage networks Literature searches for soil, meteorological, and water quality data Soil boring and well logs can be used for water table elevation determination	Problems with determining area extent of filed study	Information presented in text form plus: Surface topographic and hydrologic maps of 1:5,000–1:12,000 SCS soils maps Tabulated surface water flow, volume, and quality data

Source: After Hazelwood and Turc (1989).

clay-rich strata), anhydrite, and certain limestones (i.e., chalk). Although these rock types have all been considered at one time or another for the disposal of radioactive waste, certain rock types such as rock salt and granite have also been considered and, in some cases, used as hazardous waste geologic repositories. General characteristics of these potential host rock types are further discussed below.

Rock Salt

Rock salt has been the most studied host-rock type for radioactive-waste disposal. Worldwide, there are several thousands of salt caverns. The emphasis on rock salt in the U.S. focused on bedded, or stratified, and domal type deposits. In the U.S., bedded deposits cover a wide geographical range, while dome deposits are limited to the Gulf Coast area (Figure 12-1). Bedded deposits include those of Permian age which underlie much of the Panhandle region of western Texas (i.e., the Pala Duro Basin); bedded and flow-distorted units within the Pennsylvania Paradox Group; and within the Paradox Basin of southeastern Utah (i.e., Salt Valley, Lisbon Valley, Gibson Dome, and Elk Ridge). A typical bedded salt core from the Permian basin of western Texas is shown in Figure 12-2. Domal structures include seven salt domes in the Gulf Coast states, comprised of Rayburns and Vacherie domes in Louisiana; Cypress Creek, Lampton, and Richton domes in Mississippi, and Keechi and Oakson domes in Texas. The Gibson Dome in southeastern Utah was recommended in 1981 as the principal structure for future detailed study for radioactive-waste disposal. Major rock salt sites throughout the U.S. that have been considered for radioactive or hazardous waste disposal, including liquid brines and solids, are shown in Figure 12-1.

Salt beds originate from repeated cycles of evaporation and recharge of saltwater seas followed by cycles of sediment deposition. These salt deposits experienced a variety of temperature and pressure influences reflecting in part their depth below ground surface and geographic proximity to either interior continental or coastal areas. Within the relatively stable interior areas of the continents (or cratons), temperature and pressure influences are such that the salt is consolidated into horizontally stratified formations, termed bedded salt. In less stable areas such as along the continental margins or coastal areas, salt beds tend to exist at deeper depths; thus, temperatures as high as 400°F and pressure influences generate buoyancy differentials causing vertical movement of the salt formation, forcing it upward through the overlying sediments. Such structures are termed salt domes. The reduced viscosity of the salt, in conjunction with its relative buoyancy and upward vertical flow, also function as a mechanism which expels impurities. Other salt structures associated with such salt-dome structures include anticlines, pillows, rollers, salt walls, bulbs, stems, roots, numakier, diapiric stocks, and detached diapirs as illustrated in Figure

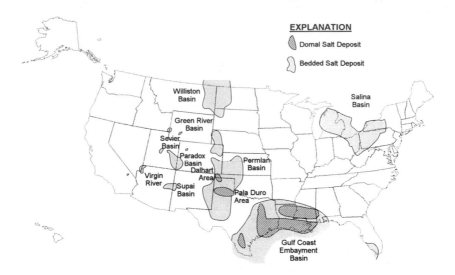

Figure 12-1 Locations of major rock salt occurrences in the U.S.

12-3. Not shown in Figure 12-3 are nappes and massifs, which are an order of magnitude larger in size.

Domal salts have numerous favorable characteristics which make them suitable as geologic repositories. These characteristics include:

- Essentially homogeneous since surrounding formation materials (i.e., clay, carbonates, and anhydrites) are selectively sorted and displaced by buoyancy and plasticity mechanism during dome development
- Intrabedded sedimentary layers which may serve as potential pathways for water to enter the body of the salt formation or as discontinuities enhancing the potential for cavern roof cave-ins are absent
- Relatively dense and impermeable to fluids due to repeated stages of recrystallization
- Geologically stable on the order of millions of years
- Typically large occupying several square miles in extent allowing for a sufficient salt buffer between the repository and any beneficial use groundwater resource(s)
- Considered a nonreactive material and not anticipated to react with solidified hazardous waste.

Nonfavorable characteristics of rock salt pertain to solubility, plasticity, and presence of naturally occurring brines. Salt is more soluble than other potential host rock by two orders of magnitude. Thus, inadvertent dissolution of a salt bed or dome via groundwater or other waters is an obvious concern. Natural dissolution can take the form of breccia pipes, collapsed chimneys, and vertical-

PERMIAN BASIN SALT CORE

DEAF SMITH COUNTY, TEXAS

LOWER SAN ANDRES FORMATION, CYCLE 4

WPP85-8-Detten #1
(2680.4-2681.0 ft)

Figure 12-2 Typical bedded rock salt core from the Permian basin of western Texas (photo courtesy of Jim Conca).

walled structures containing jumbled masses of broken rock reflecting dissolution at depth in conjunction with collapse of overlying structures. Due to salt's self-flow capacity under the appropriate pressure-temperature regime, concerns exist regarding open caverns and passageways remaining open, and renewed plastic movement within the salt dome (halokineiis) due to thermal

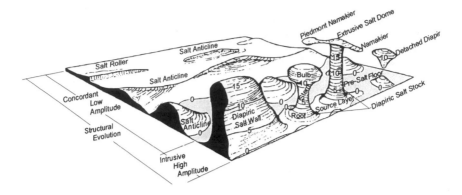

Figure 12-3 Typical rock salt structures.

stresses. All salt deposits contain brines (fluids of very high concentrations of dissolved solids) which could migrate toward waste canisters under the influence of a thermal field caused by the emplaced wastes (notable with radioactive wastes) which could corrode the canister if contact were made. Salt also has poor sorptive qualities, which affects its ability to impede radionuclides or other waste constituents. Most of these not-so-favorable characteristics can be compensated by conservative engineering design of the repository, and appropriate canister layout and storage.

Chemical compatibility of the salt host rock and the hazardous or toxic waste being stored must also be assessed. Salt deposits due to their depositional environment, contain many other dissolved and suspended substances in addition to salt (i.e., halite, NaCl). Insoluble impurities commonly encountered in salt include anhydrite ($CaSO_4$), dolomite ($CaMg(CO_3)_2$), calcite ($CaCO_3$), pyrite (FeS_2), quartz (SiO_2), and iron oxides; the common soluble impurities or ions include Ca, Mg, K, Cl, CO_3, and SO_4, with minor amounts of Ba, Sr, B, and Br. Minerals commonly associated with salt beds are summarized in Table 12-2.

Disposers of waste in salt domes do have their technical opponents. In Texas, oil and gas exploration/production have stimulated concerns regarding disposal of waste in the Boling salt dome situated south of Houston near Boling, TX. Hazardous waste is planned to be injected as a liquid and then solidified in place within leached caverns. The concern of oil and gas entities is that the plan also involves injection of waste into the Oligocene Frio sands where oil and gas exploration exists. This injection program could shut down drilling or could potentially contaminate hydrocarbon with the Frio, thus impacting the ability of the operator to acquire or maintain liability insurance.

The utilization of salt domes for the disposal of solidified hazardous waste is currently planned for the North Dayton dome, situated about 5 mi northwest of Dayton, TX, about 30 mi southwest of Houston. This salt dome actually resembles a column extending to a depth of about 45,000 ft below ground

**Table 12-2 Common Accessory Minerals
Associated with Salt Deposits**

Mineral Species	Chemical Composition
Anhydrite	$CaSO_4$
Bischofite	$MgCl_2·6H_2O$
Bloedite	$MgSO_4·NA_2SO_4·4H_2O$
Carnallite	$KMgCl_3·6H_2O$
Epsomite	$MgSO_4·7H_2O$
Glaserite	$K_3Na(SO_4)_2$
Glauberite	$CaSO_4·NA_2SO_4$
Gypsum	$CaSO_4·2H_2O$
Kainite	$MgSO_4·KCl·3H_2O$
Kiescrite	$MgSO_4·H_2O$
Krugite	$4CaSO_4·MgSO_4·K_2SO_4·2H_2O$
Langbeinite	$2MgSO_4·K_2SO_4$
Leonite	$MgSO_4·K_2SO_4·4H_2O$
Loeweite	$MgSO_4·Na_2SO_4·2\text{-}1/2H_2O$
Picromerite	$MgSO_4·K_2SO_4·6H_2O$
Polyhalite	$2CaSO_4·MgSO_4·K_2SO_4·2H_2O$
Sylvite	KCl
Syngenite	$CaSO_4·K_2SO_4·H_2O$
Tachydrite	$2MgCl_2·CaCl_2·12H_2O$
Vanthoffite	$MgSO_4·3Na_2SO_4$

surface and encompasses an area of more than 4 mi^2 at its cap. If above ground, this salt formation would dwarf Mount Everest. The North Dayton dome is only one of some 500 similar type structures in the Gulf Coast region. About 200 of these saltdome formations have an estimated storage capacity of between 500 million and several billion gallons of waste. This particular site will accommodate 44 caverns, each capable of holding about 500,000 yd^3 of solidified waste material.

The waste, mostly petroleum-based sludges, will be treated using a solidification process such that the sludge will be mixed with utility waste fly ash, cement, and other additives to produce a low-strength concrete. The liquid mixture will be pulverized and conveyed into a cavern developed using conventional solution mining techniques. The cavern will be cylindrical in shape, 125 ft in diameter, completed at depths on the order of 600 ft below the surface of the salt and extending to about 1800 ft deep. With the salt first encountered at a depth of about 900 ft, the top of the cavern will be at about 1500 ft below ground surface and extend to a depth of about 3300 ft. Upon filling, the cavern will be permanently sealed with cement grout with the borehole backfilled with

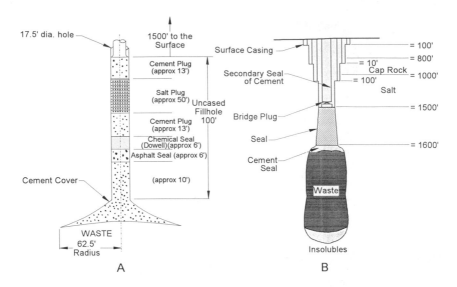

Figure 12-4 Schematics showing cavern completion (A) and closure (B) (after Klos, 1992).

salt and the cased hole cemented to ground surface. Design criteria are in compliance with UIC regulations and guidelines for well siting, construction, and monitoring. Cavern completion and closure schematics are shown in Figure 12-4.

Cavern development involves drilling and cementing concentric casings into the salt dome, then drilling an uncased hole which exposes the salt for dissolution or leaching. The concentric casings provide for the circulation of water through the well via the inner casing, or annulus, to dissolve the salt. A flow of brine is then circulated back to the surface via an outer casing. The brine is filtered for solids removal, then pumped to brine disposal wells located about 1 mi off the dome. Leaching is accomplished via two basic methods:

1. Direct circulation is initially used to enlarge the initial borehole and form the cavern chimney and sump. By injecting water near the bottom of the cavern and withdrawing brine through the annulus positioned above the injection point, maximum diameters occur near the bottom of the cavern with decreasing diameters toward the cavern top (Figure 12-5).
2. Reverse circulation is used during the final cavern stage for development of the upper portion of the cavern by injecting water down the annulus between the inner and outer casing strings, causing brine to circulate into the inner string above the injection point, thus allowing enlargement of the cavern in the top half of the cavern pending actual positioning of the suspended casing strings in the cavern.

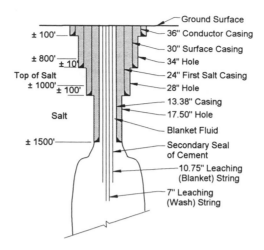

Figure 12-5 Schematic showing direct injection process for cavern development.

Granite

Granitic rocks, classified on the basis of texture and mineralogy, generally refer to the whole range of plutonic rocks. Essential minerals include varying amounts of sodic plagioclase, potassium-rich alkali plagioclase (either orthoclase or microline), and quartz (usually greater than 10% in total volume). Accessory minerals include biotite, muscovite, hornblende, and pyroxene. Granitic rocks exhibit several favorable characteristics which make them suitable for the disposal of waste materials. Along continental margins, belts of granitic rocks developed as batholiths composed of hundreds of relatively smaller sized plutons. Some of the more prominent batholiths in North America include the Coast Ranges, Boulder-Idaho, Sierra Nevada, and Baja California. The largest is more than 1500 km long and 200 km wide. The Sierra Nevada, for example, is composed of about 200 plutons separated by many smaller plutons, some only a few kilometers wide. Locations of major granitic rocks throughout the U.S. are shown in Figure 12-6.

Granitic rocks within the shallow crust away from zones of high seismicity, faulting and structural complexity. In addition, granites typically exhibit a high level of physical strength, and are mechanically stable in large underground openings, chemically stable, low content of pore water, presence of secondary minerals (i.e., clays, hydrous oxides, etc.) which can increase its sorptive qualities, and diminishing fractures with depth.

Fracturing, as with other rock types, can be problematic. It is assumed that fractures in granitic bodies can be numerous and water filled near the surface but may diminish in both number and magnitude at depth (i.e., 1000 m), yielding a relatively impermeable rock mass. However, these masses have not

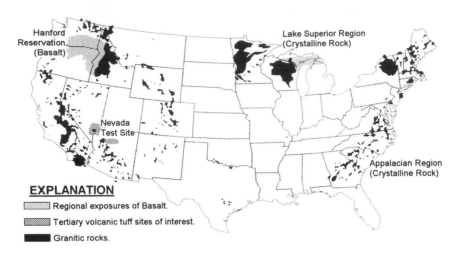

Figure 12-6 Location of major igneous rock masses in the U.S.

been studied intensively due to their relatively low abundance of ore deposits. Thus, it is assumed that, at repository depths, fractures would persist, and that any water that would be present would be free to migrate within this fracture-flow system and eventually be in contact with any emplaced canisters. Economically, DOE estimated that a granite repository may cost up to 2.6 times that of rock salt, reflecting higher excavation costs due to difficulties in excavation and potentially deeper depths of excavation. Another concern is the thermal load induced by the waste, especially in the case of radioactive waste, which could potentially generate artificial fractures. Excavation and construction-related activities may also contribute to artificial fracturing.

Storage of hazardous waste materials in a granitic repository currently exists in Mulen, Norway. Since 1928, Norway's largest zinc smelting and processing plant, Norzink AS, has generated both a zinc waste and leaching residue. The zinc waste was jarosite, a mixture of Zn, Fe, Cd, Hg, and sulfuric acid, which was pumped directly into the nearby fjord and subsequently settled on the sea bottom. The leaching residue was pumped into surface settlement ponds and essentially left there. Environmental pressures resulted in the company developing a plan to store both wastes in a granitic gneiss geologic repository. Located about 2 km from the plant, the first of several caverns commenced in 1984. Each cavern is approximately 235 m long × 17.5 m wide × 23.5 m high with an overall volume of 60,000 to 70,000 m³. The caverns are inclined and situated in a row parallel to one another, with access via tunnel aligned perpendicular to the row. The enveloping granitic gneiss has a compressive strength of 3000 kg/cm². The waste is pumped some 2 km to the repository. As the waste is piped into the cavern, about one third of the waste volume is water. Thus, once in the cavern, solid materials settle to the bottom while the water is pumped out and either reused, used as process water, or pumped to the

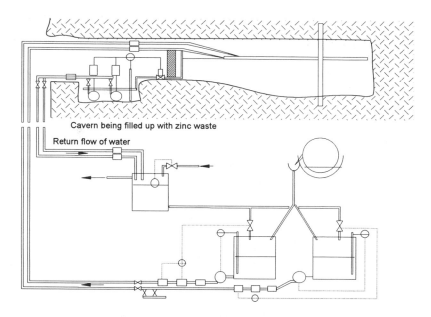

**Figure 12-7 Schematic showing caverns in a granitic gneiss host rock for
the storage of zinc waste and water recycling system.**

purification system and ultimately returned to the zinc plant in a closed-circuit
pipe system (Figure 12-7).

No actual geologic repository for the storage of hazardous waste currently
exists in the U.S. as in the case of Minnesota, Missouri, and Ohio, reflecting
economic, regulatory, and/or public opposition-related limitations. However,
arsenic dust produced through the roasting of arsenopyrite is stored in seven
separate underground chambers within the Grant Yellowknife Mines, Yukon
Territory, Canada. The chambers are located in the permafrost zone at about
75 m below ground level. About 170,000 tons of arsenic dust has been disposed
of since 1951. At the Herfa-Neurode Facility in Germany, about 270,000 tons
of waste has been stored in mines since 1972, with an annual volume ranging
between 35,000 and 40,000 tons. Furthermore, about 1000 tons has since been
retrieved and reused.

Basalts

Basalts are the most abundant of rock types, comprising the deep ocean
floor in addition to widespread occurrences on the continents, notably as flood
basalts. Extensive flood basalt provinces include the Karroo in southern Africa,
the Parana in South America, the Deccan basalts in India, and the Columbia
River basalts in the northwestern U.S. The most impressive aspect of flood

basalts are their overall dimensions. Single lava flows may exceed 500 km^3 in volume with an areal extent of 40,000 km^2 or more. Entire lava fields may have volumes on the order of 100,000 to 1 million km^3. Basalts also comprise dikes and sills which can also be extensive. The feeder dike swarms of the Grande Ronde dike swarm of Oregon, Washington, and Idaho extend for some 200 km and have a width of about 50 km. The location of the Columbia River basalt group is shown in Figure 12-6.

Basalts exhibit favorable characteristics including:

- Intermediate thermal conductivity
- Silicate mineralogy formed from cooling of high-temperature lavas
- Ability to withstand high thermal loads
- High physical rock strength
- Low primary permeability
- Structurally thick (up to 50 m) individual and nearly horizontal units
- Presence of alteration minerals (i.e., clays, zeolites, etc.) which enhance the basalt's sorptive qualities.

Characteristics not considered favorable include numerous zones of secondary permeability such as columnar joints, porous vesicular intervals, differential weathering, and sedimentary interbeds (Figure 12-8). These zones serve as potential conduits for groundwater flow, thus providing preferred pathways for contaminant migration from the geologic repository to surface water bodies (i.e., Columbia River) and beneficial-use groundwater at deeper depth. In addition, groundwater occurrence and flow path is very complex, consisting of multiple zones of saturation with varying degrees of interconnection. Principle aquifers are associated with interflow zones and sedimentary interbeds between basalt flows. The interflow zones typically have high to very high secondary permeability but low storativity, reflecting the open nature but limited volume of water-bearing joints and fractures. An aerial view of the Columbia River basalt group, typical of basalt host rock for waste produced at the Hanford, WA site, is shown in Figure 12-9.

Volcanic Tuffs

Volcanic tuffs for use as geologic repositories include primarily two types: welded tuffs and zeolitic tuffs. Welded tuffs originated as volcanic ash flows that were fused during formation. Characterized as dense, relatively low porosity and permeability, and accepting of high thermal load, welded tuffs are comparable to basalts in regards to rock strength, thermal conductivity, and other properties. Zeolitic tuffs contain zeolites, which are hydrous silicate minerals formed by the alteration of silicon- and aluminum-rich minerals (i.e., silicic volcanic glass). Characterized by open structures in which water molecules and large-radius cations occupy cavities between tetrahedral bonds,

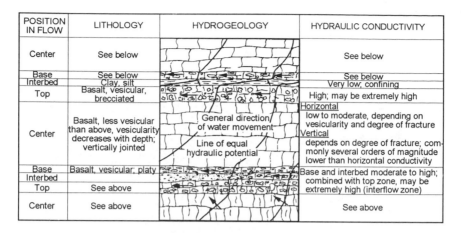

POSITION IN FLOW	LITHOLOGY	HYDROGEOLOGY	HYDRAULIC CONDUCTIVITY
Center	See below		See below
Base	See below		See below
Interbed	Clay, silt		Very low; confining
Top	Basalt, vesicular, brecciated		High; may be extremely high
Center	Basalt, less vesicular than above, vesicularity decreases with depth; vertically jointed	General direction of water movement / Line of equal hydraulic potential	Horizontal low to moderate, depending on vesicularity and degree of fracture / Vertical depends on degree of fracture; commonly several orders of magnitude lower than horizontal conductivity
Base	Basalt, vesicular; platy		Base and interbed moderate to high; combined with top zone, may be extremely high (interflow zone)
Interbed			
Top	See above		
Center	See above		See above

Figure 12-8 Intraflow structural controls on water movement in a discharge area of the Snake River (after Lindholm and Vacarro, 1988).

resulting in very high sorptive capacity which can serve to impede or filter radionuclides and other waste constituents. In contrast to welded tuffs, zeolitic tuffs exhibit low density, high porosity and permeability, and only moderate rock strength and thermal conductivity.

Ideal geologic conditions would entail welded tuff insulated above and below by zeolitic tuffs which would take advantage of those favorable characteristics of both rock types. Such geologic conditions occur at the Yucca Mountain site in Nevada. Although not selected as a repository, this is the only candidate site for the storage of radioactive waste (i.e., spent fuel from commercial nuclear reactors, high-level waste from defense activities, and small quantities of commercial high-level waste) to be characterized by the DOE. Yucca Mountain is underlain by a sequence of silicic volcanic rocks ranging from greater than 3,000 to 10,000 ft thick, dipping 5 to 10° to the east. Underlying rocks consist mainly of welded and nonwelded ash-flow and air full tuffs.

The repository would be constructed in an ash-flow unit called the "Topopah Spring Member" of the Paintbrush Tuff Formation (Figure 12-10). The Topopah Spring Member is about 1100 ft thick and is the lowermost and most extensive in the area. Underlying the Paintbrush Tuff are the tuffaceous beds of Calico Hills, Crater Flat Tuff, and older tuffs.

Sedimentary Rocks

Two sedimentary rock types, excluding salt, have received serious consideration as geologic repositories: shale and anhydrite. Shale refers to a wide

Figure 12-9 Columbia River basalt group in south-central Washington, which is typical of basalt host rock for waste produced at the Hanford Reservation.

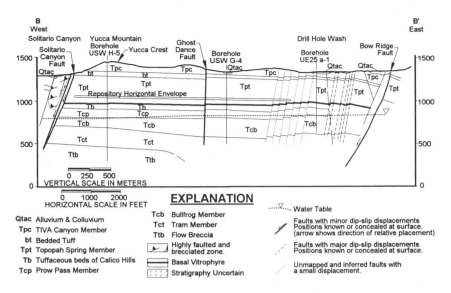

Figure 12-10 Geologic cross section showing ash-flow tuff repository at Yucca Mountain, Nevada.

spectrum of fine-grained argillaceous rock including mudstones, claystones, clays, and argillites. In the U.S., these rock types are widely distributed with a large number of homogeneous, largely unfractured, thick (greater than 75 m) units which show merit. These include the Pierre Shale of the Northern Great Plains, the Maquoketa Shale and equivalent units throughout the Midwest, and the Antrim-Ellsworth-Coldwater interval in Michigan (Figure 12-11). Favorable characteristics for shale include:

- Low porosity and permeability
- Insoluble
- Capable of plastic flow to seal fractures at depth
- High sorptive characteristics (even at temperatures greater than 100°C)
- Long-term and almost complete retention of fission-product radionuclides (about 1.8 billion years).

Shale units, however, quite often contain water, usually saline or mineralized, and commonly occur in proximity to water-bearing strata and other resources (i.e., petroleum, coal, and other sedimentary rocks of economic value). Any thermal load could cause dehydration of clay minerals, generate liquids and gases under the influence of temperature, and potentially release additional water into the repository. Furthermore, porosity and permeability could also be adversely affected. Other concerns associated with shale are its overall low physical strength in comparison to granite and basalt, and its overall stability in deep excavations.

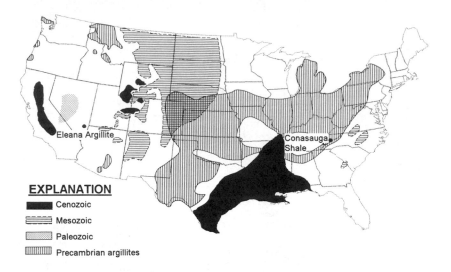

EXPLANATION
- Cenozoic
- Mesozoic
- Paleozoic
- Precambrian argillites

Figure 12-11 Location of major argillaceous strata of varying ages within 1000 m of the surface in the U.S.

Anhydrite has also received consideration as a geologic repository. Anhydrite is an evaporitic deposit of anhydrous calcium sulfate ($CaSO_4$). Associated with many salt deposits, anhydrite tends to occur in thick, impermeable, homogeneous beds. Anhydrite also has high thermal conductivity and is chemically stable at high temperatures.

Anhydrite, however, is soluble in water and can be hydrated to gypsum via the following reaction:

$$CaSO_4 + H_2O \rightleftharpoons CaSO_4 \cdot 2H_2O$$

anhydrite   water   gypsum

This process under the right conditions is reversible, and thus could contribute to future changes in porosity and permeability.

DESIGN CONSIDERATIONS

Because of the great diversity in the hazards associated with the varied types of waste materials, little attention has focused on underground disposal design for hazardous waste. Where nuclear waste has a half-life, the hazard is eventually reduced with time. This allows for development of an operational life span for a nuclear waste repository. However, certain hazardous waste (i.e., heavy metals) will remain toxic indefinitely; thus, the operational life span of the repository must be approached differently. Conceptually, the hazardous waste repository must be designed for total containment or controlled release for a finite number of years, and continuous controlled release to the environment. Thereafter, monitoring and an established protocol for remedial actions should containment fail must also be considered.

As previously discussed, a variety of repository designs exist, ranging from unlined repositories to well-designed containment systems involving multiple barriers. comprised of man-made and natural barriers. Man-made barriers include containers and chemically inert and structurally sound encapsulating matrix. For example, concrete vaults can be used for underground storage of containerized waste under unsaturated conditions, and have exhibited favorable hydraulic performance (i.e., low permeability) even under saturated conditions. This allows for consideration of repository candidates located in humid climates, rather than being restricted to arid climates.

Three types of underground storage facilities have been used or are being proposed for the storage of hazardous and toxic waste materials. In order of technological development, these are existing mines, solution-mined caverns in salt, and new mined caverns.

Existing Mines

The potential use of existing mined space in the U.S. is notable in the Midwest, throughout central and eastern Missouri. Sizable underground limestone and dolostone mines are located throughout Missouri. Some of these mines have been converted to secondary uses. Originally developed by conventional room-and-pillar techniques, these mines are typically developed at shallow depths (less than 100 ft below ground surface) within thick, horizontally bedded limestone, dolostone or sandstone units. These mines are situated beneath 30 to 60 ft of overburden to avoid freeze-thaw effects, and above the saturated zone. Although not currently used for the disposal or storage of hazardous material, hundreds of such mines exist, many of which would be excellent candidates as geologic repositories.

Underground mining methods are usually discussed in terms of the extraction process, and the means by which the walls and roofs are supported or reinforced (Table 12-3). A drift mine has a horizontal entry, whereas slope mines have inclined entries. Underground openings created by the extraction of ore are called stopes. Stoping methods are classified as to whether the openings are naturally or artificially supported, are encouraged to cave, or a combination of several methods depending upon the uniformity of the subsurface geologic conditions. Existing mines (i.e., salt, potash, gypsum, limestone, and coal) have conventionally been mined by the room-and-pillar method, resembling a checkerboard pattern resulting in a series of rooms alternating with supporting pillars (Figure 12-12). Room-and-pillar mining accounts for 60% of the total noncoal mineral production in the U.S. This method is well

Table 12-3 Underground Mining Methods[a]

Opening	Principal Method
Naturally supported	Room and pillar mining, open stoping, sublevel open stoping, longhold open stoping, shrinkage stoping
Artificially supported	Stull stoping, cut-and-fill stoping, square-set stoping, longwall mining, shortwall mining, top slicing
Caving	Block caving, sublevel caving

[a] After Peters (1978).

suited to gently dipping, relatively uniform bedded deposits. In hard rock, this method is usually limited to depths of 1000 to 2000 m due to the potential for rock bursts and other similar phenomena related to a high concentration of stress (i.e., floor heave, gas emission, etc.). Stress patterns must thus be carefully monitored, and the placement and design of pillars must be conservatively addressed with regard to height, structural defects, and creep. Unlike with mining of ore deposits where the amount of ore extracted is maximized, overly conservative pillar design for the underground storage of materials can easily be accommodated, all else being considered equal.

The Herfa-Neurode facility was established in 1972 in West Germany for the storage of hazardous materials. This repository is located in a depleted portion of a potash mine at a depth of approximately 700 m in horizontal beds of salt and potash. Regarding the storage of waste materials, once a room is filled, it is sealed off from active portions of the mine by means of a concrete block wall. The stored waste may eventually be useful for reprocessing or reuse, and can be retrieved simply by removal of the wall.

Solution-Mined Caverns

Solution-mined caverns in salt formations are currently used to store crude oil. Since 1978, for example, the DOE has stored over 750 million barrels of crude oil in such caverns for the Strategic Petroleum Reserve. Private industry in the U.S. also operates more than 1800 caverns for the storage of liquid petroleum products, petrochemicals, and natural gas. In the Dayton area, two natural gas storage salt repositories have been operational for more than four years. European countries have also used salt formation caverns as containment sites for both hazardous and nuclear wastes. Certain German states actually require disposal of irreducible solid waste in salt caverns.

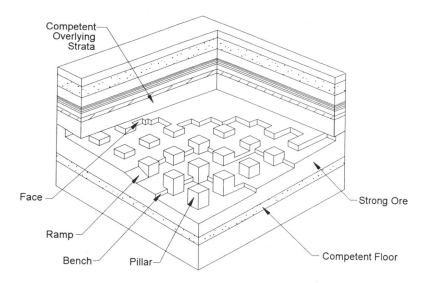

Competent Overlying Strata

Face

Ramp

Bench   Pillar

Strong Ore

Competent Floor

Figure 12-12 Schematic showing typical room-and-pillar method.

Mining Methods

Solution mining can be accomplished with several techniques, depending upon the type and character of the waste to be stored. These techniques include brine-balanced, gas-balanced with zero discharge, atmospheric cavern with controlled gas discharge, in situ solidified waste disposal, and string-of-pearls methodologies. These methods are further discussed below. The advantages and limitations for each method are summarized in Table 12-4.

The brine-balanced method is widely used for the storage of liquid hydrocarbons. In a cavern, liquid hydrocarbon floats on top of brine solutions because it is less dense. When hydrocarbon liquids are withdrawn, brine is added to the bottom of the cavern such that the cavern remains full of fluid at all times. When hydrocarbon liquids are injected into the cavern for storage, the displaced brine is forced up the casing string to a brine holding pond or disposed of via an underground injection well. In considering the appropriate number of caverns required for a given volume of waste, the physical dimensions of bedded vs dome salt deposits must be considered. Louisiana, for example, requires a minimum of 200 ft between adjacent caverns and at least 100 ft between a cavern and the property line, in addition to consideration of drilling and solutioning tolerances. In addition, unless the hazardous or toxic waste does not exceed 150°F, conventional casing and cementing procedures are adequate. As in the case with hydrocarbon liquid, the specific gravity of the liquid waste or waste slurry must be significantly lighter or greater than that of the brine solution. If the specific gravity of the hazardous waste slurry is significantly lighter than the brine, similar to hydrocarbon liquids, the slurry

Table 12-4 Summary of Solution Mining Methods, Advantages, and Limitations

Solution Mining Method	Advantages	Limitations
Brine-balanced	Established methodology Stress regime relatively constant	Potential generation of gas or vapors with heavier waste types Excludes waste types whose specific gravity and particle size approximates that of the brine Potential cross contamination of the brine solution
Gas-balanced with zero discharge	Potential for cross contamination of brine solution absent	Stringent monitoring of design pressures required Requires heightened concern regarding permeability, structural stability, well installation, and seal implacement protocol Potential closure due to salt plasticity Limited in size to maintain structural integrity
Atmospheric cavern with controlled gas discharge	Potential for cross contamination of brine solution absent	Requires a scrubber or flare Limited in size to maintain structural integrity Requires a scrubber or flare
In situ solidified waste disposal	Potential for cross contamination of brine solution absent Low risk of future migration Minimal adverse import in inadvertently drilling into cavern	Limited in size while exposed to atmospheric pressure
String-of-pearls	Incorporates solidification of waste	Constrained by the vertical dimension of the dome Limited in size to maintain structural integrity while exposed to atmospheric pressure

can be injected into the top of the cavern while displacing brine up the casing string. If the specific gravity of the waste slurry is greater, then the slurry would remain in the bottom of the cavern, while the floating brine would be displaced from the top of the cavern. In either case, no significant change in the stress regime occurs. Waste types whose specific gravity approximates that of the brine pose limitations since the cross-contamination of the brine could occur. Another consideration is the potential generation of gases or vapors which could be released by the heavier waste types.

The gas-balanced method involves the displacement of brine with an inert gas at high pressures once the cavern had been developed to capacity. The cavern would then be sealed at the minimum design pressure after which gaseous, liquid, or slurry wastes would be injected into the cavern until the maximum design pressure of the cavern is reached. The appropriate minimum and maximum pressure gradients vary for bedded vs dome deposits. A conservative maximum pressure gradient in bedded salt is on the order of 0.65 psi/ft of overburden (or depth), reflecting the invariable presence of interbedded layers of shale, anhydrite, or other sedimentary rock types. In dome deposits, the maximum operating pressure should not exceed 0.85 to 0.90 psi/ft of overburden since it is commonly assumed in the industry that a pressure gradient of 1 psi/ft of overburden will fracture the salt formation at some location. The gas-balanced method eliminates the potential for cross-contamination of brine solutions; however, concerns regarding permeability, casing installation and seals, and structural stability of the cavity are heightened.

The atmospheric cavern with the controlled gas discharge method is usually used for the storage of condensate (natural gas) and involves the collection of displaced vapors and gases, with subsequent above-ground treatment via a scrubber or burning by use of a flare following cavern completion, and brine removal. Chemically compatible liquid waste or waste slurry is injected while the open cavern is maintained at just above atmospheric pressure. The cavern size needs to be limited to maintain the structural integrity. The potential for brine contamination is absent.

In situ solidified waste disposal involves the mixing of the waste material with a cement of polymer slurry prior to injection into a cavern. The permanent solidification of the waste material reduces several risks and concerns associated with variable permeability of the host rock, seismicity, and potential migration pathways over time. However, the size of the cavern is limited to maintain the overall structural integrity of the cavern while exposed to atmospheric pressure. The solidification of the waste materials also alleviates concerns of inadvertently drilling into the cavern.

The string-of-pearls method involves the construction of a series of caverns, one above the other, from one deep solution well. Brine would be removed from the initial deeper cavern and immediately filled with a waste slurry. A cement slurry could then be mixed with the waste via in situ solidification as previously discussed. Upon solidification, an upper cavern could then be developed with each cavern being sealed by the installation of a cement plug in the top portion of the lower cavern prior to starting the upper cavern. As with the previous method, size limitations are imposed to maintain structural integrity of the caverns.

Rock Mechanics

Three phases are characteristic of a waste disposal facility in salt: construction, operation, and postclosure (Figure 12-13). During construction, no significant difference in rock mechanical properties exists between a conventional and a waste containment cavern, assuming both are under atmospheric pressure. During the operational phase, the method of injection is of minor importance. A continuous injection flow should be maintained, although a safety factor should be considered in case plugging occurs during injection or the delivery of waste to the cavern is interrupted. Assuming a net cavern volume of about 300,000 m^3 and an annual wasteflow of 60,000 to 100,000 m^3, this phase is anticipated to be operational for about 3 to 5 years.

During postclosure, however, rock mechanical behavior is dependent upon the mechanical properties of the contained waste. The waste itself may remain liquid permanently, such that the viscosity of the waste is minimal; then hydrostatic stress conditions are assumed and no shear stresses between the waste and rock exist, or a solid waste which can withstand deviatoric stresses; thus, shear stresses between the rock and waste are likely. From a rock mechanics perspective, these two cases reflect substantial differences. With a liquid waste, cavern closure can force a pressure increase in addition to the existing hydrostatic pressure as determined by the rock pressure of the lower cavern area. In addition, the specific gravity of the fluid waste is less than that of the surrounding rock. Thus, the fluid pressure in the vicinity of the cavern roof might eventually exceed the maximum allowable pressure (i.e., 80% of the lithostatic pressure). This could result in fractures in the surrounding rock, and a potential pathway to the biosphere or ground surface. With solid waste, more favorable conditions exist because of the internal strength of the solid material (i.e., monolithic block or pellets) and its ability to withstand shear forces at the cavern well. The load conditions on the waste require understanding of the mechanical properties (i.e., strength and deformation behavior) of both the rock and the contained waste. Issues that need to be addressed include:

- Adhesion of the waste to the rock
- Binder additives
- Binding of water and brine
- Bridging
- Compression behavior
- Creep behavior

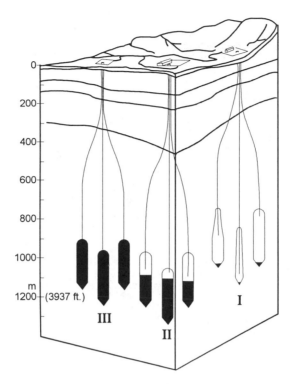

Figure 12-13 Schematic showing the various phases of operations in salt caverns.

- Density
- Formation of hydration heat
- Material feed capability
- Mixing ratio of the waste, binder, and water
- Permeability
- Physical and chemical long-term stability
- Pore space
- Segregation properties
- Solidification
- Spreading
- Stability
- Temperature stability
- Volume changes (i.e., swelling, shrinkage, etc.)

Stress-strain conditions must be considered in and around filled and partially filled caverns. These rock mechanical design issues before and after cavern abandonment are shown in Figure 12-14.

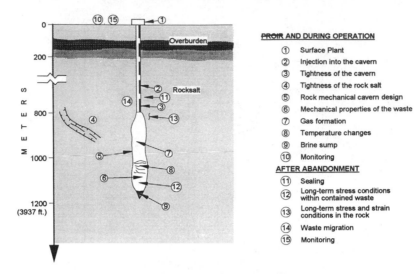

Figure 12-14 Issues related to rock mechanical design.

Newly Mined Space

Although economics are such that it is easier to utilize or modify an existing repository or mined space than to generate a new one, the future use of newly mined space has merit. With consideration for the long-term storage of waste material, site selection can incorporate the most suitable geologic host rock within a reasonable distance to urban centers. Site-specific design specifications can also incorporate the most efficient layout to accommodate the type of material being stored and handled, as well as storage criteria.

BIBLIOGRAPHY

Alyar, G.S. and Hsin, F.J., 1987, Design of Saltstone Vaults: *American Nuclear Society Transactions,* Vol. 54, p. 79.

Auld, F.A., 1983, Design of Underground Plugs: *International Journal of Mining Engineering,* pp. 189-228.

Bredehoeft, J.D. and Maini, T., 1981, Strategy for Radioactive Waste Disposal in Crystalline Rocks: *Science*, Vol. 213, pp. 293-296.

Bredehoeft, J.D., England, A.W., Stewart, D.B., Trask, N.J. and Winograd, I.J., 1978, Geologic Disposal of High-level Radioactive Wastes - Earth Science Perspectives: U.S. Geological Survey Circular 779.

Brookins, D.G., 1976, Shale as a Repository for Radioactive Waste - The Evidence from Oklo: *Environmental Geology,* Vol. 1, pp. 255-259.

Chapman, N.A. and McKinley, I.G., 1987, *The Geological Disposal of Nuclear Waste:* John Wiley & Sons, New York, 280 p.

Economics of Mined Repositories for Acutely Hazardous Wastes: *The Hazardous Waste Consultant,* Sept./Oct., 1983, pp. 32-36.

Ekren, E.B., Dinwidelie, G.A., Mytton, J.W., Thordarson, W., Weir, J.E., Hinrichs, E.N. and Schroder, L.J., 1974, Geologic and Hydrologic Considerations for Various Concepts of High-level Radioactive-Waste Disposal in Coterminous United States: U.S. Geological Survey Open-File Report 74-158.

Esposito, M.P., Thompson, W.E., Gerber, J.S. and Ponder, J.C., 1985, Mined Space - A Viable Alternative to Landfilling for the Long-Term Retention of Contaminated Soils and Other Hazardous Wastes in Missouri: In Proceedings of the Sixth National Conference on Management of Uncontrolled Hazardous Waste Sites, Washington, D.C., pp. 387-392.

Forsberg, C.W., 1984, Disposal of Hazardous Elemental Wastes: *Environmental Science and Technology,* 18, 56A-62A.

Freeman, H.M., 1989, *Standard Handbook of Hazardous Waste Treatment and Disposal:* McGraw-Hill, New York, 1,417 p.

Funderburk, R.E., 1985, Salt Cavern Repositories for Hazardous Waste: In Proceedings of the Society of Mining Engineers Conference on Salt and Brines (Edited by Smith, W.J.), pp. 47-62.

Ghosh, D.K., 1981, Subsurface Space for Waste Disposal Around Korba, Madhya Pradesh, India: In *Proceedings of Rockstore 80,* Pergamon Press, New York, pp. 285-294.

Gnirk, P.F. et al., 1973, Analysis and Evaluation of the Rock Mechanics Aspects of the Proposed Salt Mine Repository, Summary Progress Report: NTIS (1972), Washington, D.C.

Gonzales, S., 1982, Host Rocks for Radioactive-Waste Disposal: *American Scientist,* Vol. 70, No. 2, pp. 191-200.

Hazelwood, D. and Turi, J., 1989, Hazardous Waste Storage and Disposal in Geologic Repositories: In Proceedings of the Hazardous Materials Research Control Institute Conference on Hazardous Wastes and Hazardous Materials, New Orleans, 1989, pp. 387-392.

Hooper, M.W., Geeselman, J.N. and Noel, T.E., 1983, Mined Cavities in Salt - A Land Disposal Alternative: In Proceedings of the Fourth National Conference on Management of Uncontrolled Hazardous Waste Sites, Washington, D.C., pp. 266-269.

Jackson, M.P.A. and Talbot, C.J., 1986, External Shapes, Strain Rates, and Dynamics of Salt Structures: *Geological Society of America Bulletin,* Vol. 97, pp. 305-323.

Jefferies, S.A., 1991, The Design of Waste Containment Systems: *Environmental Pollution,* 1-1 CEP.1, pp. 299-306.

Johnsson, G., 1978, Untertage-Deponie Herfa-Neurode, Hazardous Waste Emplacement in Mined Openings - Four Years Experience in the Disposal of Special Industrial Wastes in a Mined-out Section of a Potash Mine in the Werra Basin (Summary): In *Proceedings of First International Symposium , Rockstore 77* (Edited by Bergman, M.), Pergamon Press, New York, NY.

Johnsson, G., 1978, Storage in Excavated Rock Caverns: In *Proceedings of First International Symposium, Rockstore 77* (Edited by Bergman, M.), Pergamon Press, New York, pp. 173-174.

Johnsson, G., 1981, Underground Disposal at Herfa-Neurode: In *Proceedings of the NATO CCMS Symposium on Hazardous Waste Disposal,* Plenum Press, New York, 1983.

Karably, L.S., Jr., 1983, High-Integrity Isolation of Industrial Waste in Salt: In *Sixth International Symposium on Salt, Vol. 2* (Edited by Schreiber, B.C. and Harmer, H.L.), The Salt Institute, Alexandria, VA, pp. 211-215.

Klingsberg, C. and Duguid, J., 1982, Isolating Radioactive Wastes: *American Scientist,* Vol. 70, No. 2, pp. 182-190.

Klos, T.R., 1992, A Salt Dome Repository for Hazardous Waste: In Proceedings of the Hazardous Materials Control Research Institute HMC-South '92 Conference, New Orleans, pp. 161-175.

Kown, B.T., et al., 1977, Cost Assessment for the Emplacement of Hazardous Materials in a Salt Mine: EPA-600/2-77/215; Prepared by Bechtel Corp. for U.S. EPA, Cincinnati.

Krauskopf, K.B., 1988, Radioactive Waste Disposal and Geology: *Topics in the Earth Sciences, Vol. 1,* Chapman and Hall, New York, 145 p.

Lefond, S.J., 1969, Handbook of World Salt Resources: *Monographs in Geoscience,* Plenum Press, New York.

Martin, D., 1990, Bottling Up Zinc Waste in Secure Mountain Caverns: *Tunnels and Tunnelling,* Vol. 22, No. 3, pp. 52-53.

Merewether, E.A., Sharps, J.A., Gill, J.R. and Cooley, M.E., 1973, Shale, Mudstone, and Claystone as Potential Host Rocks for Underground Emplacement of Waste: U.S. Geological Survey Open-File Report 4339-5.

Oblath, S.B., 1989, Leaching from Solidified Waste Forms Under Saturated and Unsaturated Conditions: *Environmental Science and Technology,* Vol. 23, pp. 1098-1102.

Pendery, E.C., 1969, Distribution of Salt and Potash Deposits - Present and Potential Effect on Potash Economics and Exploration: In Proceedings of the Northern Ohio Geological Society Third Symposium on Salt, Vol. II, pp. 85-95.

Peters, W.C., 1978, *Exploration and Mining Geology:* John Wiley & Sons, New York, 696 p.

Rahman, M.S. and Booker, J.R., 1989, Pollutant Migration from Deeply Buried Repositories: *International Journal for Numerical and Analytical Methods in Geomechanics:* Vol. 13, pp. 37-51.

Ringwood, T., 1982, Immobilization of Radioactive Wastes in SYNROC: *American Scientist,* Vol. 70, No. 2, pp. 201-207.

Rodgers, Z.W., 1974, Process for Refuse Disposal in Solution-Mined Salt Cavities: In Coogan's Fourth Symposium on Salt, Vol. 11.

Roxburgh, I.S., 1987, *Geology of High-Level Nuclear Waste Disposal - An Introduction:* Chapman and Hall, New York, 229 p.

Schneider, H.J., 1988, Safe, Zero - Immission of Ultimate Disposal of Solid Hazardous Wastes in Salt Caverns: In *Hazardous Waste - Detection, Control, Treatment* (Edited by Abbou, R.), Elsevier Science Publishers, The Netherlands, pp. 1505-1512.

Siegenthaler, C., 1987, Hydraulic Fracturing - A Potential Risk for the Safety of Clay-Sealed Underground Repositories for Hazardous Wastes: *Hazardous Waste and Hazardous Materials,* Vol. 4, No. 2, pp. 111-117.

Skinner, B.J. and Walker, C.A., 1982, Radioactive Wastes: *American Scientist,* Vol. 70, No. 2, pp. 180-181.

Smedes, H.E., 1980, Rationale for Geologic Isolation of High-level Radioactive Waste and Assessment of the Suitability of Crystalline Rock: U.S. Geological Survey Open-File Report 80-165.

Stauffer, T.P., Sr., 1978, Kansas City's Use of Limestone Mines for Business, Industry, and Storage: In *Proceedings of First International Symposium, Rockstore 77,* (Edited by Bergman, M.), Pergamon Press, New York.

Stauffer, T., Jr., 1981, Grain, Seed, Food Storage and Farm Machinery Manufacturing - Kansas City's Use of Underground Space in the U.S. Agricultural Midwest: In *Proceedings of International Symposium on Subsurface Space, Rockstore 80, Pergamon Press, New York.*

Stone, R.B., et al., 1975, Evaluation of Hazardous Wastes Emplacement in Mined Openings: EPA-600/2-75-040; Prepared by Fenix and Scisson, Inc., for U.S. EPA, Cincinnati.

Stone, R.B., Moran, T.R., Weyand, L.W. and Sparkman, C.U., 1985, Using Mined Space for Long-Term Retention of Non-radioactive Hazardous Waste, Vol. 1, Conventional Mines: U.S. Environmental Protection Agency Report No. PB85-177111, 38 pp.

Stone, R.B., Covell, K.A. and Weyand, L.W., 1985, Using Mined Space for Long-Term Retention of Non-radioactive Hazardous Waste, Vol. 2, Solution Mined Salt Caverns: U.S. Environmental Protection Agency Report No. PB85-177129, 53 p.

Stone, R.B., 1987, Underground Storage of Hazardous Waste: *Journal of Hazardous Materials,* Vol. 14, pp. 23-37.

Swanson, D.A. and Wright, T.L., 1978, Bedrock Geology of the Southern Columbia Plateau and Adjacent Areas: In *The Channeled Scabland* (Edited by V.R. Baker and D. Nommedal), National Aeronautical and Space Administration Planetary Geology Program, Washington, D.C., pp. 37-57.

Testa, S.M., 1988, Hazardous Waste Disposal Site Hydrogeologic Characterization: In Proceedings of the Second International Conference on Case Histories in Geotechnical Engineering, St. Louis, Paper No. 1.26, pp. 97-106.

Testa, S.M., 1990, Igneous Rocks: Granitic: In *Magill's Survey of Science, Earth Science Series,* (Edited by Magill, F.N.), Vol. 3, Salem Press, Pasadena, CA, pp. 1180-1185.

U.S. Department of Energy, 1988, Site Characterization Plan Overview - Yucca Mountain Site, Nevada Research and Development Area, Nevada: DOE Office of Civilian Radioactive Waste Management Report DOE/RW-0198, 164 p.

U.S. Department of Energy, 1991, Site Characterization Progress Report - Yucca Mountain Site, Nevada: DOE Office of Civilian Radioactive Waste Management Report DOE/RW-0307P.

Walton, J.C. and Sugar, B., 1990, Aspects of Fluid Flow Through Small Flaws in Membrane Liners, *Environmental Science and Technology,* Vol. 24, pp. 920-924.

Walton, J.C., 1991, Fluid Flow and Placement of Concrete Vaults in the Saturated or Unsaturated Zone: *Waste Management,* Vol. 11, pp. 3-10.

Whitfield, J.W., 1981, Underground Space Resources in Missouri: Missouri Department of Natural Resources, Division of Geology and Land Survey, Rolla, MO, 1981.

Winograd, I.J., 1974, Radioactive-Waste Storage in the Arid Zone: *American Geophysical Union Transactions,* Vol. 55, pp. 84-94.

Winograd, I.J., 1981, Radioactive-Waste Disposal in Thick Unsaturated Zones: *Science*, Vol. 212, pp. 1457-1464.

Witherspoon, P.A., Cook, N.G.W. and Gale, J.E., 1981, Geologic Storage of Radioactive Waste - Field Studies in Sweden: *Science*, Vol. 211, pp. 894-900.

13   OCEAN DISPOSAL

"It is a curious situation that the sea, from which life first arose, should now be threatened by the activities of one form of that life. But the sea, though changed in a sinister way, will continue to exist; the threat is rather to life itself."
(Rachel Carson, 1961)

INTRODUCTION

Ocean waste disposal typically occurs on the continental shelves where abundant sea life is concentrated. In addition, approximately half of the 30,000 land-based hazardous waste sites that have been identified are situated in coastal areas. Add to this 90% of the world's marine fish catch (by weight) and more than one fifth of the world's population living in coastal areas by the turn of the century, and population without concurrent control on waste management may be the single most important factor in coastal pollution. Contaminants derived from these sites contribute to some degree either directly or indirectly to adverse water quality and biological effects (with correlation of upstream activities to coastal impact very difficult to establish). Although the cause-and-effect relationship is difficult to demonstrate conclusively, data has been generated from chemical testing of sediments and testing of fish and mussels indicating the occurrence of relatively high levels of contamination in many of the major harbors of the U.S. — Boston Harbor, Corpus Christi Bay, parts of Long Island Sound, New York Harbor, Puget Sound, San Antonio Bay, San Diego Harbor, San Francisco Bay, and Santa Monica Bay, in addition to other areas such as Manila Bay, the North Sea, and the Mediterranean Sea.

The National Oceanic and Atmospheric Administration (NOAA) defines coastal ocean as beginning at the head of tidal waters and extending to the edge of the continental shelf. The Hudson River, for example, as far upstream as Albany is considered part of the coastal ocean. And although public policy has historically treated estuaries and marine waters separately, these areas are now perceived (including adjacent terrestrial areas) as a very complex interconnected inseparable system. The estuaries eventually receive most of the waste in some form that is generated in coastal areas, and actually inadvertently serve as a repository for it. This condition is exacerbated when small water volumes and minimal tidal flushing conditions exist.

The past 15 years have witnessed a substantial decrease in ocean dumping of industrial waste in the U.S., although the volume of municipal wastewater treatment sludge has increased. Although most wastes are no longer literally dumped into the ocean, contaminants do enter the coastal areas via other pathways including river and coastal runoff, atmospheric fallout and precipitation, and point and nonpoint source discharges.

Only a few permits currently exist for industrial waste disposal in the ocean. Most of this type of waste is referred to as diluted acid waste which is quickly neutralized by the ocean's enormous buffering capacity. Since 1981, there has been at least a ten fold reduction in the volume of industrial waste being dumped in the ocean. In fact, disposal of industrial waste was ordered to cease in all waters under the jurisdiction of the U.S. by 1991. Disposal of domestic sewage sludge in the coastal environment was ordered to cease after 1992. Although these actions are consistent with the thoughts of Jacques Cousteau and other environmentalists, some ocean waste disposal programs have been successful in the past and some argue that parts of the ocean can accept some wastes if introduced properly. So, although some would give the ocean preferential treatment and believe it needs to be left unspoiled and unaffected by man's intervention, others feel that the ocean can be used in a prudent manner for the assimilation of certain wastes.

In the U.S., four deep ocean sites have received various wastes; one of them is still active. These sites include the active Deepwater Dumpsite (DWD)-106 and other sites in the Mid-Atlantic Bight situated southeast of Long Island; one that is now inactive north of Arecibo, Puerto Rico; and two inactive sites in the Gulf of Mexico (Figure 13-1). Waste types and volumes per location are summarized in Table 13-1. Nearshore dump sites historically used between 1947 and 1973 in state and federal waters of the Southern California Bight is shown in Figure 13-2. Based on these volumes, the amount of waste disposed nearshore (i.e., Mid-Atlantic Bight) significantly exceeds that of deep ocean sites combined. In Europe, three deep ocean dumpsites situated west of the U.K. in the eastern Atlantic Ocean, with depths of about 4000 m, have been used for the disposal of various wastes including industrial, sewage sludge, and low-level nuclear wastes (Figure 13-1).

Dredged materials are predominantly disposed of nearshore. In 1980, for example, about 41% of all dredged materials were disposed of within 5.5 km of the shoreline with all but one site on the continental shelf. The only deep ocean site for dredged materials is south of Honolulu, Hawaii, at a depth of about 410 m where no continental shelf exists. Most of the materials incinerated

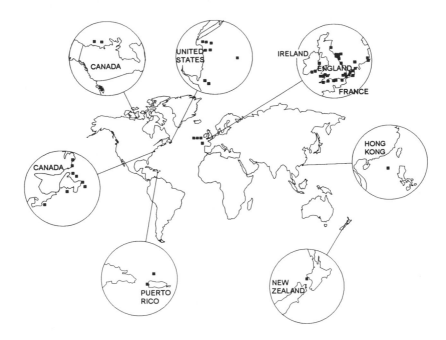

Figure 13-1 Location of waste dumpsites.

at sea are primarily in the North Sea. The overall extent of past and recent unreported disposal of obsolete ships, munitions, and other military material is uncertain; however, in one case, a potential mining operation for phosphorite on the continental shelf of California had to be abandoned due to the presence of a munitions dumpsite, including unexploded shells overlying the phosphorite deposits. Major dredged materials dumpsites are shown in Figure 13-3.

Presented in this chapter is discussion of types of ocean wastes, the ocean environment, ocean processes and geologic considerations.

TYPES OF OCEAN WASTE

Several types of waste are disposed of in the ocean. The primary types of waste in descending volumes include dredged materials, sewage sludge, industrial waste, construction debris, and incineration residue. Other types of waste materials include solid waste, munitions and explosives, medical wastes, coal wastes, and low-radioactive wastes. These wastes exhibit a variety of physical and chemical properties. Information on the volume of these materials as of 1983 is presented in Table 13-2 and illustrated in Figure 13-4. These waste types and their respective characteristics are further discussed below.

Table 13-1 Summary of Location, Waste Types, and Volumes for Nearshore and Deep Ocean Dumpsites

Disposal Site	Ocean Environment	Depth (m)	Location	Waste Types
Deepwater Dumpsite-106	Deep	2000	196 km southeast of New York City	Industrial Sewage sludge Coal ash Low-level nuclear
Puerto Rico	Deep	6000	70 km north of Arecibo, Puerto Rico	Industrial
Gulf of Mexico	Deep		220 km south of Galveston, Texas	Industrial
	Deep		90 km south of the Mississippi River	Industrial
New York Bight	Nearshore	2000		Industrial Sewage sludge
Denmark	Deep	4000	Eastern Atlantic Ocean	Industrial (liquid)
Germany	Deep	4000	Eastern Atlantic Ocean	Sewage sludge
U.K.	Deep	4000	Eastern Atlantic Ocean	Industrial (solid)
Nuclear Energy Agency	Deep	4000	Eastern Atlantic Ocean	Low-level nuclear

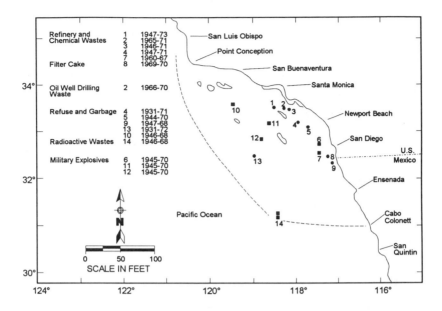

Figure 13-2 Ocean dump sites designated and used between 1947 and 1973 in the Southern California Bight (after Chartrand, et al., 1985).

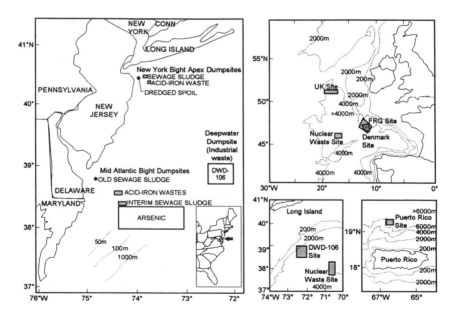

Figure 13-3 Location of dredged materials nearshore dumpsites.

Table 13-2 Primary Classes of Ocean Particulate Wastes in U.S.
Coastal Waters[a,b]

Year	Industrial Waste ($\times 10^6$ tons)	Sewage Waste ($\times 10^6$ tons)	Demolition Construction ($\times 10^6$ tons)	Dredged Material ($\times 10^6$ tons)	Incineration at Sea ($\times 10^3$ tons)
1968	4.2	4.5	0.52	ND[c]	ND
1973	5.0	4.9	0.88	51	9.8
1974	4.1	4.5	0.70	75	13.3
1975	3.1	4.5	0.36	66	5.6
1976	2.4	4.8	0.28	50	7.9
1977	1.6	4.6	0.35	31	13.6
1978	2.2	5.0	0.22	40	16.2
1979	ND	ND	ND	61	ND

[a] Modified after Duedall, et al., 1983.
[b] Wasteloadings are approximate metric tons, except as noted.
[c] ND = No data available.

Ocean dumping in the United States

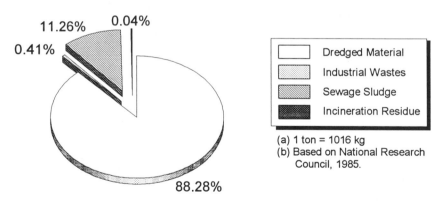

Figure 13-4 Volume of waste materials disposed via ocean dumping in the
U.S. (modified after National Research Council, 1985).

Dredged Materials

Most dredged materials in the U.S., approximately 200 million m^3 of
sediment per year, is generated during routine operations associated with
maintenance of federal navigation projects. Dredged materials are typically
sedimentary in origin with bulk densities varying from 1.2 to 2.0 g/cm^3 as a
function of mineral composition, grain size, age, and handling technique (i.e.,

dredging, transport, and disposal). Texturally, these materials encompass the entire grain size spectrum from clays and silts to fine-grained sands to gravels. Contaminant loads are also highly variable, with the higher loads associated with the finer-grained sediments and organic content. Approximately 10% of dredged materials in total volume is considered highly contaminated, whereas approximately 10 to 20% is considered clean. The remaining volume is comprised of moderate concentrations of a variety of organic and inorganic contaminants.

Sewage Sludge

Sewage sludge is a concentrated semisolid material and is the principal waste produced at municipal wastewater treatment facilities. Sewage sludge is characterized as a liquid in which up to 10% is a heterogeneous mixture of solids including microorganisms, organic detritus or aggregates, fibers, mineral grains, and food residues. With no well-defined morphology, the size of the particulates varies from readily visible to microscopic to colloidal. Bulk densities are on the order of 1.01 g/cm^{-3}, which is slightly less than the density of seawater, and a solid density of approximately 1.5 g/cm^{-3}. Sewage sludge when mined with industrial waste may also contain significant amounts of heavy metals.

Environmental concerns pertaining to disposal of sewage sludge to the ocean include chemical changes which may enhance the mobility of some elements, notably metals, by means of oxidation due to precipitation of oxides, increased microbial breakdown of organic matter releasing organically bound heavy metals, or displacement of adsorbed metals by ion exchange processes, reflecting the large amounts of magnesium and calcium present in seawater. Other concerns pertain to changes in turbidity, burial of benthic organisms, and lethal toxicity to biota. Most important to predicting the fate of sewage sludge after discharge into the ocean are the settling speeds of its constituents, which depend upon solid content and grain size. The New York Bight apex area is bordered by an industrialized metropolitan region encompassing approximately 2000 km^2, and is intensively used for commerce, recreation, and disposal of acid wastes, construction rubble, dredged spoils, and sewage sludge. Contaminant monitoring of sediments from the New York Bight revealed trace metal-contaminated surface sediments, notably lead, that were reported at levels 0.5-fold higher than preindustrial levels. Although sewage sludge dumping contributed far less mass to the apex area, approximately eight times as much area in and around the dump has been impacted relative to dredged spoils.

One alternative under consideration to the direct discharge of sewage sludge to the ocean is the stabilization of this material into block form. This is accomplished by the use of fly ash, gypsum, lime, and Portland cement. This approach allows the sludge to be reclassified as nonhazardous. Most importantly, preliminary tests have shown that the structural integrity of the blocks following submersion into seawater is maintained. As an additional benefit, these blocks may serve as artificial reef construction materials.

Industrial Wastes

Industrial wastes are mainly dissolved in an acidic (pH <1.0) and highly concentrated iron (approximately 0.8 M) solution which forms an acid-iron waste plume. The acid in acid iron waste is either sulfuric acid (H_2SO_4) or hydrochloric acid (HCl), or a combination of the two. Acid iron waste is comprised of suspended solids ranging from 0.01 to 10 g/L^{-1}, and specific gravities ranging from 1.084 to 1.35 g cm^{-3}.

Typical waste disposed of at the Deepwater Dumpsite-106, for example, has pH and bulk density of about 0.01 and 1.5 g cm^3, respectively, in comparison to seawater at approximately 8.2 and 1.02 g mL^{-1}, respectively. Upon disposal, two major chemical processes occur: neutralization and hydrous ferric oxide precipitation. Neutralization occurs rapidly due to the abundance of carbonate and borate systems in seawater. The formation of hydrous ferric oxide precipitate occurs in accordance with the following reaction:

$$Fe^{3+} + (n+2)H_2O \rightarrow FeOOH \cdot nH_2O + 3H^+$$

where n represents the variable extent of hydration of the solid phase. Rapid dilution of FeOOH · nH$_2$O on the order of 10^4 to 10^5 occurs within a few hours of mixing. Other particulates of concern which can vary seasonally include copper, cadmium, and lead.

Industrial waste disposed of in the Mid-Atlantic Bight have included byproducts of the manufacturing of titanium dioxide, caustic, and white-water oil emulsions. Titanium dioxide is a white pigment used as a replacement of the highly toxic lead-based white pigment in the manufacturing of paints, paper, printing inks, plastics, textile fibers, food additives, and cosmetics. Wastes derived from the sulfate and chloride processes used for production of titanium dioxide include iron and mother liquor, and metal chlorides, respectively. White water refers to the stable 0.1-on-water emulsions consisting of approximately 85% water, 15% oil, and small amounts of sodium antioxidants, organic acids, alcohol, and heavy metals with a pH usually between 8 and 13. Spent caustic refers to process wastewater containing approximately 0.1 to 7.5% (by weight) free sodium hydroxide and possibly trace amounts of sulfides, phenols, and heavy metals with a pH between 9.5 to 14.

Britain contributes the largest volume of industrial waste to the North Sea relative to other European countries. Three quarters of this material is characterized as fly ash waste derived from power stations and colliery wastes as discussed later. Other wastes include acids and small quantities of toxic waste mainly derived from the scrubbing of gases in industrial incinerators to prevent air pollution. These waste types are derived from the pharmaceutical, chemical, and textiles industries. Other industrial wastes disposed of in the North Sea include compounds of organotin used in pesticides, and wastes from offshore oil operations and the titanium oxide industry. Although water currents off the east coast of Britain are sufficiently strong to disperse waste and sediment into the main body of the North Sea, a significant quantity tends to accumulate in the Wodden Sea off the west coast of Germany.

Construction Debris

Construction debris, as discussed in Chapter 7 (Waste Characterization), includes primarily debris generated during building renovation or demolition activities, but may also include discarded equipment, broken drums, and containers, among other items. Construction debris has long been disposed of offshore and favorably utilized for artificial reef structures as well as erosion abatement. Some of these materials may contain hazardous or toxic constituents, depending on their source and historical usage.

Incineration Residue

Incineration of combustible wastes at sea is a viable method of disposal of certain types of waste including toxic organic chemicals, which are converted to carbon dioxide, water, acidic gases (mainly HCl), and a small amount of residual material. Incinerated waste can include ashes, slag, and dust. Soluble constituents of dissolved dust derived from waste incineration are comprised mainly of calcium, sodium, chlorides, and sulfates with subordinate amounts of heavy metals. Typical eluate analysis of dust derived from the waste incineration process is presented in Table 13-3.

Other Waste Types

Other waste types include pharmaceutical waste and fly ash-type water, which includes consolidated fly ash and scrubber sludge. Pharmaceuticals include medical wastes and may be generated by hospitals, clinical laboratories, blood banks, and doctors' offices. Types of wastes include infectious, pathological, bulk blood, carcasses of pathogen-exposed animals, needles and syringes, and tainted medical care equipment, among others. Suspended solids and specific gravity for pharmaceutical-type wastes are on the order of 0.5 g l^{-1} and 1.025 g cm^{-3}, respectively.

**Table 13-3 Eluate Analysis of Dust
From Waste Incineration Process**

Constituent	Concentration (mg/L)
Calcium	1023.0
Magnesium	0.3
Sodium	803.0
Chloride	2800.0
Fluoride	7.04
Sulfate	2244.0
Zinc	0.19
Cadmium	0.03
Mercury	0.002
Lead	0.25
Thallium	0.002
Molybdenum	0.26
Arsenic	0.001

Fly ash is denser than water and characterized as a solid produced in large quantities from the combustion of coal; it consists mostly of small, solid spheres with diameters ranging from 0.5 to approximately 200 μm. The settling behavior of fly ash is complex. Obviously, smaller sized particles (i.e., <10 μm) will have very low settling rates (on the order of 10^{-3} cm/s^{-1}) relative to larger particles. Some of the fly ash will float, that being the cenospheres. Cenospheres are glassy spheres ranging in diameter from 20 to 200 μm that are filled with nitrogen and carbon dioxide gases, comprising up to 20% in total volume of fly ash. The variability in fly ash composition and physical properties reflect the origin of the coal, degree of pulverization before the coal is burned, and the type of boiler unit utilized. Scrubber sludge is principally gypsum and calcium sulfite hemihydrate. Solids density for fly ash ranges from 1.5 to 1.7 g cm^{-3}.

THE OCEAN ENVIRONMENT

The nearshore ocean environment comprises about 10% of all oceanic areas and include estuaries, bays, and continental shelves (Figure 13-5). The nearshore waters also support more than 99% of the commercial fisheries. Deep ocean disposal is that which occurs in waters deeper than 200 m. Although arbitrary, it excludes most of the continental shelves and includes waters that have a permanent pycnocline. The exception to this is in the polar regions where deep and bottom waters are formed at the sea surface. A summary of nearshore and deep ocean characteristics is presented in Table 13-4.

Figure 13-5 Extent of continental shelves of the oceans.

The ocean water is stratified due to variations in density, temperatures, and salinity with depth. These variations reflect latitudinal differences in surface temperature, precipitation, zonal winds, and cold and dense bottom water in polar regions. Additional zones may be ascertained based upon penetration of sunlight (photic zone), effective wave base, variations in the amount of dissolved oxygen, among other factors. Each of these zones or water masses approaches the continental margins at an approximate known depth range forming bathymetric horizons demarcated by distinct variations in faunal and sedimentologic parameters (Figure 13-6). Significant microfaunal and sedimentologic trends across major physical oceanographic boundaries based on patterns evident off southern California and in the Gulf of California are shown in Figure 13-7.

The characteristics of the circulation of oceanic waters and associated forces must be understood since distribution of sediments and organisms, many broad-scale geologic features and geochemical fractionation processes are controlled by such circulation patterns through the water column. The primary driving force of surface currents and of heat transport is the prevailing wind system. The most important properties of seawater are temperature, density, and salinity. With depth, water is also set in motion via density differences in seawater as a result of temperature and salinity differences; warmer water is less dense than colder water and high salinity water is more dense than low salinity water, respectively. The wind-driven system of currents has little connection with deeper (thermohaline processes). Density-driven circulation, or thermohaline circulation, is affected by the topographic configuration of the ocean floor and may have a significant effect on the geology.

Density differences through the water column are vertically stratified with currents at varying levels flowing in different directions. Temperatures gener-

Table 13-4 Summary of Physical Characteristics Between Nearshore and Deep Ocean Environments[a]

Parameter	Nearshore (Continental Shelf)				Deep Ocean		
	Deltas and Fans	Submarine Canyons	Rise	Slope	Abyssal Plains	Abyssal Hills	Trenches
Water depth (km)	0.1 to 5	NA[b]	3 to 5	0.2 to 3	4 to 6	5 to 6	7 to 11
Regional slope (degrees)	0.5 to 5	NA	<1	4 to 10	<0.1	<1	2 to 20
Sediment type	Sand, silt, clay	Sand, silt, clay	Silt, clay	Silt, clay	Sand, silt, clay	Clay	Sand, silt, clay
Sedimentation (cm/1000 years)	10 to 500	NA	5 to 50	<0 to 3	2 to 5	<1 to 10	1 to 10
Currents (cm/s)	5 to 30	NA	5 to 30	5 to 20	<5	2 to 10	2 to 10
Depositional environment (erosion or deposition)	Deposition	Erosion	Deposition, erosion	Deposition, erosion	Deposition	Deposition	Deposition
Seismicity	Low to high[c]	NA	Low to high[c]	Low to high[c]	Very low	Very low	Very high
Biological activity	High	NA	Moderate	High	Low	Very low	Moderate
Sediment failure frequency	Very high	Very high	Moderate	Very high	Low, some faulting	Very low	High
Geologic stability (predictability)	Very low	Very low	Moderate	Very low	Moderate	High	Very low
Accessibility	Moderate	Moderate	Low	Moderate	Very low	Very low	Very low

[a] Modified after Silva, 1990.
[b] NA = Information not available.
[c] Dependent upon location relative to plate boundary.

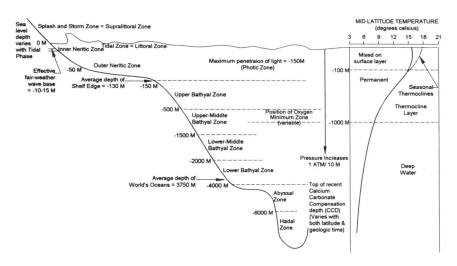

Figure 13-6 Classification of benthonic marine environments in terms of depth and positions of critical oceanographic boundaries or transitional zones (modified after Ingle, 1980).

ally decrease with depth. Zones of rapidly changing temperature or density are referred to as the thermocline or pycnocline, respectively. The water column can essentially be divided into three zones based on changes in temperature and density: surface or mixed layer, permanent thermocline, and intermediate and bottom waters zone. The surface or mixed layer, as previously stated, is wind driven with little interaction with deeper circulation and is directly affected by winds, wave action, and seasonal cooling. The salinity increases as a result of evaporation. This zone also contains the primary source of food via photosynthesis by marine plants with no replacement by photosynthesis or mixing since this zone lies below the thermocline. Low oxygen values can also be indicative of older, deeper water.

Salinity distribution is roughly zonal and typically more diffused in comparison to temperature patterns. Salinity values normally occupy a small range from about 33 to 37%, averaging about 35%. Processes affecting development of salinity patterns include evaporation, precipitation, and ice melting. High values of about 41% are encountered in the Red Sea, whereas values less than 33.5% are encountered near the Equator in the east.

Dissolved oxygen also decreases from the surface to deeper depths. At intermediate depths of about 150 to 1000 m, relatively low levels of dissolved oxygen exist. This level is referred to as the oxygen minimum layer. Geologically, this zone of reduced oxidation is important because it leads to the formation of organic-rich sediments. At deeper depths, oxygen is consumed by living organisms and the oxidation of detritus.

The two important oceanographic considerations regarding the ocean as a waste repository are the permanent pycnocline and the oxygen minimum zone.

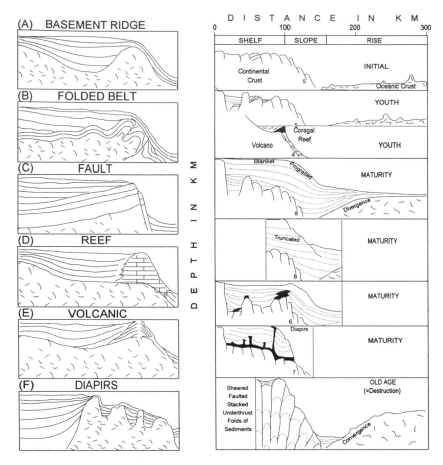

Figure 13-7 **Significant microfaunal and sedimentologic trends across major physical oceanographic boundaries based on pattern off Southern California and in the Gulf of California (after Ingel, 1980).**

If waste dumped into the deep ocean reaches the permanent pycnocline, the possibility of this material spreading and thus undergoing more dispersion before ultimately reaching the ocean floor exists. Conditions that encourage this include waste properties, discharge procedures, and water-column stratification.

The oxygen minimum zone on the other hand is generated by biochemical oxidation. Oxygen-deficient water masses in the oxygen minimum layer commonly intersect the continental margins on the upper slope. Although oxygen levels beneath this zone are large, oxygen is also renewed very slowly. Given the slow rate of oxygen renewal, any waste-disposal strategy needs to consider potential effects on oxygen concentrations, particularly in oxygen depletion zones which occur in topographic lows.

Oxygen-deficiency basins may also develop from the bottom upward in offshore basins if water within the basin becomes density stratified due to the formation and preservation of either a thermal or salinity layer. In such oxygen-deficient basins, oxygen is irreplaceably consumed in the water column and on the seafloor by the destruction of organic matter. Recognition of such basins is important in the disposal of organic-rich waste since biological activity will be nil and will remain for a long period of time unless transported via sediment processes.

Continental Margins

Continental margins are comprised of the continental shelf, continental slope, shelf break, and continental rise. The inclination of the continental shelf averages 7' while the slope exhibits an average inclination of 4°; the break is considered to occur at a depth of about 200 m. Continental shelves and slopes receive most of the waste disposed of at sea. The characterization of these areas regarding morphology, geologic history, depositional and erosional patterns, and sedimentation cycles is essential to addressing stability and predictability of the waste over time. In most cases, for example, the continental slope exhibits well-defined depressions, ridges, step-like features, isolated mounts, submarine canyons, and faults.

Continental margins can be distinguished and characterized by five morphological types: shelf-rise type, bordering-depression type, marginal plateaus, continental borderlands, and steep rift continental margins. These morphological types are illustrated in Figure 13-8 and characterized as follows:

Shelf-rise types: A broad, well-developed, continental shelf, slope, and rise. Dams are present internally below the continental shelf and rise. Examples include the east coast of the U.S. and Gulf of Mexico.

Bordering-depression types: Deep-sea trenches exist with some filled with sediment deposits of varying thicknesses. Examples include the Peru-Chile Trench (i.e., greater than 1000 m of trench-filled sediments) and the Japan Trench (i.e., thin deposits due to lack of sediment supply).

Marginal plateaus: Plateau-like structures occur on the continental slope. Dam structures may exist on the outer margin of the plateau. An examples is the Blake Plateau.

Continental borderlands: A series of basins and ridges exist due to repeated folding and faulting. Some ridges may reach above sea level as islands. Basins are filled with sediments of which some may be deformed from continuous tectonism, and probably deposited via low-velocity, low-density turbidity currents with the basin closest to land being filled first. Examples include off Southern California and northern Baja California.

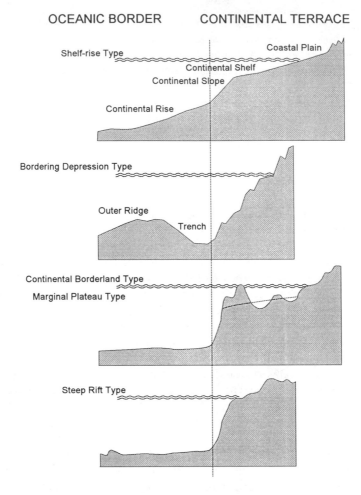

Figure 13-8 Various morphological types of continental margins (after Curray, 1969)

Steep rift continental margins: Continental rise is poorly developed with no trench or depression present at the base of the continental slope. With time, material will accumulate at the base of the continental slope. An example is the west coast of the Iberian Peninsula.

Downslope processes that play an important role in modifying the sea floor range from very slow creep-type deformation to slumps and slides which may involve large volumes of saturated sediments which can travel long distances. Other processes, as illustrated in the Gulf of Mexico (Figure 13-9), include gas seeps, salt diapirs, mud lumps (i.e., soft sediments that are extruded upward through surficial layers), and collapse features.

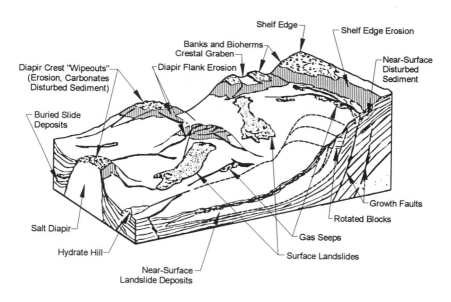

Figure 13-9 Some observed features and inferred processes in the Gulf of Mexico (after Campbell, et al., 1986).

Continental Shelf

The continental shelves are essentially shallow platforms adjacent to land masses. Continental shelves are characterized by an irregular topography comprised of banks, basins, and valleys. The average width is about 75 km with total relief on the order of 40 m or greater. The seaward limit is usually defined as the shelf break encountered at depths ranging from about 100 to 200 m, which may be gradual or sharp.

Continental Slope

The continental slope comprises the central part of the continental margin and is the most consistent and significant topographic discontinuity in the ocean basin. Less studied relative to continental shelves and abyssal plains, continental slopes cover 10 to 15% of the earth's surface. Width of the continental slope ranges from 20 to 100 km. Most information available on sediments of the continental slope are based upon seismic reflecting profiling. The upper boundary of the continental slope is the shelfbreak. The lower boundary is gradual and located at depths ranging from 1400 to 3200 m or greater, notably at the continental rise where the gradient decreases to 1:40. This lower boundary is not always located due to the presence of deep-sea fan deposits, or is found in depressions (i.e., a deep-sea trench where the slope may terminate at a depth of 10,000 m).

Slopes act as sites of permanent deposition comprising part of the accretion-ary wedges flanking stable cratons, or as temporary storage areas for sediments in transit to deeper water. Submarine canyons are the main conduits for transport of terrigenous materials into the deep ocean.

Continental slopes may be smooth, slightly convex, or irregular on a large or small scale. In the western Gulf of Alaska above the Aleutian Trench, the middle portion of the slope is terraced; whereas, an inner and outer slope characterizes the southern California borderland, with banks and islands sepa-rating deep basins. The Blake Plateau is a marginal plateau with a relatively smooth surface. Small-scale irregularities reflect mass movement, folding and faulting, diapirism, and slope erosion. Dominant morphological features are incisions which cut more or less transversely, and are referred to as submarine canyons. Submarine canyons initiate near the top of the slope or on the continental shelf. As previously mentioned, submarine canyons are conduits for sediment migration to deep-sea fans, continental rises, and abyssal plains. A modern slope undergoing numerous small-scale mass-movement events poses certain potential geologic hazards which could adversely influence areas utilized for ocean waste disposal and influence the selection of future sites. Mass wasting can take place on slopes as low as 0.8°. Large slumps 10 to 35 m in thickness have been reported to have traveled 7 to 10 km over slopes 3 to 5°.

Shelfbreak

The shelfbreak is a distinct interface on continental margins and delineates the major physiographic boundary between the continental shelf and continen-tal slope. Defined as the first major change in the gradient at the outermost edge of the continental shelf, it is highly variable in depth and distance from the shore. Diversity of shelf-to-slope configurations emphasizes geomorphology (i.e., genetic-descriptive classification), although the main controlling factors are geological framework and substrate mobility. Several schemes showing shelf-to-slope configuration emphasizing structural framework and modifying effects of sedimentary processes are shown in Figure 13-10. Different types of dams and thus subsequent effects on the shelfbreak are also shown in Figure 13-10.

Depositional processes include river-supplied sediment, carbonate forma-tion and varying rates of reef build-up, influence of ice transport in higher altitudes, and interactions between fluid-driven and gravitative processes at and adjacent to the shelfbreak. Physical, biological, and chemical processes also assist in modifying the outer edge of the shelf, notably during lower enstatic sea level stands when the shelfbreak is closer to the shoreline.

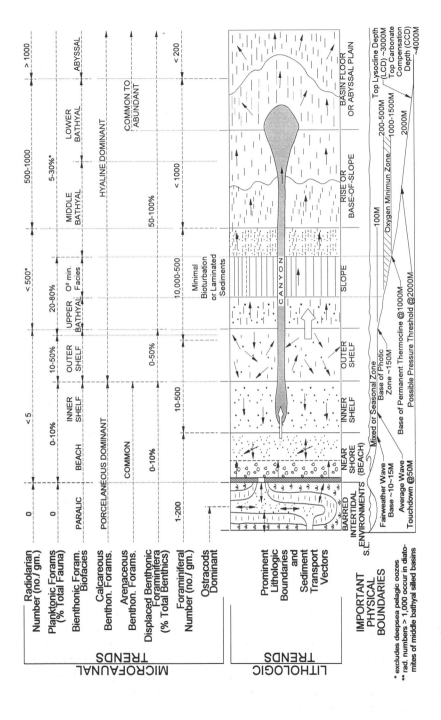

Figure 13-10 Several schemes showing shelf-to-slope configurations (modified after Hedberg, 1970 and Emery, 1980)

Continental Rise

Seaward of the continental slope is the continental rise. The gradient along the rise is quite variable, ranging from 1:50 to 1:800. It is typically composed of a long wedge of sediments, usually several kilometers in thickness. These sediments accumulate through mass movement (i.e., sliding and slumping processes) and/or sediment transport via turbidity currents down the slope. Rises along the margins of the Atlantic and Indian Ocean are wide relative to the Pacific Ocean, where rises are narrow or even absent.

Deep Ocean Regimes

Deep ocean morphological features include abyssal plains and abyssal hills, and, to a lesser degree, trenches, fracture zones, fan areas seaward of large deltas, flanks, crests, and rift valleys of mid-oceanic ridges. Most pertinent to ocean disposal are abyssal plains and abyssal hills.

Abyssal Plains

Abyssal plains are characterized as large, nearly flat areas of the deep ocean floor adjacent to the outer margins of the continental rise. Sedimentary deposits are comprised of varying amounts of pelagic clays, hemipelagic muds, calcareous oozes, and turbidite deposits, depending upon proximity to the continental margin sediment sources (Table 13-5). Stratification is quite variable. Pelagic zones of relatively low permeability occur interstratified with turbidite layers which generally grade from coarse at the base to relatively fine at the top of the sequence. Thus, permeability varies with grain size. Depositional patterns are also complex. Deep-seated faults or other structural elements also exist, reflecting tectonic related activities or differential consolidation.

Abyssal Hills

Abyssal hills are characterized as broad sedimentary deposits of relatively low relief. Slopes range from about 1 to 15% and commonly undergo slumping on the deeper slopes and localized erosional processes. These regions cover vast areas of some ocean basins (i.e., 80 to 85% of the Pacific Ocean floor) and are considered seismically passive. Sediments are primarily comprised of pelagic and hemipelagic clays of relatively low permeability with varying amounts of carbonate and siliceous material.

Table 13-5 Classification of Deep-Sea Sediments[a]

Sediment Type	Sediment Subtype	Description
Pelagic deposits (oozes and clays) <25% of fraction >5 μm is of terrigenous, volcanogenic, and/or neritic origin. Median grain size <5 μm (except authigenic minerals and pelagic organisms)	A. Pelagic clays; $CaCO_3$ and siliceous fossils <30%	1. $CaCO_3$ 1-10%; (slightly) calcareous clay 2. $CaCO_3$ 10-30%; very calcareous (or marl) clay 3. Siliceous fossils 1-10%; (slightly) siliceous clay 4. Siliceous fossils 10-30%; very siliceous clay
	B. Oozes; $CaCO_3$ or siliceous fossils >30%	1. $CaCO_3$ > 30% < 2/3$CaCO_3$; marl ooze >2/3$CaCO_3$; chalk ooze 2. $CaCO_3$ < 30% < 30% siliceous fossils: diatom or radiolarian ooze
Hemipelagic deposits (muds) > 25% of fraction >5 μm is of terrigenous, volcanogenic, and/or neritic origin; median grain size >5 μm (except authigenic minerals and pelagic organisms)	A. Calcareous muds; $CaCO_3$ > 30%	1. <2/3$CaCO_3$; marl mud. >2/3$CaCO_3$: chalk mud 2. Skeletal $CaCO_3$ > 30%: foram ~, nanno~, coquina~
	B. Terrigenous muds; $CaCO_3$ < 30%; quartz, feldspar, mica dominant. Prefixes: quartzose, arkosic, micaceous	
	C. Volcanogenic muds. $CaCO_3$ < 30%, ash, palagonite, etc., dominant	
Pelagic and/or hemipelagic deposits		1. Dolomite-sapropelite cycles 2. Black (carbonaceous) clay and mud: sapropelites 3. Silicified claystones and mudstones: chert 4. Limestone

[a] Modified after Berger, 1974.

OCEAN PROCESSES

Wastes disposed of in the ocean are complex mixtures varying widely in composition and concentrations. These mixtures obviously undergo dispersion and advection, but are also altered by physical, geochemical, and biological processes. In considering the short- and long-term effects of ocean disposal, the ultimate fate of a particular waste reflects the integration of these processes. For example, a waste material is initially influenced by physical processes which continuously act to dilute the waste until it reaches the seafloor, where other processes serve to concentrate it. Chemical and biological processes are, for the most part, short term in nature until the waste reaches the seafloor where these processes are then more long term in nature. Once the waste reaches the seafloor, sedimentological processes (i.e., settling, sedimentation, and bioaccumulation processes) also serve to concentrate it, although geochemical processes can affect the mobility of some wastes.

Physical Processes

Physical processes affecting the disposal of wastes in the ocean include dilution, advection, sedimentation, and the overall physical and chemical properties of the waste itself. Materials discharged by either mechanical or hydraulic dredging operations tend to settle through the water column as a function of settling velocity and local hydrographic conditions. Settling velocities vary as a function of sediment type and properties, mass concentration, and velocity field structure. The overall process is illustrated in Figure 13-11.

Chemical Processes

Chemical processes affecting wastes include acid-base neutralization, dissolution, precipitation or flocculation (the agglomeration of small particles to form larger ones), surface exchange reactions between particles and seawater (i.e., adsorption, desorption, etc.), volatilization at the sea surface, and changes in oxidation state. These processes are important factors which determine the chemical speciation of constituents in the waste material and in seawater and their respective chemical reactivity. For example, acid-iron dumping, although rapidly neutralized by the buffering system of seawater, precipitates iron oxides forming a turbid plume in the wake of the discharging vessel. Iron is thus converted from a dissolved to a particulate phase which includes the oxidation reaction of Fe^{2+} to Fe^{3+} for those wastes where iron is in a reduced state. The scavenging ability of iron oxide particles for trace metals (i.e., copper, lead, and cadmium) via adsorption allows for this process to occur. Flocculation may also occur.

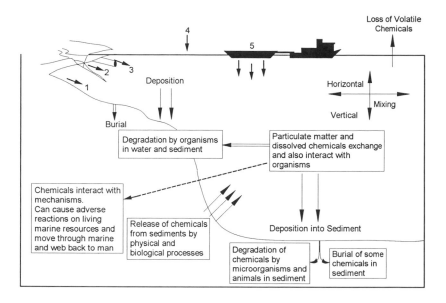

Figure 13-11 Physical processes affecting dispersal of waste materials (after Bohlen, 1990).

How a chemical undergoes speciation requires consideration of a particular element's oxidation state and its ability to form ion-pairs and organic complexes, and to incorporate into solid phases. Speciation-classification based on size is presented in Figure 13-12. The chemical form of certain elements in waste can also be changed by undergoing oxidation-reduction reactions. This is accomplished through a change in valency and is dependent upon the system's redox potential (Eh[1] and pH).

Once waste reaches the seafloor, complex geochemical and biological processes occur. Some trace elements may become mobile in organic-rich sediments. As the sediment is also sulfide ion enriched, a fraction of the trace elements may become immobilized through precipitation as insoluble metal sulfides.

Waste constituents may also be ingested by some organisms or mixed deeper in the sediment column by burrowing organisms in the benthic zone, a process referred to as bioturbation. Physiochemical processes and properties may further affect the mobility of these constituents.

Classification of elements in relation to residence times and removal mechanisms can also prove useful in predicting time scales and perturbations to chemical cycles that can result from the disposal of waste into the ocean. Residence time for a particular element is the amount of an element in the ocean divided by the rate of its input to or removal from the ocean:

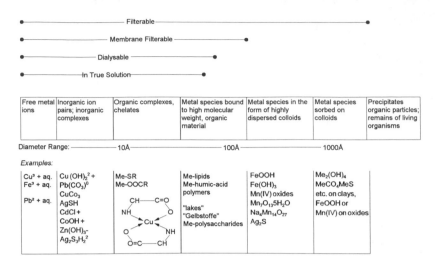

Figure 13-12 Examples of metal speciation-classification based on size (after Stumn and Morgan, 1981).

$$r = \frac{A}{r_i} = \frac{A}{r_o}$$

where r = residence time, A = total amount of the element dissolved or in suspension in the ocean, r_i = input rate, and r_o = output rate.

Residence time assumes that chemical cycles are in steady state; thus, the rate of input and rate of output are equal for a particular element. Based in part on oceanic concentrations of the elements in seawater, a distinct periodicity is evident with the alkali, alkaline earth, and halide ions having long residence times and mid-period elements having short residence times. Residence time in relation to fundamental controlling mechanism is shown in Figure 13-13. Long-term geochemical cycles such as weathering and subduction processes control residence times for certain elements (i.e., B, Na, Mg, Cl, K, Br, Rb, Sr, and U) in the range of 10^7 to 10^8 years. Biological cycles in value Si, P, and Ca and certain trace metals (i.e., Cu, Ni, Zn, Fe, Mn, V, Ba, Hg, I, and As) and control residence times in the range of 10^4 to 10^6 years. Insoluble solid phases, that being those elements that have an affinity for solid particulate and insoluble hydroxide phases (i.e., Al, Fe, La, Pb, and Th), control residence times in the range of 1000 years.

Biological Processes

Understanding of biological processes is obviously of importance to avoid potential adverse effects not only on marine organisms and the marine ecosys-

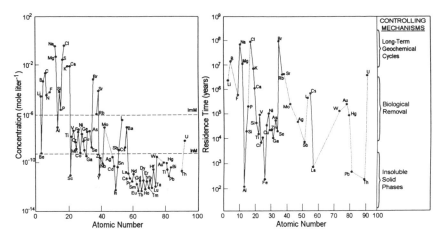

Figure 13-13 Graphs showing (left) concentration of elements and (right) residence time as a periodic function of atomic number (modified after Turner and Whitfield, 1983 and Brewer, 1975).

tem, but also on human health by the ingestion of contaminated seafood. Certainly, change in the marine environment has the potential to adversely affect migration routes and feeding and breeding behavior, or influence critical stages of the life cycle or chemical signals. Biological processes thus are viewed with regard to the response of marine organisms to waste constituents, incorporation of certain constituents within the living organism, and modification of waste constituents by marine organisms.

Since certain constituents in a waste material are toxic (i.e., mercury, cadmium, organohalogen compounds, etc.), the response of an organism to such constituents must be evaluated. Toxicity is a function of (1) the concentration of the toxic constituent and (2) the time of exposure. Bioassays which are performed in the laboratory typically are expressed by a notation such as "96-hr LC_{50}", identifying the concentration of the toxic substance required to kill half the population of the test organism in a period of 96 hr. Chronic, sublethal effects of such toxic constituents are more difficult to evaluate under laboratory conditions. Such effects are evaluated either in long-term laboratory tests or direct observation of contaminated environments over long periods of time. Biological processes may also be enhanced by introduction of certain nutrients such as nitrogen and phosphorous compounds, both common in sewage effluents and sludge. Enhanced photosynthesis (eutrophication) increases the total productivity of the population and, when excessive, may destroy the normal phytoplankton assemblage and replace it with a nuisance species. The nuisance species in turn may not be as suitable as food and could result in a collapse of the entire natural ecosystem.

Organisms may also incorporate certain toxic constituents primarily by bioconcentration, bioaccumulation, or biomagnification. Bioconcentration occurs when substances enter the aquatic organism through the gills or integument directly from solution in the water, allowing equilibrium concentration either less than that of water or greater than it by several orders of magnitude. Bioaccumulation occurs when the organism stores substances in various tissues or organs of the body either through direct exchange with substances dissolved in water or in the food consumed. With bioaccumulation, different substances will be accumulated at varying degrees in different tissues (i.e., liver, muscle, digestive tract, fatty tissue, bone) or organisms. Biomagnification is the increase in concentration of bioaccumulated toxic constituents as they pass up the food chain through two or more trophic levels. Biological processes involving incorporation are best studied by selecting the most critical substance in a waste material for evaluation with a suitable organism.

Modification of waste occurs by metabolic processes which result in either a change in the character of the compound or complete oxidation to carbon dioxide and water. Most natural organic materials in waste are easily metabolized with a half-life on the order of days or weeks. Some natural organic substances, however, are more resistant to decomposition (i.e., cellulose, lignins, and tannins of plants). Crude oil, for example, is a natural product formed under anoxic conditions. When released by natural seeps or by man's activities, these substances are slowly oxidized in the marine environment under aerobic conditions with a half-life ranging from years to many decades, depending on the constituents. However, when incorporated in anoxic sediments, crude oil can persist indefinitely. Synthetic organisms are even more resistant to decomposition in the marine environment with a half-life ranging from decades to centuries. Another major concern is the complete decomposition of an organic material which may produce anoxic conditions and subsequent release of hydrogen sulfide by the removal of oxygen from the water. Historically, this process was poorly evaluated due to a lack of understanding of the processes involved. However, acceptable environmental conditions can be monitored and environmental impact evaluated successfully despite economic constraints frequently being a limitation.

GEOLOGIC CONSIDERATIONS

As earlier stated, the majority of waste disposed of offshore occurs on the continental shelf. The primary geologic characteristic that distinguishes the nearshore environment from that of the deep ocean is topography. Disposal onto the continental shelf can adversely affect the bottom topography by creating an uneven surface in which benthic organisms are buried or marine fishing resources are influenced by the damage of trawler nets. Dredged materials disposed of on the seafloor in the New York Bight apex, as previously discussed, have produced a 6 to 8-m-high mound causing changes in water circulation and interfering with commerce. Specific considerations include factors such as stability and predictability, and sediment properties and processes. These factors are discussed below.

Stability and Predictability

The continental shelf can be considered an extension of the continent, but is certainly not featureless, and often comprises submarine canyons. These canyons have steep walls and can extend into the continental slope and rise. Large volumes of sediments occur in such areas (both natural and anthropogenic) in the form of turbidity currents that can rapidly reshape the configuration of the canyon and sea floor. Certain portions of the continental shelf are considered areas of instability and thus are not suitable for the permanent disposal of waste at a particular location (i.e., radioactive and toxic wastes).

Conversely, certain portions of the deep ocean environments are considered very stable with a long-established geologic history. Abyssal-clay regions have, for example, been considered as potential disposal sites, notably for radioactive and toxic waste, in lieu of several geologic factors. These considerations include:

- Adequate distance away from seismically and tectonically active lithospheric plate boundaries
- Adequate distance away from active or young volcanoes
- Comprised of thick layers of very uniform fine-grained clay
- Devoid of exploitable natural resources
- Geologic and oceanographic processes in this particular environment are well understood and have remained unchanged during past oceanic and climatic change
- Contains sedimentary records of tens of millions of years of slow, uninterrupted deposition of fine-grained clay, thus allowing for predictions of future stability.

Sediment Properties

Sources and particulate (including contaminants) fates are not assessed solely by concentration measurements but in regards to some contaminants, notably metals, are also influenced by sediment grain size. Finer-grained fractions typically predominate at deeper depths relative to coarser-grained soils which predominate at shallower depths. Since elevated metals concentrations are naturally associated with finer-grained soil, they may thus reflect grain-size distribution rather than indicating contamination.

Sediment Processes

Understanding of the natural behavior of particulate materials can serve as guidance in comprehension of the behavior of waste materials. Sedimentation rates vary considerably in the open area, although they are generally orders of magnitude smaller than that experienced nearshore. Much more information is available on nearshore sedimentation processes vs deep ocean due to the numerous studies conducted in association with the shipping and fisheries industries.

Particles also tend to seek certain preferred sites of deposition within the coastal zone regardless of their point of introduction, with exception to the immediate area of the dump site. The migration pathway of particulates tend to follow natural transport processes toward depositional sinks. These sinks largely occur within the intracoastal zone which includes marshes, estuaries, and lagoons. Thus, the assimilation capacity becomes of concern vs the open ocean. Other processes requiring consideration include erosional processes and those characterized as downslope displacement-type processes including creep, slumping, and sediment flow.

BIBLIOGRAPHY

Basta, D., 1989, On the State of Oceans and the Nature of Doing Business: *Journal Water Pollution Control Federation,* Vol. 61, No. 4, pp. 440-445.

Berger, W.H., 1974, Deep Sea Sedimentation: In *Geology of the Continental Margins* (Edited by Burke, C.A. and Drake, C.L.), Springer-Verlag, New York.

Bishop, W.P. and Hollister, C.D., 1974, Seabed Disposal—Where to Look: *Nuclear Technology,* Vol. 24, pp. 245-443.

Bohlen, W.F., 1990, Ocean Disposal of Particulate Wastes - Practices, Properties, and Processes: In *American Society for Testing and Materials, WTP 1087* (Edited by Demars, K.R. and Chaney, R.C.), Philadelphia, pp. 21-49.

Booth, J.S., 1979, Recent History of Mass-Wasting on the Upper Continental Slope, Northern Gulf of Mexico, as interpreted from the Consolidation States of the Sediments: In *Society of Economic Paleontologists and Mineralogists Special Publication* (Edited by Doyle, L.J. and Pilkey, O.H.), No. 27, pp. 153-164.

Bouma, P.H., 1979, Continental Slopes: In *Society of Economic Paleontologists and Mineralogists Special Publication* (Edited by Doyle, L.J. and Pilkey, O.H.), No. 27, pp. 1-15.

Brennan, B., 1991, Chemical Partitioning and Remobilization of Heavy Metals from Sewage Sludge Dumped in Dublin Bay: *Water Resources,* Vol. 25, No. 10, pp. 1193-1198.

Brewer, P.G., 1975, Minor Elements in Sea Water: In *Chemical Oceanography,* (Edited by Riley, J.P. and Skirrow, G.) Second Edition, Vol. 8, Academic Press, London, pp. 415-496.

Brown, M.F., Kester, D.R. and Dowd, J.M., 1983, Automated Iron Measurements after Acid-Iron Waste Disposal: In *Wastes in the Ocean, Vol. 1* (Edited by Duedall, I.W., et al.), John Wiley & Sons, New York, pp. 157-169.

Buelow, R.W., 1968, Ocean Disposal of Waste Material: In *Transactions: National Symposium on Ocean Sciences and Engineering of the Atlantic Shelf*, Marine Technology Society, Washington, D.C., pp. 311-337.

Byrne, C.D., Law, R.J., Hudson, P.M., Thain, J.E. and Fileman, T.W., 1988, Measurement of the Dispersion of Liquid Industrial Waste Discharged into the Wake of a Dumping Vessel: *Water Resources,* Vol. 22, No. 12, pp. 1577-1584.

Campbell, K.J., Hooper, J.R. and Prior, D.B., 1986, Engineering Implications of Deepwater Geologic and Soil Conditions, Texas-Louisiana Slope: In Proceedings of the 18th OTC, Paper 5015, Houston.

Carson, R.L., 1961, *The Sea Around Us:* Oxford University Press, New York, 250 p.

Champ, M.A., 1973, A Survey of Three Atlantic Ocean Disposal Sites: The City of Philadelphia Sewage Sludge Dump Site, the DuPont Acid Waste Dump Site, and the New York (DWD-106) Toxic Industrial Waste Site: Program Publication No. 1, Center for Earth Resources and Environmental Studies, American University, Washington, D.C., 169 p.

Champ, M.A. and Park, P.K., 1981, Ocean Dumping of Sewage Sludge: A Global Review: *Sea Technology,* Vol. 22, pp. 18-24.

Chaney, R.C. and Fang, H.Y., 1986, Static and Dynamic Properties of Marine Sediments - A State of the Art: In *Marine Geotechnology and Nearshore/Offshore Structures,* American Society for Testing and Materials, pp. 74-111.

Cook, H.E., 1979, Ancient Continental Slope Sequences and Their Value in Understanding Modern Slope Development: In *Society of Economic Paleontologists and Mineralogists Special Publication* (Edited by Doyle, L.J. and Pilkey, O.H.), No. 27, pp. 287-305.

Curray, J.R., 1969, Shallow Structure of the Continental Margin: In *The New Concepts of Continental Margin Sedimentation* (Edited by Stanley, D.J.); American Geological Institute, JC-12, pp. 1-22.

Doyle, L.J., Pilkey, O.H. and Woo, C.C., 1979, Sedimentation on the Eastern U.S. Continental Slope: In *Society of Economic Paleontologists and Mineralogists Special Publication* (Edited by Doyle, L.J. and Pilkey, O.H.), No. 27, pp. 119-129.

Duedall, I.W., O'Connors, H.B., Oakley, S.A. and Stanford, H.M., 1977, Short-term Water Column Perturbations Caused by Wastewater Sludge Dumping in the New York Bight Apex: *Journal Water Pollution Control Federation,* Vol. 49, pp. 2074-2080.

Duedall, I.W., Roethel, F.J., Seligman, J.D., O'Connors, H.B., Parker, J.H., Woodhead, P.M.J., Dayal, R., Chezar, B., Roberts, B.K. and Mullen, H., 1981, Stabilized Power Plant Scrubber Sludge and Fly Ash in the Marine Environment. In: *Ocean Dumping of Industrial Wastes* (Edited by Ketchum, B.H., Kester, D.R. and Parks, P.K.), Plenum Press, New York, pp. 315-346.

Duedall, I.W., Ketchum, B.H., Park, P.K. and Kester, D.R., 1983, Global Inputs, Characteristics, and Fates of Ocean-Dumped Industrial and Sewage Wastes - An Overview: In *Wastes in the Ocean, Vol. 1* (Edited by Duedall, I.W., et al.), John Wiley & Sons, New York, pp. 3-45.

Dutz, R.S., 1964, Origin of Continental Slopes: *American Scientist,* Vol. 52, pp. 50-69.

Emery, K.O., 1980, Continental Margins - Classification and Petroleum Prospects: *American Association of Petroleum Geologists Bulletin,* Vol. 64, pp. 297-315.

Falk, L.L., Meyers, T.D. and Thomann, R.V., 1972, Waste Dispersion Characteristics and Effects in an Oceanic Environment: Technical Report EAW-12020, U.S. Environmental Protection Agency, Washington, D.C., 393 p.

Farrington, J.W., Capuzzo, J.M., Leschnic, T.M. and Champ, M.A., 1982, Ocean Dumping: *Oceanus,* Vol. 25, No. 4, pp. 39-50.

Goldberg, E.D., 1981, The Oceans as Waste Space: The Argument: *Oceanus,* Vol. 24, pp. 2-9.

Goldberg, E.D., 1990, Protecting the Wet Commons: *Environmental Science and Technology,* Vol. 24, No. 4, pp. 450-454.

Gordon, R.B., 1974, Dispersion of Dredge Spoil Dumped in Near Shore Waters: *Estuarine and Coastal Marine Science,* Vol. 2, pp. 349-358.

Gorsline, D.S. and Emery, K.O., 1959, Turbidity Current Deposits in San Pedro and Santa Monica Basins off Southern California: *Geologic Society of America Bulletin,* Vol. 70, pp. 279-290.

Hagen, A.A., 1983, History of Radioactive Waste Disposal into the Ocean: In: *Wastes in the Ocean, Vol. 3* (Edited by Park, P.K., Kester, D.R., Duedall, I.W. and Ketchum, B.H.); John Wiley & Sons, New York.

Heath, G.R., Hollister, C.D., Anderson, D.R., and Leinen, M., 1983, Why Consider Subseabed Disposal of High-Level Nuclear Wastes? In: *Wastes in the Ocean, Vol. 3* (Edited by Park, P.K., Kester, D.R., Duedall, I.W. and Ketchum, B.H.); John Wiley & Sons, New York, pp. 303-325.

Hedberg, H.D., 1970, Continental Margins from Viewpoint of the Petroleum Geologist: *American Association of Petroleum Geologists Bulletin,* Vol. 54, pp. 3-43.

Hollister, C.D., Anderson, D.R. and Heath, G.R., 1981, Subseabed Disposal of Nuclear Wastes: *Science,* Vol. 213, pp. 1321-1326.

Ingle, J.C., Jr., 1980, Cenozoic Paleobathymetry and Depositional History of Selected Sequences within the Southern California Continental Borderland: Cushman Foundation Special Publication No. 19, pp. 163-195.

Jackson, G.A., 1982, Sludge Disposal in Southern California Basins: *Environmental Science and Technology,* Vol. 16, pp. 746-757.

Jones, R.W., 1983, Organic Matter Characteristics near the Shelf-Slope Boundary: In *Society of Economic Paleontologists and Mineralogists Special Publication* (Edited by Stanley, D.J. and Moore, G.T.), No. 33, pp. 391-405.

Keller, G.H., Lambert, D.N. and Bennett, R.H., 1979, Geotechnical Properties of Continental Slope Deposits - Cape Hatteras to Hydrographer Canyon: In *Society of Economic Paleontologists and Mineralogists Special Publication* (Edited by Doyle, L.J. and Pilkey, O.H.), No. 27, pp. 131-151.

Kennett, J.P., 1982, *Marine Geology:* Prentice-Hall, Englewood Cliffs, N.J., 813 p.

Kester, D.R., Hittinger, R.C. and Mukherji, P., 1981, Transition and Heavy Metals Associated with Acid-Iron Waste Disposal at Deep Water Dumpsite 106: In: *Ocean Dumping of Industrial Wastes* (Edited by Ketchum, B.H., Kester, D.R. and Park, P.K.), Plenum Press, New York, pp. 215-232.

Kester, D.R., Ketchum, B.H., Duedall, I.W. and Park, P.D., Editors, 1983, *Dredged Material Disposal in the Ocean:* John Wiley & Sons, New York.

Kester, D.R., Burt, W.V., Capuzzo, J.M., Park, P.K. and Duedall, I.W., 1985, Disposal of Wastes in the Deep Sea - Oceanographic Considerations: In *Wastes in the Ocean, Vol. 5* (Edited by Kester, D.R., et al.), John Wiley & Sons, New York, pp. 319-335.

Ketchum, B.H. and Ford, W.L., 1952, Rate of Dispersion in the Wake of a Barge at Sea: *Transactions, American Geophysical Union,* Vol. 33, p. 680-684.

Ketchum, B.H., 1980, Marine Industrial Pollution: In *Oceanography: The Past,* (Edited by Sears, M. and Merriman, D.); Springer-Verlag, New York, pp. 397-413.

Lavelle, J.W., Ozturgut, E., Baker, E.T., Tennant, D.A. and Walker, S.L., 1988, Settling Speeds of Sewage Sludge in Seawater: *Environmental Science and Technology,* Vol. 22, No. 10, pp. 1201-1207.

Lee, H.J., Demars, K.R. and Chaney, R.C., 1990, Geotechnical Engineering for Ocean Waste Disposal - An Introduction: In *American Society for Testing and Materials, WTP 1087* (Edited by Demars, K.R. and Chaney, R.C.), Philadelphia, pp. 3-117.

Lowe, D.R., 1979, Sediment Gravity Flows: Their Classification and Some Problems of Application to Natural Flows and Deposits: In *Society of Economic Paleontologists and Mineralogists Special Publication* (Edited by Doyle, L.J. and Pilkey, O.H.), No. 27, pp. 75-82.

MacKenzie, L.D. and Chou, G.P., 1991, RCRA Land Disposal Restrictions for Contaminated Debris: In Proceedings of the Hazardous Materials Control Research Institute Superfund '91 Conference, Washington, D.C., pp. 376-379.

Manheim, F.T., 1983, Who is Doing What in Marine Dumping: In *Wastes in the Ocean, Vol. 1* (Edited by Duedall, I.W., et al.), John Wiley & Sons, New York, pp. 47-65.

Muir, W.C., 1983, History of Ocean Disposal in the Mid-Atlantic Bight: In *Wastes in the Ocean, Vol. 1* (Edited by Duedall, I.W., et al.), John Wiley & Sons, New York, pp. 273-291.

Mukherji, P. and Kester, D.R., 1983, Acid-Iron Disposal Experiments in Summer and Winter at Deepwater Dumpsite-106: In *Wastes in the Ocean, Vol. 1* (Edited by Duedall, I.W., et al.), John Wiley & Sons, New York, pp. 141-155.

Nardin, T.R., Hein, F.J., Gorsline, D.S. and Edwards, B.D., 1979, A Review of Mass Movement Processes, Sediment and Acoustic Characteristics, and Contrasts in Slope and Base-of-Slope Systems vs Canyon-Fan-Basin Floor Systems: In *Society of Economic Paleontologists and Mineralogists Special Publication* (Edited by Doyle, L.J. and Pilkey, O.H.), No. 27, pp. 61-73.

O'Connor, T.P., Kester, D.R., Burt, W.V., Capuzzo, J.M., Park, P.K. and Duedall, I.W., 1985, Waste Disposal in the Deep Ocean - An Overview: In *Wastes in the Ocean, Vol. 5* (Edited by Kester, D.R., et al.), John Wiley & Sons, New York, pp. 3-30.

Palmer, H.D., 1979, Man's Activities on the Continental Slope: In *Society of Economic Paleontologists and Mineralogists Special Publication* (Edited by Doyle, L.J. and Pilkey. O.H.), No. 27, pp. 17-24.

Presley, B.J., Schofield, J.S. and Trefry, J., 1981, Waste Material Behavior and Inorganic Geochemistry at the Puerto Rico Waste Dumpsite: In *Ocean Dumping of Industrial Wastes,* (Edited by Ketchum, B.H., Kester, D.R. and Park, P.K.), Plenum Press, New York, pp. 233-246.

Proni, J.R., Newman, F.C., Sellers, R.L. and Parker, C., 1976, Acoustic Tracking of Ocean Dumped Sewage Sludge: *Science,* Vol. 193, pp. 1005-1007.

Reineck, H.E. and Singh, I.B., 1975, *Depositional Sedimentary Environments:* Springer-Verlag, New York, 439 p.

Roxburgh, I.S., 1987, *Geology of High-Level Nuclear Waste Disposal - An Introduction:* Chapman and Hall, New York, 229 p.

Shepard, F.P., 1973, *Submarine Geology:* Harper and Row, New York, 517 p.

Shepard, F.P., 1977, *Geological Oceanography: Evolution of Coasts, Continental Margins, and the Deep-Sea Floor:* Crane, Russak & Company, New York, 214 pp.

Shleh, C.S. and Roethel, F.J., 1988, Physical and Chemical Behavior of Stabilized Sewage Sludge Blocks in Seawater: *Environmental Science and Technology,* Vol. 23, No. 1, pp. 121-125.

Silva, A.J. and Hollister, C.D., 1974, Geotechnical Properties and Processes of Deposition Determined from Abyssal Atlantic Sediments Recovered with the GPC: *American Geophysical Union Transactions,* Vol. 55, No. 4.

Silva, A.J., 1990, Geotechnical Properties and Processes of Deep Ocean Sediments as Related to Disposal of Toxic Wastes - An Overview: In *American Society for Testing and Materials, WTP 1087* (Edited by Demars, K.R. and Chaney, R.C.), Philadelphia, pp. 113-155.

Stumm, W. and Morgan, J.J., 1981, *Aquatic Chemistry,* 2nd Ed., Wiley-Interscience, New York, 780 p.

Stuvier, M., Quay, P.D. and Ostlund, H.G., 1983, Abyssal Water Carbon-14 Distribution and the Age of the World Ocean: *Science,* Vol. 219, pp. 849-851.

Torrey, S., 1978, *Coal Ash Utilization:* Noyes Data Corporation, Park Ridge, NJ, 370 p.

Turner, D.R. and Whitfield, M., 1983, Inorganic Controls in the Biogeochemical Cycle of the Elements in the Ocean: In *Environmental Biochemistry* (Edited by Hallberg, R.), *Ecological Bulletin,* No. 35, Stockholm, pp. 9-37.

Vanney, J.R. and Stanley, D.J., 1983, Shelfbreak Physiography - An Overview: In *Society of Economic Paleontologists and Mineralogists Special Publication* (Edited by Stanley, D.J. and Moore, G.T.), No. 33, pp. 1-24.

Water Pollution Control Federation, Marine Water Quality Committee, 1987, Ocean's Role in a Multi-Media Disposal Environment: *Journal WPCF,* Vol. 59, No. 11, pp. 927-931.

Young, R.A., Swift, D.J.P., Clarke, T.L., Harvey, G.R. and Betzer, P.R., 1985, Dispersal Pathways for Particle Associated Pollutants: *Science,* Vol. 229, No. 4712, pp. 431-435.

Zdanowicz, V.S., 1991, Determining the Fates of Contaminated Wastes Dumped in the New York Bight by Use of Metal Enrichment Factors: *Environmental Science and Technology,* Vol. 25, No. 10, pp. 1760-1766.

APPENDICES

To convert from	To	Multiply by
Area		
square millimeters	square centimeters (cm^2)	0.01
	square inches (in^2)	1.550×10^{-3}
square centimeters	square millimeters (mm^2)	100.0
	square meters (m^2)	1.0×10^{-4}
	square inches	0.1550
	square feet (ft^2)	1.07639×10^{-3}
square inches	square millimeters	645.16
	square centimeters	6.4516
	square meters	6.4516×10^{-4}
	square feet	69.444×10^{-4}
square feet	square meters	0.0929
	hectares (ha)	9.2903×10^{-6}
	square inches	144.0
	acres	2.29568×10^{-5}
square yards	square meters	0.836 13
	hectares	8.3613×10^{-5}
	square feet	9.0
	acres	2.06612×10^{-4}
square meters	hectares	1.0×10^{-4}
	square feet	10.763 91
	acres	2.471×10^{-4}
	square yards (yd^2)	1.19599
acres	square meters	4046.8564
	hectares	0.404 69
	square feet	4.356×10^4
hectares	square meters	1.0×10^4
	acres	2.471

To convert from	To	Multiply by
square kilometers	square meters	1.0×10^6
	hectares	100.0
	square feet	107.6391×10^5
	acres	247.10538
	square miles (mi^2)	0.3861
square miles	meters	258.998
	hectares	81×10^4
	square kilometers (km^2)	258.998 81
	square feet	2.589 99
	acres	2.78784×10^7
		640.0

Force Per Unit Area
Pressure-Stress

To convert from	To	Multiply by
pounds per square inch	kilopascals (kPa)	6.89476
	meters-head[a]	0.70309
	mm of Hg[b]	51.7151
	feet of water[a]	2.3067
	pounds per square foot (lb/ft^2)	144.0
	std. atmospheres	68.046×10^{-3}
pounds per square foot	kilopascals	0.047 88
	meters-head[a]	4.8826×10^{-3}
	mm of Hg[b]	0.35913
	feet of water[a]	16.0189×10^{-3}
	pounds per square inch	6.9444×10^{-3}
	std. atmospheres	0.47254×10^{-3}
short tons per square foot	kilopascals	95.76052
	pounds per square inch (lb/in^2)	13.88889
meters-head[a]	kilopascals	9.80636
	mm of Hg[b]	73.554
	feet of water[a]	3.28084
	pounds per square inch	1.42229
	pounds per square foot	204.81

To convert from	To	Multiply by
feet of water[a]	kilopascals	2.98898
	meters-head[a]	0.3048
	mm of Hg[b]	22.4193
	inches of Hg[b]	0.88265
	pounds per square inch	0.43351
	pounds per square foot	62.4261
kilopascals	newtons per square meter (N/m²)	1.0×10^3
	mm of Hg[b]	7.50064
	meters-head[a]	0.10197
	inches of Hg[b]	0.2953
	pounds per square foot	20.8854
	pounds per square inch	0.14504
	std. atmospheres	9.8692×10^{-3}
kilograms (f) per square meter	kilopascals	$9.806\ 65 \times 10^{-3}$
	mm of Hg[b]	73.556×10^{-3}
	pounds per square inch	1.4223×10^{-3}
millibars (mbars)	kilopascals	0.10
bars	kilopascals	100.0
std. atmospheres	kilopascals	101.325
	mm of Hg[b]	760.0
	pounds per square inch	14.70
	feet of water[a]	33.90

Length

micrometers	millimeters	1.0×1^{-3}
	meters	1.0×1^{-6}
	angstrom units (A)	1.0×10^{-4}
	mils	.03937
	inches	3.93701×10^{-5}
millimeters	micrometers	1.0×10^3
	centimeters (cm)	0.1
	meters	1.0×10^{-3}
	mils	39.37008
	inches	.039 37
	feet(ft)	3.28084×10^{-3}

To convert from	To	Multiply by
centimeters	millimeters	10.0
	meters	0.01
	mils	0.3937×10^3
	inches	0.3937
	feet	0.03281
inches	millimeters	25.40
	meters	0.0254
	mils	1.0×10^3
	feet	0.08333
feet	millimeters	304.8
	meters	0.3048
	inches	12.0
	yards (yd)	0.33333
yards	meters	0.9144
	inches	36.0
	feet	3.0
meters	millimeters	1.0×10^3
	kilometers (km)	1.0×10^{-3}
	inches	39.37008
	yards	1.09361
	miles (mi)	6.21371×10^{-4}
kilometers	meters	1.0×10^3
	feet	3.28084×10^3
	miles	0.62137
miles	meters	1.60934×10^3
	kilometers	1.60934
	feet	5280.0
	yards	1760.0
Mass		
grams	kilograms (kg)	1.0×10^{-3}
	ounces (avdp)	0.03527
ounces (avdp)	grams (g)	28.34952
	kilograms	0.02835
	pounds (avdp)	0.0625

To convert from	To		Multiply by
pounds (avdp)	kilograms		0.45359
	ounces (avdp)		16.00
kilograms	kilograms (force)-second squared per	meter	0.10197
	(kgf · s²/m)		
	pounds		2.20462
	slugs		0.06852
slugs	kilograms		14.5939
short tons	kilograms		907.1847
	metric tons (t)		0.90718
	pounds (avdp)		2000.0
metric tons (tonne or	kilograms		1.0×10^3
megagram)	pounds (avdp)		2.20462×10^3
	short tons		1.10231
long tons	kilograms		1016.047
	metric tons		1.01605
	pounds (avdp)		2240.0
	short tons		1.120

Mass Per Unit Volume, Density and Mass Capacity

		Multiply by
pounds per cubic foot		
	kilogram per cubic meter (kg/m³)	16.01846
	slugs per cubic foot (slug/ft³)	0.03108
	pounds per gallon (lb/gal)	0.13368
pounds per gallon	kilograms per cubic meter (kg/m³)	119.8264
	slugs per cubic foot	0.2325
pounds per cubic yard	kilograms per cubic meter	0.59328
	pounds per cubic foot (lb/ft³)	0.03704
grams per cubic centimeter	kilograms per cubic meter	1.0×10^3
ounces per gallon (oz/gal)	grams per liter (g/l)	7.48915

To convert from	To	Multiply by
kilograms per cubic meter	grams per cubic centimeter (g/cm^3)	1.0×10^{-3}
	pounds per cubic foot (lb/ft^3)	1.0×10^{-3}
	pounds per gallon	62.4297×10^{-3}
	pounds per cubic yard	1.68556
long tons per cubic yard	kilograms per cubic meter	1328.939
ounces per cubic inch (oz/in^3)	kilograms per cubic meter	1729.994
slugs per cubic foot	kilograms per cubic meter	515.3788
Velocity		
feet per second	centimeter per square meter (cm^2)	3.11×10^{-4}
	square feet (ft$_2$)	3.35×10^{-7}
	darcy	3.15×10^4
	meters per second (m/s)	0.3048
	gallons per day per square feet (gal/d/ft^2)	6.46×10^5
	kilometers per hour (km/h)	1.09728
	miles per hour (mi/h)	0.68182
meters per second	centimeters pere square meter (cm^2)	1.02×10^{-3}
	square feet (ft^2)	1.10×10^{-6}
	darcy	1.04×10^5
	feet per second (ft/s)	3.28
	gallons per day per square feet (gal/d/ft^2)	2.12×10^6
	kilometers per hour	3.60
	feet per second (ft/s)	3.28084
	miles per hour	2.23694
centimeters per squared meter (cm^2)	square feet (ft^2)	1.08×10^{-3}
	darcy	1.01×10^8
	meters per second (m/s)	9.80×10^2
	feet per second (ft/sec)	3.22×10^3
	gallons per day per square feet (gal/d/ft^2)	1.85×10^9
square feet (ft^2)	centimeters per squared meter (cm^2)	9.29×10^2
	darcy	9.42×10^{10}
	meters per second (m/s)	9.11×10^5
	feet per second (ft/s)	2.99×10^6
	gallon per day per square feet (gal/d/ft^2)	1.71×10^{12}

To convert from	To	Multiply by
kilometers per hour	meters per second	0.27778
	feet per second	0.91134
	miles per hour	0.62147
miles per hour	kilometers per hour	1.690934
	meters per second	0.44704
	feet per second	1.46667

Viscosity

To convert from	To	Multiply by
centipoise	pascal-second (Pas)	1.0×10^{-3}
	poise	0.01
	pound per foot-hour (lb/ft · h)	2.41909
	pound per foot-section (lb/ft · s)	6.71969×10^{-4}
	slug per foot-second (slug/ft · s)	2.08854×10^{-5}
pascal-second	centipoise	1000.0
	pound per foot-hour	2.41990×10^{3}
	pound per foot-second	0.67197
	slug per foot-second	20.8854×10^{-3}
pounds per foot-hour	pascal-second	4.13379×10^{-4}
	pound per foot-second	2.77778×10^{-4}
	centipoise	0.41338
pounds per foot-second	pascal-second	1.48816
	slug per foot-second	31.0809×10^{-3}
	centipoise	1.48816×10^{3}
centistokes	square meters per second (m²/s)	1.0×10^{-6}
	square feet per second (ft²/s)	10.76391×10^{-6}
	stokes	0.01
square feet per second	square meters per second	9.2903×10^{-2}
	centistokes	0.2903×10^{4}
stokes	square meters per second	1.0×10^{-4}
rhe	1 per pascal-second (1/Pa s)	10.0

To convert from	To	Multiply by
Volume-Capacity		
cubic millimeters	cubic centimeters (cm^3)	1.0×10^{-3}
	liters (l)	1.0×10^{-6}
	cubic inches (in^3)	61.02374×10^{-6}
cubic centimeters	liters	1.0×10^{-3}
	milliliters (ml)	1.0
	cubic inches	61.02374×10^{-3}
	fluid ounces (fl.oz)	33.814×10^{-3}
milliliters	liters	1.0×10^{-3}
	cubic centimeters	1.0
cubic inches	milliliters	16.38706
	cubic feet (ft^3)	57.87037×10^{-5}
liters	cubic meters	1.0×10^{-3}
	cubic feet	0.03531
	gallons	0.26417
	fluid ounces	33.814
gallons	liters	3.78541
	cubic meters	3.78541×10^{-3}
	fluid ounces	128.0
	cubic feet	0.13368
cubic feet	liters	28.31685
	cubic meters (m^3)	28.31685×10^{-3}
	cubic dekameters (dam^3)	28.31685×10^{-6}
	cubic inches	1728.0
	cubic yards (yd^3)	37.03704×10^{-3}
	gallons (gal)	7.48052
	acre-feet (acre-ft)	22.95684×10^{-6}
cubic miles	cubic dekameters	4.16818×10^6
	cubic kilometers (km^3)	4.16818
	acre-feet	3.3792×10^6
cubic yards	cubic meters	0.76455
	cubic feet	27.0

To convert from	To	Multiply by
cubic meters	liters	1.0×10^3
	cubic dekameters	1.0×10^{-3}
	gallons	264.1721
	cubic feet	35.31467
	cubic yards	1.30795
	acre-feet	8.107×10^{-4}
acre-feet	cubic meters	1233.482
	cubic dekameters	1.23348
	cubic feet	43.560×10^3
	gallons	325.8514×10^3
cubic dekameters	cubic meters	1.0×10^3
	cubic feet	35.31467×10^3
	acre-feet	0.81071
	gallons	26.41721×10^4
cubic kilometers	cubic dekameters	1.0×10^6
	acre-feet	0.81071×10^6
	cubic miles (mi^3)	0.23991

Volume Per Cross Sectional Area Per Unit Time: Transmissivity[a]

To convert from	To	Multiply by
cubic feet per foot per day $ft^3/(ft \cdot d))$	cubic meters per meter per day ($m^3/(m \cdot d)$)	0.0929
	gallons per foot per day ($gal/(ft \cdot d)$)	7.48052
	liters per meter per day ($l/(m \cdot d)$)	92.903
gallons per foot per day	cubic meters per meter per day ($m^3/(m \cdot d)$)	0.01242
	cubic feet per foot per day ($ft^3/(ft \cdot d)$)	0.13368

To convert from	To	Multiply by
Volume Per Unit Area Per Unit Time: Hydraulic Conductivity (Permeability)[c]		
cubic feet per square foot per day	cubic meters per square meter per day ($m^3/m^2/d$)	0.3048
	cubic feet per square foot per minute ($ft^3/ft^2/min$)	0.6944×10^{-3}
	liters per square meter per day ($l/m^2/d$)	304.8
	gallons per square foot per day ($gal/ft^2/d$)	7.48052
	cubic millimeters per square millimeter per day ($mm^3/mm^2/d$)	304.8
	cubic millimeters per square millimeter per hour ($mm^3/mm^2/h$)	25.4
	cubic inches per square inch per hour ($in^3/in^2/h$)	0.5
gallons per square foot per day	centimeter per square meter (cm^2)	5.42×10^{-10}
	darcy	5.49×10^{-2}
	feet per second (ft/s)	1.55×10^{-6}
	meters per second (m/s)	4.72×10^{-7}
	square feet (ft^2)	5.83×10^{-13}
	cubic meters per square meter per day ($m^3/m^2/d$)	40.7458×10^{-3}
	liters per square meter per day ($l/(m^2/d)$)	40.7458
	cubic feet per square foot per day ($ft^3/(ft^2/d)$)	0.13368
darcy	centimeter per square meter (cm^2)	9.87×10^{-9}
	square feet (ft^2)	1.06×10^{-11}
	meters per second (m/s)	9.66×10^{-6}
	feet per second (ft/s)	3.17×10^{-5}
	gallons per day per squared feet ($gal/d/ft^2$)	1.71×10^{12}
Volume Per Unit Time Flow		
	liters per second (l/s)	28.31685
cubic feet per second	cubic meters per second (m^3/s)	0.02832
	cubic dekameters per day (dam^3/d)	2.44657
	gallons per minute (gal/min)	448.83117
	acre-feet per day (acre-ft/d)	1.98347
	cubic feet per minute (ft^3/min)	60.0

To convert from	To	Multiply by
gallons per minute	cubic meters per second	0.631×10^{-4}
	liters per second	0.0631
	cubic dekameters per day	5.451×10^{-3}
	cubic feet per second (ft^3/s)	2.228×10^{-3}
	acre-feet per day	4.4192×10^{-3}
acre-feet per day	cubic meters per second	0.01428
	cubic dekameters per day	1.23348
	cubic feet per second	0.50417
cubic dekameters per day	cubic meters per second	0.01157
	cubic feet per second	0.40874
	acre-feet per day	0.81071

(a) Column of H_2O (water) measured at 4 °C.
(b) Column of Hg (mercury) measured at 0 °C.
(c) Many of these units can be dimensionally simplified: for example, m^3/(m • d) can also be written m^2/d.

Provided below is a glossary of regulatory terms of geologic significance. Terms are listed in alphabetical order, followed by reference to each term's respective location as contained in the Codes of Federal Regulations by part or parts in parentheses, followed by each term's definition as given in the Codes of Federal Regulations.

Abandoned well (146.3): means a well of which use has been permanently discontinued or which is in a state of disrepair such that it cannot be used for its intended purpose or for observation purposes.

Above ground tank (260.10): means a device meeting the definition of "tank" in § 260.10 and that is situated in such a way that the entire surface area of the tank is completely above the plane of the adjacent surrounding surface and the entire surface area of the tank (including the tank bottom) is able to be visually inspected.

Aboveground release (280.12): means any release to the surface of the land or to surface water. This includes, but is not limited to, releases from the aboveground portion of a UST system and aboveground releases associated with overfills and transfer operations as the regulated substance moves to or from a UST system.

Accessible environment (191.12): means (1) the atmosphere; (2) land surfaces; (3); surface waters; (4) oceans; and (5) all of the lithosphere that is beyond the controlled area.

Action level (141.2): is the concentration of lead or copper in water specified in § 141.80(c) which determines, in some cases, the treatment requirements contained in subpart I of this part that a water system is required to complete.

Active institutional control (191.12): means (1) controlling access to a disposal site by any means other than passive site by any means other than passive institutional controls; (2) performing maintenance operations or remedial action at a site, (3) controlling or cleaning up releases from a site, or (4) monitoring parameters related to disposal system performance.

Active portion (260.10): means that portion of a facility where treatment storage, or disposal operations are being or have been conducted after the effective date of part 261 of this chapter and which is not a closed portion. (See also "closed portion" and "inactive portion".)

Active life (260.10): of a facility means the period from the initial receipt of hazardous waste at the facility until the Regional Administrator receives certification of final closure.

Agency (142.1): means the U.S. Environmental Protection Agency.

Air emissions (129.2): means the release or discharge of a toxic pollutant by an owner or operator into the ambient air either (1) by means of a stack or (2) as a fugitive dust, mist, or vapor as a result inherent to the manufacturing or formulating process.

Altered discharge (125.58): means any discharge other than a current discharge or improved discharge, as defined in this regulation.

Ambient water criterion (129.2): means that concentration of a toxic pollutant in a navigable water that, based upon available data, will not result in adverse impact on important aquatic life, or on consumers of such aquatic life, after exposure of that aquatic life for periods of time exceeding 96 hours and continuing at least through one reproductive cycle; and will not result in a significant risk of adverse health effects in a large human population based on available information such as mammalian laboratory toxicity data, epidemiological studies of human occupational exposures, or human exposure data, or any other relevant data.

Animals (116.4): means appropriately sensitive animals which carry out respiration by means of a lung structure permitting gaseous exchange between air and the circulatory system.

Approved State primacy program (142.1): consists of those program elements listed in § 142.11(a) that were submitted with the initial State application for primary enforcement authority and approved by the EPA Administrator and all State program revisions thereafter that were approved by the EPA Administrator.

Aquatic flora (116.4): means plant life associated with the aquatic ecosystem including, but not limited to, algae and higher plants.

Aquatic animals (116.4): means appropriately sensitive wholly aquatic animals which carry out respiration by means of a gill structure permitting gaseous exchange between water and the circulatory system.

Aquifer (146.3, 191.12, 257.3-4, 260.10, 144.3): means a geological formation, group of formations, or part of a formation that is capable of yielding a significant amount of water to a well or spring.

Area of review (146.3, 144.3): means the area surrounding an injection well described according to the criteria set forth in § 146.06 or in the case of an area permit, the project area plus a circumscribing area the width of which is either 1/4 of a mile or a number calculated according to the criteria set forth in § 146.06.

Areawide agency: an agency designated under section 208 of the Act, which has responsibilities for WQM planning within a specified area of a State.

Attainability analysis (131.3): is a structured scientific assessment of the factors affecting the attainment of the use which may include physical, chemical, biological, and economic factors as described in § 131.10(g).

Background soil pH (257.3-5): means the pH of the soil prior to the addition of substances that alter the hydrogen ion concentration.

Barrel (113.1): means 42 U.S. gallons at 60 degrees Fahrenheit.

Barrier (191.12): means any material or structure that prevents or substantially delays movement of water or radionuclides toward the accessible environment. For example, a barrier may be a geologic structure, a canister, a waste form with physical and chemical characteristics that significantly decrease the mobility of radionuclides, or a material placed over and around waste, provided that the material or structure substantially delays movement of water or radionuclides.

Based flood (257.3-1): means a flood that has a 1 percent or greater chance of recurring in any year, or a flood of a magnitude equaled or exceeded once in 100 years on the average over a significantly long period.

Belowground release (280.12): means any release to the subsurface of the land and to ground water. This includes, but is not limited to, releases from the belowground portions of an underground storage tank system and belowground releases associated with overfills and transfer operations as the regulated substance moves to or from a UST system.

Beneath the surface of the ground (280.12): means beneath the ground surface or otherwise covered with earthen materials.

Best Management Practice or BMP (130.2): methods, measures, or practices selected by an agency to meet its nonpoint source control needs. BMPs include but are not limited to structural and nonstructural controls and operation and maintenance procedures. BMPs can be applied before, during, and after pollution-producing activities to reduce or eliminate the introduction of pollutants into receiving waters.

Blowdown (401.11): means the minimum discharge of recirculating water for the purpose of discharging materials combined in the water, the further buildup of which would cause concentration in amounts exceeding limits established by best engineering practice.

Casing (146.3): means a pipe or tubing of appropriate material, of varying diameter and weight, lowered into a borehole during or after drilling in order to support the sides of the hole and thus prevent the walls from caving, to prevent loss of drilling mud into porous ground, or to prevent water, gas, or other fluid from entering or leaving the hole.

Catastrophic collapse (146.3): means the sudden and utter failure of overlying "strata" caused by removal of underlying materials.

Cation exchange capacity (257.3-6): means the sum of exchangeable cations a soil can absorb expressed in milliequivalents per 100 grams of soil as determined by sampling the soil to the depth of cultivation or solid waste placement, whichever is greater, and analyzing by the summation method for distinctly acid soils or the sodium acetate method for neutral, calcareous, or saline soils (*Methods of Soil Analysis, Agronomy Monograph No. 9,* C.A. Black, Ed., American Society of Agronomy, Madison, WI, pp. 891-901, 1965).

Cementing (146.3): means the operation whereby a cement slurry is pumped into a drilled hole and/or forced behind the casing.

Coagulation (141.2): means a process using coagulant chemicals and mixing by which colloidal and suspended materials are destabilized and agglomerated into flocs.

Coastal (435.41): means any body of water landward of the territorial seas as defined in 40 CFR.

Community water system (191.12): means a system for the provision to the public of piped water for human consumption, if such system has at least 15 service connections used by year round residents or regularly serves at lease 25 year-round residents.

Community water system (141.2): means a public water system which serves at least 15 service connections used by year-round residents or regularly serves at least 25 year-round residents.

Compliance period (141.2): means a three-year calendar year period within a compliance cycle. Each compliance cycle has three-year compliance periods. Within the first compliance cycle, the first compliance period runs from January 1, 1993 to December 31, 1995.

Compliance cycle (141.2): means the nine-year calendar year cycle during which public water systems must monitor. Each compliance cycle consists of three-year compliance periods. The first calendar year cycle begins January 1, 1993 and ends December 31, 2001; the second begins January 1, 2002 and ends December 31, 2010; the third begins January 1, 2011 and ends December 31, 2019.

Confined aquifer (260.10): means an aquifer bounded above and below by impermeable beds or by beds of distinctly lower permeability than that of the aquifer itself; an aquifer containing confined ground water.

Confining bed (146.3): means a body of impermeable or distinctly less permeable material stratigraphically adjacent to one or more aquifers.

Confining zone (146.3): means a geological formation, group of formations, or part of a formation that is capable of limiting fluid movement above an injection zone.

Confluent growth (141.2): means a continuous bacterial growth covering the entire filtration area of a membrane filter, or a portion thereof, in which bacterial colonies are not discrete.

Container (260.10): means any portable device in which a material is stored, transported, treated, disposed of, or otherwise handled.

Contaminant (141.2, 143.3, 144.3, 146.3): means any physical, chemical, biological or radiological substance or matter in water.

Contaminate (257.3-4): means introduce a substance that would cause (i) the concentration of that substance in the ground water to exceed the maximum contaminant level specified in Appendix I, or (ii) an increase in the concentration of that substance in the ground water where the existing concentration of that substance exceeds the maximum contaminant level specified in Appendix I.

Contiguous zone (110.1, 116.3, 122.2): means the entire zone established or to be established by the U.S. under article 24 of the Convention on the Territorial Sea and the Contiguous Zone.

Continuous discharge (122.2): means a "discharge" which occurs without interruption throughout the operating hours of the facility, except for infrequent shutdowns for maintenance, process changes, or other similar activities.

Controlled area (191.12): means (1) a surface location, to be identified by passive institutional controls, that encompasses no more than 100 square kilometers and extends horizontally no more than five kilometers in any direction from the outer boundary of the original location of the radioactive wastes in a disposal system; and (2) the subsurface underlying such a surface location.

Conventional mine means an open pit or underground excavation for the production of minerals.

Conventional filtration (141.2): means treatment by a series of processes including coagulation, flocculation, sedimentation, and filtration resulting in substantial particulate removal.

Corrosion inhibitor (141.2): means a substance capable of reducing the corrosivity of water toward metal plumbing materials, especially lead and copper, by forming a protective film on the interior surface of those materials.

Curie (Ci) (190.02): means that quantity of radioactive material producing 37 billion nuclear transformations per second. (One millicurie (mCi)=0.001 Ci.)

Current discharge (125.58): means the volume, composition, and location of an applicant's discharge as of anytime between December 27, 1977 and December 29, 1982, as designated by the applicant.

Daily discharge (122.2): means the "discharge" of a pollutant" measured during a calendar day or any 24-hour period that reasonably represents the calendar day for purposes of sampling. For pollutants, limitations are expressed in units of mass.

Deepwater port (110.1): means an offshore facility as defined in section (3)(10) of the Deepwater Port Act of 1974 (33 U.S.C. 1502(10).

Designated facility (260.10): means a hazardous waste treatment, storage, or disposal facility which (1) has received a permit (or interim status) in accordance with the requirements of parts 270 and 124 of this chapter, (2) has received a permit (or interim status) from a State authorized in accordance with part 271 of this chapter, or (3) is regulated under § 261.6(c)(2) or subpart F of part 266 of this chapter, and (4) that has been designated on the manifest by the generator pursuant to § 260.20. If a waste is destined to a facility in an authorized State which has not yet obtained authorization to regulate that particular waste as hazardous, then the designated facility must be a facility allowed by the receiving State to accept such waste.

Diatomaceous earth filtration (141.2): means a process resulting in substantial particulate removal in which (1) a precoat cake of diatomaceous earth filter media is deposited on a support membrance (septum), and (2) while the water is filtered by passing through the cake on the septum, additional filter media known as body feed is continuously added to the feed water to maintain the permeability of the filter cake.

Dike (260.10): means an embankment or ridge of either natural or man-made materials used to prevent the movement of liquids, sludges, solids, or other materials.

Direct filtration (141.2): means a series of processes including coagulation and filtration but excluding sedimentation resulting in substantial particulate removal.

Direct discharge (122.2): means the "discharge of a pollutant."

Discharge or hazardous waste discharge (260.10): means the accidental or intentional spilling, leaking, pumping, pouring, emitting, emptying, or dumping of hazardous waste into or on any land or water.

Discharge (109.2, 110.1, 116.3, 122.2): includes but is not limited to any spilling, leaking, pumping, pouring, emitting, emptying, or dumping. When used in relation to section 311 of the Act, includes but is not limited to any spilling, leaking, pumping, pouring, emitting, emptying, or dumping, but excludes (a) discharges in compliance with a permit under section 402 of the Act, (b) discharges resulting from circumstances identified and review and made a part of the public record with respect to a permit issued or modified under section 402 or the Act, and subject to a condition in such permit, and (c) continuous or anticipated intermittent discharges from a point source, identified in a permit or permit application under section 402 of the Act, that are caused by events occurring within the scope of relevant operating or treatment systems.

Discharge of pollutant(s) (401.11): (1) the addition of any pollutant to navigable waters from any point source and (2) any addition of any pollutant to the waters of the contiguous zone or the ocean from any point source, other than from a vessel or other floating craft. The term "discharge" includes wither the discharge of a single pollutant or the discharge of multiple pollutant.

Disinfectant contact time or "T" in CT calculations (141.2): means the time in minutes that it takes for water to move from the point of disinfectant application or the previous point of disinfectant residual measurement to a point before or at the point where residual disinfectant concentration ("C") is measured. Where only one "C" is measured, "T" is the time in minutes that it takes for water to move from the point of disinfectant application to a point before or at where residual disinfectant concentration ("C") is measured, "T" is (a) for the first measurement of "C", the time in minutes that it takes for water to move from the first or only point of disinfectant application to a point to the "C" measurement point for which the particular "T" is being calculated. Disinfectant contact time in pipelines must be calculated based on "plug flow" by dividing the internal volume of the pipe by the maximum hourly flow rate through that pipe. Disinfectant contact time within mixing basins and storage reservoirs must be determined by tracer studies or an equivalent demonstration.

Disinfectant means any oxidant, including but not limited to chlorine, chlorine dioxide, chloramines, and ozone added to water in any part of the treatment or distribution process, that is intended to kill or inactivate pathogenic microorganisms.

Disinfection (141.2): means a process which inactivates pathogenic organisms in water by chemical oxidants or equivalent agents.

Disposal (260.10): means the discharge, deposit, injection, dumping, spilling, leaking, or placing of any solid waste or hazardous waste or any constituent thereof may enter the environment or be emitted into the air or discharged into any waters, including ground water.

Disposal system (191.12): means any combination of engineered and natural barriers that isolate spent nuclear fuel or radioactive waste after disposal.

Disposal well (146.3): means a well used for the disposal of waste into a subsurface stratum.

Disposal facility (260.10): means a facility or part of a facility at which hazardous waste is intentionally placed into or on any land or water, and at which waste will remain after closure.

Dose equivalent (190.02): means the product of absorbed dose and appropriate factors to account for differences in biological effectiveness due to the quality of radiation and its spatial distribution in the body. The unit of dose equivalent is the "rem". (1 millirem =0.001 rem.)

Dose equivalent (141.2): means the product of the absorbed dose and appropriate factors to account for differences in biological effectiveness due to the type of radiation and its distribution in the body as specified by the International Commission on Radiological Units and Measurements (ICRU).

Drilling mud (144.3): means a heavy suspension used in drilling an "injection well," introduced down the drill pipe and through the drill bit.

Effective corrosion inhibitor residual (141.2): for the purpose of subpart I of this part only, means a concentration sufficient to form a passivating film on the interior walls of a pipe.

Effluent limitation (122.2): means any restriction imposed by the Director on quantities, discharge rates, and concentrations of "pollutants" which are "discharged" from "point sources" into "waters of the United States," the waters of the "contiguous zone," or the ocean.

Effluent limitations guidelines (401.11): means any effluent limitations guidelines issued by the Administrator pursuant to section 304(b) of the Act.

Effluent limitation (401.11): means any restriction established by the Administrator on quantities, rates, and concentrations of chemical, physical, biological, and other constituents which are discharged from point sources, other than new sources, into navigable waters, the waters of the contiguous zone or the ocean.

Effluent standard (129.2): means for purposes of section 307, the equivalent of effluent limitation as that term is defined in section 502(11) of the Act with the exception that it does not include a schedule of compliance.

Exempted aquifer (144.3): means an "aquifer" or its portion that meets the criteria in the definition of "underground source of drinking water" but which has been exempted according to the procedures in § 144.7, 144.8(b).

Existing source (129.2): means any source which is not a new source as defined above.

Existing hazardous waste management or HWM (260.10): facility or existing facilitation or for which construction commenced on or before November 19, 1980.

Existing injection well (146.3): means an "injection well" other than a "new injection well."

Experimental technology (146.3): means a technology which has not been proven feasible under the conditions in which it is being tested.

Facility or activity (124.2, 144.3): means any "HWM facility," UIC "injection well," NPDES "point source," or "treatment works treating domestic sewage" or State 404 dredge or fill activity, or any other facility or activity, or any other facility or activity (including land or appurtenances thereto) that is subject to regulation under the RCRA, UIC, NPDES, or 404 programs, any UIC "injection well," or an other facility or activity that is subject to regulation under UIC program.

Facility (260.10): means all contiguous land, and structures, other appurtenances, and improvements on the land, used for treating, storing, or disposing of hazardous waste. A facility may consist of several treatment, storage, or disposal operation units (e.g., one or more landfills, surface impoundments, or combinations of them).

Fault (146.3): means a surface or zone of rock fracture along which there has been displacement.

Federal agency (142.1): means any department, agency, or instrumentality of the U.S.

First draw sample (141.2): means a one-liter sample of tap water, collected in accordance with § 141.86(b)(2), that has been standing in plumbing pipes at lease 6 hours and is collected without flushing the tap.

Floodplain (257.3-1): means the lowland and relatively flat areas adjoining inland and coastal waters, including flood-prone areas of offshore islands, which are inundated by the base flood.

Floodplains (257.3-1): facilities or practices in floodplains shall not restrict the flow of the base flood, reduce the temporary water storage capacity of the floodplain, or result in washout of solid waste, so as to pose a hazard to human life, wildlife, or land or water resources.

Flow rate (146.3): means the volume per time unit given to the flow of gases or other fluid substance which emerges from an orifice, pump, turbine or passes along a conduit or channel.

Fluid (144.3, 146.3): means material or substance which flows or moves whether in a semisolid, liquid, sludge, gas, or any other form or state.

Formation fluid (144.3, 270.2): means "fluid" present in a "formation" under natural conditions as opposed to introduced fluids, such as drilling mud.

Formation (144.3, 146.3): means a body of consolidated or unconsolidated rock characterized by a degree of lithologic homogeneity which is prevailingly, but not necessarily, tabular and is mappable on the earth's surface or traceable in the subsurface.

Free product (280.12): refers to a regulated substance that is present as a nonaqueous phase liquid (e.g., liquid not dissolved in water).

Fugitive dust, mist, or vapor (129.2): means dust, mist, or vapor containing a toxic pollutant regulated under this part which is emitted from any source other than through a stack.

General environment (190.02): means the total terrestrial, atmospheric, and aquatic environments outside sites upon which any operation which is part of a nuclear fuel cycle is conducted.

Generator (261): means any person, by site location, whose act or process produces hazardous waste identified.

Ground water under direct influence of surface water (141.2): means any water beneath the surface of the ground with (1) significant occurrence of insects or other microorganisms, algae, or large-diameter pathogens such as *Giardia lamblia,* or (2) significant and relatively rapid shifts in water characteristics such as turbidity, temperature, conductivity, or pH which closely correlate to climatological or surface water conditions. Direct influence point is not subject to recontamination by surface water runoff.

Ground water (144.3, 191.12, 257.3-4, 260.10, 270.2): means water below the land surface in a zone of saturation.

Hazardous substance (122.2): means any substance designated under 40 CFR part 116 pursuant to section 311 of CWA.

Hazardous Waste Management facility or **"HWM facility"** (146.3): means all contiguous land, and structures, other appurtenances, and improvements on the land used for treating, storing, or disposing of hazardous waste. A facility may consist of several treatment, storage, or disposal operational units (for example, one or more landfills, surface impoundments, or combination of them).

Hazardous waste management unit (260.10): is a continuous area of land on or in which hazardous waste is placed, or the largest area in which there is significant likelihood of mixing hazardous waste constituents in the same area. Examples of hazardous waste management units include a surface impoundment, a waste pile a land treatment area, a landfill cell, an incinerator, a tank and its associated piping and underlying container storage area. A container alone does not constitute a unit; the unit includes containers and the land or pad upon which they are placed.

Heavy metal (191.12): means all uranium, plutonium, or thorium placed into a nuclear reactor.

In operation (270.2): means a facility which is treating, storing, or disposing of hazardous waste.

Incorporated into the soil (257.3-6): means the injection of solid waste beneath the surface of the soil or the mixing of solid waste with the surface soil.

Indirect discharger (122.2): means a nondomestic discharger introducing "pollutants".

Industrial source (125.58): means any source of nondomestic pollutants regulated under section 307 (b) or (c) of the Clean Water Act which discharges into a POTW.

Inground tank (260.10): means a device meeting the definition of "tank" in § 260.10 whereby a portion of the tank wall is situated to any degree within the ground, thereby preventing visual inspection of that external surface area of the tank that is in the ground.

Injection zone (144.3, 146.3): means a geological "formation" group of formations, or part of a formation receiving fluids through a "well".

Injection well (144.3, 146.3, 260.10): means a well into which fluids are injected. (See also "underground injection".)

Inner liner (260.10): means a continuous layer of material placed inside a tank or container which protects the construction materials of the tank or container from the contained waste or reagents used to treat the waste.

Interstate Agency (124.2, 142.1): means an agency of two or more States established by or under an agreement or compact approved by the Congress, or any other agency of two or more States or Indian Tribes having substantial powers or duties pertaining to the control of pollution as determined and approved by the Administrator.

Land (192.11): means any surface or subsurface land that is not part of a disposal site and is not covered by an occupiable building.

Land disposal (270.2): means placement in or on the land and includes, but is not limited to, placement in a landfill, surface impoundment, waste pile, injection well, land treatment facility, salt dome formation, salt bed formation, underground mine or cave, or placement in a concrete vault or bunker intended for disposal purposes.

Land treatment facility (260.10): means a facility or part of a facility at which hazardous waste is applied onto or incorporated into the soil surface; such facilities are disposal facilities if the waste will remain after closure.

Landfill (260.10): means a disposal facility or part of a facility where hazardous waste is placed in or on land and which is not a pile, a land treatment facility, a surface impoundment, an underground injection well, a salt dome formation, a salt bed formation an underground mine, or a cave.

Large water system (141.2): for the purpose of subpart I of this part only, means a water system that serves more than 50,000 persons.

LC50 (116.4): means that concentration of material which is lethal to one half of the test population of aquatic animals upon continuous exposure for 96 hours or less.

Leachate (257.2, 260.10): means any liquid, including any suspended components in the liquid, that has percolated through or drained from hazardous waste.

Leak-detection system (260.10): means a system capable of detecting the failure of either the primary or secondary containment structure or the presence of a release of hazardous waste or accumulated liquid in the secondary containment structure. Such a system must employ operational controls (e.g., daily visual inspections for releases into the secondary containment system of aboveground tanks) or consist of an interstitial monitoring device designed to detect continuously and automatically the failure of the primary or secondary containment structure or the presence of a release of hazardous waste into the secondary containment structure.

Liner (260.10): means a continuous layer of natural or man-made materials, beneath or on the sides of a surface impoundment, landfill, or landfill cell, which restricts the downward or lateral escape of hazardous waste constituents, or leachate.

Lithology (146.3): means the description of rocks on the basis of their physical and chemical characteristics.

Lithosphere (191.12): means the solid part of the Earth below the surface, including any ground water contained within it.

Load or loading: an amount of matter or thermal energy that is introduced into a receiving water; to introduce matter or thermal energy into a receiving water. Loading may be either man-caused (pollutant loading) or natural (natural background loading).

Loading capacity (130.2): the greatest amount of loading that a water can receive without violating water quality standards.

Major facility (124.2): means any RCRA, UIC, NPDES, or 404 "facility or activity" classified as such by the Regional Administrator, or, in the case of "approved State programs," the Regional Administrator in conjunction with the State Director.

Maximum contaminant level (141.2): means the maximum permissible level of a contaminant in water which is delivered to the free flowing outlet of the ultimate user of a public water system, except in the case of turbidity where the maximum permissible level is measured at the point of entry to the distribution system. Contaminants added to the water under circumstances controlled by the user, except those resulting from corrosion of piping and plumbing caused by water quality, are excluded from this definition.

Maximum contaminant level goal or **MCLG** (141.2): means the maximum level of a contaminant in drinking water at which no known or anticipated adverse effect on the health of persons would occur, and which allows an adequate margin of safety.

Medium-size water system (141.2): for the purpose of subpart I of this part only, means a water system that serves greater than 3,300 and less than or equal to 50,000 persons.

Mining overburden returned to the mine site (260.10): means any material overlying an economic mineral deposit which is removed to gain access to that deposit and is then used for reclamation of a surface mine.

Mixture (116.4): means any combination of two or more elements and/or compounds in solid, liquid, or gaseous form except where such substances have undergone a chemical reaction so as to become inseparable by physical means.

Modified discharge (125.58): means the volume, composition, and location of the discharge proposed by the applicant for which a modification under § 301(h) of the Act is requested. A modified discharge may be a current discharge, improved discharge, or altered discharge.

Municipality (142.1): means a city, town, or other public body created by or pursuant to State law, or an Indian Tribe which does not meet the requirements of subpart H of this part.

National primary drinking water regulation (142.1): means any primary drinking water regulation contained in part 141 of this chapter.

Navigable waters (401.11): includes all navigable waters of the U.S.; tributaries of navigable waters of the U.S.; interstate waters; intrastate lakes, rivers, and streams which are utilized by interstate travelers for recreational or other purposes; intrastate lakes, rivers, and streams from which fish or shellfish are taken and sold in interstate commence; and intrastate lakes, rivers and streams which are utilized for industrial purposes by industries in interstate commence.

New injection wells (144.3): means an "injection well" which began injection after a UIC program for the State applicable to the well is approved or prescribed.

New source (129.2): means any source discharging a toxic pollutant, the construction of which is commenced after proposal of an effluent standard or prohibition applicable to such source if such effluent standard or prohibition is thereafter promulgated in accordance with § 307.

New source (401.11): means any building, structure, facility, or installation from which there is or may be the discharge of pollutants, the construction of which is commenced after the publication of proposed regulations prescribing a standard of performance under § 306 of the Act which will be applicable to such source if such standard is thereafter promulgated in accordance with section 306 of the Act.

No discharge of free oil (435.11): shall mean that a discharge does not cause a film or sheen upon or a discoloration on the surface of the water or adjoining shorelines or cause a sludge or emulsion to be deposited beneath the surface of the water or upon adjoining shorelines.

Noncontact cooling water pollutants (401.11): means pollutants present in noncontact cooling waters.

Noncontact cooling water (401.11): means water used for cooling which does not come into direct contact with any raw material, intermediate product, waste product, or finished product.

Nonindustrial source (125.58): means any source of pollutants which is not an industrial source.

Nuclear fuel cycle (190.02): means the operations defined to be associated with the production of electrical power for public use by any fuel cycle through utilization of nuclear energy.

Ocean waters (125.58): means those coastal waters landward of the baseline of the territorial seas, the deep waters of the territorial seas, or the waters of the contiguous zone.

Oil (Parts 109.2, 110.6, 110.1, 113.1) means oil of any kind or in any form, including, but not limited to, petroleum, fuel oil, sludge, oil refuse, and oil mixed with wastes other than dredged spoil.

On-site (270.2): means on the same or geographically contiguous property which may be divided by public or private right(s)-of-way, provided the entrance and exit between the properties is at a cross-roads intersection, and access is by crossing as opposed to going along, the right(s)-of-way. Noncontiguous properties owned by the same person but connected by a right-of-way which the person controls and to which the public does not have access, is also considered on-site property.

Open dump (257.2): means a facility for the disposal of solid waste which does not comply with this part.

Passive institutional control (191.12): means (1) permanent markers placed at a disposal site, (2) public records and archives, (3) government ownership and regulations regarding land or resource use, and (4) other methods of preserving knowledge about the location, design, and contents of a disposal system.

Percent removal (133.100): a percentage expression of the removal efficiency across a treatment plant for a given pollutant parameter, as determined from the 30-day average values of the raw wastewater influent pollutant concentrations of the facility and the 30-day average values of the effluent pollutant concentrations for a given time period.

Permit (124.2, 129.2): means an authorization, license, or equivalent control document issued by EPA or an "approved State" to implement the requirements of this part and parts 122, 123, 144, 145, 233, 270, and 271. "Permit" includes RCRA "permit by rule" (§ 270.60), UIC area permit (§ 144.33), NPDES or 404 "general permit" (§§ 270.61, 144.34, and 233.38). Permit does not include RCRA interim status (§ 270.70), UIC authorization by rule (§ 144.21), or any permit which has not yet been the subject of final agency action, such as a "draft permit" or a "proposed permit."

Person (142.1): means an individual; corporation; company; association; partnership; municipality, or State, federal, or Tribal agency.

Pesticides (125.58): means demeton, guthion, malathion, mirex, methoxychlor, and parathion.

Petroleum UST system (280.12): means an underground storage tank system that contains petroleum or a mixture of petroleum with *de minimis* quantities of other regulated substances. Such systems include those containing motor fuels, jet fuels, distillate fuel oils, residual fuel oils, lubricants, petroleum solvents, and used oils.

Pile (260.10): means any noncontainerized accumulation of solid, nonflowing hazardous waste that is used for treatment or storage.

Plugging (144.3): means the act or process of stopping the flow of water, oil, or gas into or out of a formation through a borehole or well penetrating that formation.

Point source (122.2, 401.11): means any discernible, confined, and discrete conveyance, including but not limited to, any pipe, ditch, channel, tunnel, conduit, well discrete fissure, container, rolling stock, concentrated animal feeding operation, landfill leachate collection system, vessel or other floating craft from which pollutants are or may be discharged. This term does not include return flows from irrigated agriculture or agricultural storm water runoff. (See § 122.3).

Point-of-entry treatment device (141.2): is a treatment device applied to the drinking water entering a house or building for the purpose of reducing contaminants in the drinking water distributed throughout the house or building.

Point-of-use treatment device (141.2): is a treatment device applied to a single tap used for the purpose of reducing contaminants in drinking water at that one tap.

Pollutant (122.2, 401.11): means dredged spoil, solid waste, incinerator residue, filter backwash, sewage, garbage, sewage sludge, munitions, chemical wastes, biological materials, radioactive materials (except those regulated under the Atomic Energy Act of 1954, as amended (42 U.S.C. 2011 et seq.), heat, wrecked or discarded equipment, rock, sand, cellar dirt and industrial, municipal, and agricultural waste discharged into water. It **does not** mean (a) sewage from vessels; or (b) water, gas, or other material which is injected into a well to facilitate production of oil or gas, or water derived in association with oil and gas production and disposed of in a well, if the well used either to facilitate production or for disposal purposes is approved by authority of the State in which the well is located, and if the State determines that the injection or disposal will not result in the degradation of ground or surface water resources.

Pollution (130.2, 401.11): the man-made or man-induced alteration of the chemical, physical, biological, and radiological integrity of water.

Practice (257.2): means the act of disposal of solid waste.

Primary enforcement responsibility (142.1): means the primary responsibility for administration and enforcement of primary drinking water regulations and related requirements applicable to public water systems within a State.

Process wastes (129.2): means any designated toxic pollutant, whether in wastewater or otherwise present, which is inherent to or otherwise present, which is inherent to or unavoidably resulting from any manufacturing process, including that which comes into direct contact with or results from the production or use of any raw material, intermediate product, finished product, by-product, or waste product and is discharged into the navigable waters.

Process wastewater (122.2, 401.11): means any water which, during manufacturing or processing, comes into direct contact with or results from the production or use of any raw material, intermediate product, finished product, by-product, or waste product.

Process waste water pollutants (401.11): means pollutants present in process waste water.

Prohibited (129.2): means that the constituent shall be absent in any discharge subject to these standards, as charge subject to these standards, as determined by any analytical method.

Public water supplies (125.58): means water distributed from a public water system.

Public water system (125.58, 141.2): means a system for the provision to the public of piped water for human consumption if such system has at least 15 service connections or regularly serves at least 25 individuals. This term includes (1) any collection, treatment, storage and distribution facilities under the control of the operator of the system and used primarily in connection with the system, and (2) any collection or pretreatment storage facilities not under the control of the operator of the system and used primarily in connection with the system.

Publicly Owned Treatment Works or POTW (260.10, 403.3): means a treatment works as defined by section 212 of the Act, which is owned by a State or municipality (as defined by section 502(4) of the Act). This definition includes any devices and systems used in the storage, treatment, recycling and reclamation of municipal sewage or industrial wastes of a liquid nature. It also includes sewers, pipes and other conveyances only if they convey wastewater to a POTW Treatment Plant. The term also means the municipality as defined in section 502(4) of the Act, which has jurisdiction over the indirect discharges to and the discharges from such a treatment works. Any device or system used in the treatment (including recycling and reclamation) of municipal sewage or industrial wastes of a liquid nature which is owned by a "State" or "municipality" (as defined by section 502(4) of the CWA). This definition includes sewers, pipes, or other conveyances only if they convey wastewater to a POTW providing treatment.

Radiation (190.02): means any or all of the following: alpha, beta, gamma, or X-rays; neutrons; and high-energy electrons, protons, or other atomic particles, but not sound or radio waves, nor visible, infrared, or ultraviolet light.

Radioactive material (190.02): means any material which spontaneously emits radiation.

Radioactive waste (144.3): means any waste which contains radioactive material in concentrations which exceed those listed in 10 CFR part 20, appendix B, table II, column 2.

Recharge area (149.2): means an area in which water reaches the zone of saturation (ground water) by surface infiltration; in addition, a major recharge area is an area where a major part of the recharge to an aquifer occurs.

Recharge (149.2): means a process natural or artificial, by which water is added to the saturated zone of an aquifer.

Recommencing discharger (122.2): means a source which recommences discharge after terminating operations.

Regulated substance: means (a) any substance defined in section 101(14) of the Comprehensive Environmental Response, Compensation and Liability Act (CERCLA) of 1980 (but not including any substance regulated as a hazardous waste under subtitle C), and (b) petroleum, including crude oil or any fraction thereof that is liquid at standard conditions of temperature and pressure (60 degrees Fahrenheit and 14.7 pounds per square inch absolute).

Release detection (280.12): means determining whether a release of a regulated substance has occurred from the UST system into the environment or into the interstitial space between the UST.

Release (280.12): means any spilling, leaking, emitting, discharging, escaping, leaching, or disposing from an UST into ground water, surface water or subsurface soils.

Representative sample (260.10): means a sample of a universe or whole (e.g., waste pile, lagoon, ground water) which can be expected to exhibit the average properties of the universe or whole.

Run-off (260.10): means any rainwater, leachate, or other liquid that drains over land from any part of a facility.

Run-on (260.10): means any rainwater, leachate, or other liquid that drains over land onto any part of a facility.

Saline estuarine waters (125.58): means those semienclosed coastal waters which have a free connection to the territorial sea, undergo net seaward exchange with ocean waters, and have salinities comparable to those of the ocean. Generally, these waters are near the mouth of estuaries and have cross-sectional annual means salinities greater than 25 parts per thousand.

Sanitary landfill (257.2): means a facility for the disposal of solid waste which complies with this part.

Saturated zone or zone of saturation (260.10): means that part of the earth's crust in which all voids are filled with water.

Schedule of compliance (144.3): means a schedule of remedial measures included in a "permit", including an enforceable sequence of interim requirements (e.g., actions, operations, or milestone events) leading to compliance with the "appropriate Act and regulations."

Secondary maximum contaminant levels (143.2): means SMCLs which apply to public water systems and which, in the judgement of the Administrator, are requisite to protect the public welfare. The SMCL means the maximum permissible level of a contaminant in water which is delivered to the free flowing outlet of the ultimate user of public water system. Contaminants added to the water under circumstances controlled by the user, except those resulting from corrosion of piping and plumbing caused by water quality, are excluded from the definition.

Sedimentation (141.2, 142.1): means a process for removal of solids before filtration by gravity or separation.

Septage (122.2): means the liquid and solid material pumped from a septic tank, cesspool, or similar domestic sewage treatment system, or a holding tank when the system is cleaned or maintained.

Service line sample (141.2): means a one-liter sample of water collected in accordance with § 141.86 (b)(3), that has been standing for at lease 6 hours in a service line.

Service line (141.2): means a service line of lead which connects the water main to the building inlet and any lead pigtail, gooseneck, or other fitting which is connected to such lead line.

Sewage sludge (122.2): means any solid, semisolid, or liquid residue removed during the treatment of municipal waste water or domestic sewage. Sewage sludge includes, but is not limited to, solids removed during primary, secondary, or advanced waste water treatment, scum, septage, portable toilet pumpings, type III marine sanitation device pumpings (33 SFR part 159), and sewage sludge products. Sewage sludge does not include grit or screenings, or ash generated during the incineration of sewage sludge.

Sheen (110.1): means an iridescent appearance on the surface of water.

Shellfish, fish, and wildlife (125.58): means any biological population or community that might be adversely affected by the applicant's modified discharge.

Significant source of ground water (191.12): as used in this part, means (1) an aquifer that (i) is saturated with water having less than 10,000 milligrams per liter of total dissolved solids; (ii) is within 2,500 feet of the land surface; (iii) has a transmissivity greater than 200 gallons per day per square foot; and (iv) is capable of continuously yielding at least 10,000 gallons per day to a pumped or flowing well for a period of at least a year; or (2) an aquifer that provides the primary source of water for a community water system as of the effective date of this subpart.

Significant biological treatment (133.100): the use of an aerobic or anaerobic biological treatment process in a treatment works to consistently achieve a 30-day average of at least 65 percent removal of BOD_5.

Site (122.2, 144.3): means the land or water area where any "facility or activity" is physically located or conducted, including adjacent land used in connection with the facility or activity.

Site (190.02): means the area contained within the boundary of a location under the control of persons possessing or using radioactive material on which is conducted one or more operations covered by this part.

Slow sand filtration (141.2): means a process involving passage of raw water involving passage of raw water through a bed of sand at low velocity (generally less than 0.4 m/h) resulting in substantial particulate removal by physical and biological mechanisms.

Sludge (110.1, 260.10): means any solid, semisolid, or liquid waste generated from a municipal, commercial, or industrial wastewater treatment plant, water supply treatment plant, or air pollution control facility exclusive of the treated effluent from a wastewater treatment plant.

Small water system (141.2): for the purpose of subpart I of this part only means a water system that serves 3300 persons or fewer.

Soil (192.11): means all unconsolidated materials normally found on or near the surface of the earth including, but not limited to, silts, clays, sands, gravel, and small rocks.

Sole or principal source aquifer (SSA) (149.2): means an aquifer which is designated as an SSA under section 1424(e) of the SDWA.

Solid waste boundary (257.3-4): means the outermost perimeter of the solid waste (projected in the horizontal plane) as it would exist at completion of the disposal activity.

Solid waste (257.2): means any garbage, refuse, sludge from a waste treatment plant, water supply treatment plant or air pollution control facility and other discarded material, including solid, liquid, semisolid, or contained gaseous material resulting from industrial, commercial, mining, and agricultural operations, and from community activities, but does not include solid or dissolved materials in domestic sewage, or solid or dissolved material in irrigation return flows or industrial discharges which are point sources subject to permits under section 402 of the Federal Water Pollution Control Act, as amended (86 Stat. 880), or source, special nuclear, or byproduct material as defined by the Atomic Energy Act of 1954, as amended (68 Stat. 923).

Source (129.2): means any building, structure, facility, or installation from which there is or may be the discharge of toxic pollutants designated as such by the Administration under section 307(a)(1) or the Act.

Special source of ground water (191.12): as used in this part, means those Class I ground waters identified in accordance with the Agency's Ground-Water Protection Strategy published in August 1984 that: (1) are within the controlled area encompassing a disposal system or are less than five kilometers beyond the controlled area; (2) are supplying drinking water for thousands of persons as of the date that the Department chooses a location within that area for detailed characterization as a potential site for a disposal system (e.g., in accordance with section 112(b)(1)(B) of the NWPA); and (3) are irreplaceable in that no reasonable alternative source of drinking water is available to that population.

Standard sample (141.2): means the aliquot of finished drinking water that is examined for the presence of coliform bacteria.

Standard of performance (401.11): means any restriction established by the Administrator pursuant to section 306 of the Act on quantities, rates, and concentrations of chemical, physical, biological, and other constituents which are or may be discharged from new sources into navigable waters, the waters of the contiguous zone or the ocean.

State program revision (142.1): means a change in an approved State primacy program.

State (141.2, 257.2): means the agency of the State or Tribal government which has jurisdiction over public water systems. During any period when a State or Tribal government does not have primary enforcement responsibility pursuant to Section 1413 of the Act, the term "State" means the Regional Administrator, U.S. Environmental Protection Agency.

State primary drinking water regulation (142.1): means a drinking water regulation of a State which is comparable to a national primary drinking water regulation.

States (131.3, 142.1): the 50 States, the District of Columbia, Guam, the Commonwealth of Puerto Rico, Virgin Islands, American Samoa, the Trust Territory of the Pacific Islands, and the Commonwealth of the Northern Mariana Islands, or an Indian Tribe treated as a State.

Storage (260.10): means the holding or hazardous waste for a temporary period, at the end of which the hazardous waste is treated, disposed of, or stored elsewhere.

Stratum (plural strata) (144.3): means a single sedimentary bed or layer, regardless of thickness, that consists of generally the same kind of rock material.

Stressed waters (125.58): means those receiving environments in which an applicant can demonstrate to the satisfaction of the administrator, that the absence of a balanced, indigenous population is caused solely by human perturbations other than the applicant's modified discharge.

Sump (260.10): means any pit or reservoir that meets the definition of tank and those troughs/trenches connected to it that serves to collect hazardous waste for transport to hazardous waste storage, treatment, or disposal facilities.

Supplier of water (141.2): means any person who owns or operates a public water system.

Surface impoundment or impoundment (260.10): means a facility or part of a facility which is a natural topographic depression, man-made excavation, or diked area formed primarily of earthen materials (although it may be lined with man-made materials), which is designed to hold an accumulation of liquid wastes or wastes containing free liquids, and which is not an injection well. Examples of surface impoundments are holding, storage, setting,and aeration pits, ponds and lagoons.

Surface water (141.2): means all water which is open to the atmosphere and subject to surface runoff.

Territorial seas (116.3): means the belt of the seas measured from the line of ordinary low water along that portion of the coast which is in direct contact with the open sea and the line marking the seaward limit of inland waters, and extending seaward a distance of 3 miles. The term includes but is not limited to petroleum and petroleum-based substances comprised of a complex blend of hydrocarbons derived from crude oil though processes of separation, conversion, upgrading, and finishing, such as motor fuels, jet fuels, distillate fuel oils, residual fuel oils, lubricants, petroleum solvents, and used oils.

Total maximum daily load or (TMDL) (130.2): the sum of the individual WLAs for point sources and LAs for nonpoint sources and natural background. If a receiving water has only one point source discharger, the TMDL is the sum of that point source WLA plus the LAs for any nonpoint sources of pollution and natural background sources, tributaries, or adjacent segments. TMDLs can be expressed in terms of either mass per time, toxicity, or other appropriate measure. If Best Management Practices (BMPs) or other nonpoint source pollution controls make more stringent load allocations practicable, then wasteload allocations can be made less stringent. Thus the TMDL process provides for nonpoint source control tradeoffs.

Total dissolved solids (144.3): means the total dissolved (filterable) solids as determined by use of the method specified in 40 CFR part 136.

Toxic pollutant (131.3): means any pollutant listed as toxic under section 307(a)(1) or, in the case of "sludge use or disposal practices," any pollutant identified in regulations implementing section 405(d) of the CWA.

Transmissivity (191.12): means the hydraulic conductivity integrated over the saturated thickness of an underground formation. The transmissivity of a series of formations is the sum of the individual transmissivities of each formation comprising the series.

Treatability study (260.10): means a study in which a hazardous waste is subjected to a treatment process to determine: (1) whether the waste is amenable to the treatment process, (2) what pretreatment (if any) is required, (3) the optional process conditions needed to achieve the desired treatment, (4) the efficiency of a treatment process for a specific waste or wastes, or (5) the characteristics and volumes of residuals from a particular treatment process. Also included in this definition for the purpose of the § 261.4 (e) and (f) exemptions are liner compatibility, corrosion, and other material compatibility studies and toxicological and health effects studies. A "treatability study" is not a means to commercially treat or dispose of hazardous waste.

Treatment zone (260.10): means a soil area of the unsaturated zone of a land treatment unit within which hazardous constituents are degraded, transformed, or immobilized.

Treatment technique requirement (142.1): means a requirement of the national primary drinking water regulations.

Treatment (260.10): means any method technique, or process, including neutralization, designed to change the physical chemical, or biological character or composition of any hazardous waste so as to neutralize such waste, or so as to recover energy or material resources from the waste, or so as to render such waste nonhazardous or less hazardous; safer to transport, store, or dispose of; or amenable for recovery, amenable for storage, or reduced in volume.

UIC (144.4): means the Underground Injection Control program under Part C of the Safe Drinking Water Act, including an "approved State program."

Underground tank (260.10): means a device meeting the definition of "tank" in § 260.10 whose entire surface area is totally below the surface of and covered by the ground.

Underground storage tank or UST (280.12): means any one or combination of tanks (including underground pipes connected thereto) that is used.

Underground release (280.12): means any belowground release.

Underground area (280.10): means an underground room, such as a basement, cellar, shaft, or vault providing enough space for physical inspection of the exterior of the tank situated on or above the surface of the floor.

Underground drinking water source (257.3-4): means (i) an aquifer supplying drinking water for human consumption or (ii) an aquifer in which the ground water contains less than 10,000 mg/1 total dissolved solids.

Underground source of drinking water or USDW (270.2): means an aquifer or its portion: (a) (1) which supplies any public water system or (2) which contains a sufficient quantity of ground water to supply a public water system; and (i) currently supplies drinking water for human consumption or (ii) contains fewer than 10,000 mg/1 total dissolved solids; and (b) which is not an exempted aquifer.

Underground injection (260.10, 270.2): means the subsurface emplacement of fluids through a bored, drilled or driven well; or through a dug well, where the depth of the dug well is greater than the largest surface dimension. (See also "injection well").

Unsaturated zone or zone of aeration (260.10): means the zone between the land surface and the water table.

Uppermost aquifer (260.10): means the geologic formation nearest the natural logic formation nearest the natural ground surface that is an aquifer, as well as lower aquifers that are hydraulically interconnected within the facility's property boundary.

Uranium fuel cycle (190.02): means the operations of milling of uranium, isotopic enrichment of uranium, fabrication of uranium fuel, generation of electricity by a light-water-cooled nuclear power plant using uranium fuel, and reprocessing of spent uranium fuel, to the extent that these directly support the production of electrical power of public use utilizing nuclear energy, but excludes mining operations, operations at waste disposal sites, transportation of any radioactive material in support of these operations, and the reuse of recovered nonuranium special nuclear and by-product materials from the cycle.

USDW (144.4): means "underground source of drinking water."

Washout (257.3-1): means the carrying away of solid waste by waters of the base flood.

Waste (191.12): as used in this subpart means any spent nuclear fuel or radioactive waste isolated in a disposal system.

Waste form (191.12): means the materials comprising the radioactive components of waste and any encapsulating or stabilizing matrix.

Wasteload allocation or WLA (130.2): the portion of a receiving waters loading capacity that is allocated to one of its existing or future point sources of pollution. WLAs constitute a type of water quality-based effluent limitation.

Water quality limited segment (130.2, 131.3): means any segment where it is known that water quality does not meet applicable water quality standards, even after the application of the technology-bases effluent limitations required by sections 301(b) and 306 of the Act.

Water quality standards or WQS (131.3): are provisions of State or Federal law which consist of a designated use or uses for the waters of the U.S. and water quality criteria for such waters based upon such uses. Water quality standards are to protect the public health or welfare, enhance the quality of water and serve the purposes of the Act.

Water quality standards (WQS) (130.2): means provisions of State or Federal law which consist of a designated use or uses for the waters of the U.S. and water quality criteria for such waters based upon such uses. Water quality standards are to protect the public health or welfare, enhance the quality of water and serve the purposes of the Act.

Water quality standards (125.58): means applicable water quality standards which have been approved, left in effect, or promulgated under section 303 of the Clean Water Act.

Water quality management or WQM plan (130.2): a State or areawide waste treatment management plan developed and updated in accordance with the provisions of sections 205(j), 208 and 303 of the Act and this regulation.

Well injection (144.3): means the subsurface emplacement of "fluids" through a bored, drilled, or driven "well;" or through a dug well is greater than.

Well (144.4): means a bored, drilled, or driven shaft, or a dug hole, whose depth is greater than the largest surface dimension.

Wetlands (110.1, 122.2, 435.41): means those areas that are inundated or saturated by surface or ground water at a frequency or duration sufficient to support, and that under normal circumstances do support, a prevalence of vegetation typically dated for life in saturated soil conditions. Wetlands generally include playa lakes, swamps, marshes, bogs, and similar areas such as sloughs, prairie potholes, wet meadows, prairie river overflows, mudflats, and natural ponds.

Whole effluent toxicity (122.2): means the aggregate toxic effect of an effluent measured directly by a toxicity test.

Working level (WL) (192.11): means any combination of short-lived radon decay products in one liter of air that will result in the ultimate emission of alpha particles with a total energy of 130 billion electrons volts.

Zone of engineering control (260.10): means an area under the control of the owner/operator that, upon detection of a hazardous waste release, can be readily cleaned up prior to the release of hazardous waste or hazardous constituents to ground water or surface water.

Zone of initial dilution or ZID (125.58): means the region of initial mixing surrounding or adjacent to the end of the outfall pipe or diffuser ports, provided that the ZID may not be larger than allowed by mixing zone restrictions in applicable water quality standards.

Appendix C Analysis Reference Chart

Analysis	Container For Water	Preservative For Water (Chill to 4°C)	Container For Soil (Chill to 4°C)[a]	Holding Time (From Sampling Date)	
				Water	Soil
INORGANIC CHEMISTRY					
Alkalinity	4 oz. Plastic	Unpreserved	N/A	14 days	N/A
Ammonia (NH_3)	4 oz. Plastic	.25 ml H_2SO_4[A]	4 oz. jar	28 days	28 days
BOD	16 oz. Plastic (Headspace Free)	Unpreserved	N/A	48 hr.	N/A
Boron	4 oz. Plastic	Unpreserved	4 oz. jar	28 days	28 days
Bromide	16 oz. Plastic	Unpreserved	8 oz. jar	28 days	28 days
Chloride	4 oz. Plastic	Unpreserved	8 oz. jar	28 days	28 days
COD	4 oz. Plastic	.25 ml H_2SO_4[A]	4 oz. jar	28 days	28 days
Color	4 oz. Plastic	Unpreserved	N/A	48 hr.	N/A
Cyanide (total and/or amenable)	4 Plastic[B]	2 ml 1.5N NaOH[C]	4 oz. jar	14 days	No Specified Time
EC (Electrical Conductivity)	4 oz. Plastic	Unpreserved	4 oz. jar	28 days	28 days
Fish Toxicity	5 × 1 Gal. Plastic	Unpreserved	4 oz. jar	48 hr.	No Specified Time
Flashpoint	8 oz. Amber Glass (Glass only) w/Septum (Headspace Free)	Unpreserved	8 oz. jar	28 days	28 days
Fluoride	4 oz. Plastic	Unpreserved	4 oz. jar	28 days	28 days
Formaldehyde	1 L Glass	1% Methanol	4 oz. jar	28 days-Pres. 7 days-Unp.	28 days

Appendix C Analysis Reference Chart (continued)

Analysis	Container For Water	Preservative For Water (Chill to 4°C)	Container For Soil (Chill to 4°C)[a]	Holding Time (From Sampling Date) Water	Holding Time (From Sampling Date) Soil
General Minerals					
- General Minerals	1 L Plastic	Unpreserved	16 oz. jar	28 days	28 days
- NO_3	4 oz. Plastic	.25 ml H_2SO_4[A]	4 oz. jar		
- Metals	16 oz. Plastic[B]	1 ml HNO_3[A]	N/A		
Gross Alpha/Beta	1 L Plastic	2 ml HNO_3[A D]	4 oz. jar	6 mo.	6 mo.
Hardness	4 oz. Plastic	Unpreserved	N/A	28 days	N/A
Hexavalent Chromium ($Cr + ^6$)	16 oz. Plastic	Unpreserved	4 oz. jar	24 hr.	28 days
Iodide	4 oz Plastic	Unpreserved	4 oz. jar	24 hr.	28 days
Nitrate/Nitrite (NO_3/NO_2)	4 oz. Plastic	.25 ml H_2SO_4[A]	4 oz. jar	28 days	28 days
- NO_3	4 oz. Plastic	Unpreserved	4 oz. jar	48 hr.	28 days
Odor	4 oz. Glass	Unpreserved	N/A	48 hr.	N/A
Oil & Grease	1 L Glass (Glass Only)[B]	2 ml H_2SO_4[A]	4 oz. jar	28 days	28 days
418.1 (TPH by IR)	1 L Glass (Glass only)[B]	2 ml H_2SO_4[A]	4.oz. jar	28 days	28 days
pH	4 oz. Plastic	Unpreserved	4 oz. jar	immediately	14 days
Phenolics	4 oz. Amber Glass (Glass Only)[B]	.25 ml H_2SO_4[A]	4 oz. jar	28 days	28 days
Phosphorus - Total (P)	4 oz./8 oz. Plastic	.25 ml/.5 H_2SO_4[A]	4 oz. jar	28 days	28 days
Phosphorus - Ortho (PO_4)	4 oz./8 oz. Plastic	Unpreserved	4 oz. jar	48 hr.	28 days

Parameter	Container	Preservative	Container	Holding Time	Holding Time
Silica	4 oz. (Plastic Only)	Unpreserved	4 oz. jar	28 days	28 days
Solids (Residue) - Total dissolved	16 oz. Plastic	Unpreserved	N/A	7 days	N/A
Solids (Residue) - Total suspended	16 oz. Plastic	Unpreserved	N/A	7 days	N/A
Solids (Residue) - Total settleable	1 L Plastic	Unpreserved	N/A	48 hr.	N/A
Solids (Residue) - Total solids	16 oz. Plastic	Unpreserved	N/A	7 days	N/A
Specific Gravity	4 oz. Plastic	Unpreserved	4 oz. jar	28 days	28 days
Sulfate	4 oz. Plastic	Unpreserved	4 oz. jar	28 days	28 days
Sulfide	4 oz. Plastic[B]	6 drops-2N Zn acetate & 8 drops 6N NaOH[E]	N/A	7 days	N/A
Sulfite	4 oz. Plastic	1 ml EDTA	N/A	28 days-Pres 6 hr.-Unp.	N/A
Surfactants (MBAS)	1 L Plastic	Unpreserved	N/A	48 hr.	N/A
Total Coliform	8 oz. Glass or Nalgene	Unpreserved	N/A	6-8 hr.	N/A
Total Organic Carbon (TOC)	4 oz. Amber Glass (Glass Only) w/Septum (Headspace Free)	.25 ml H_2SO_4[A]	4 oz. jar	28 days	28 days
Total Organic Halide (TOX)	8 oz. Amber Glass (Glass Only) w/Septum (Headspace Free)	.5 ml H_2SO_4[A]	4 oz. jar	7 days	No Specified Time
Total Radium	1 L Plastic	2 ml HNO_3[A D]	4 oz. jar	6 mo.	6 mo.
Turbidity	4 oz. Plastic	Unpreserved	N/A	48 hr.	N/A

Appendix C Analysis Reference Chart (continued)

Analysis	Container For Water	Preservative For Water (Chill to 4°C)	Container For Soil (Chill to 4°C)[a]	Holding Time (From Sampling Date)	
				Water	Soil
ORGANIC CHEMISTRY					
8010/8020	2X VOA (Headspace Free)	3 drops HCl[A]	4 oz. jar	14 days-Pres., 7 days-Unp.	14 days until Analysis
- 8010	2X VOA (Headspace Free)	3 drops HCl[A]	4 oz. jar	14 days	14 days until Analysis
- 8020	2X VOA (Headspace Free)	3 drops HCl[A]	4 oz. jar	14 days-Pres., 7 days-Unp.	14 days until Analysis
- BTXE	2X VOA (Headspace Free)	3 drops HCl[A]	4 oz. jar	14 days-Pres., 7 days-Unp.	14 days until Analysis
Modified 8015 (TPH)	4 oz. Amber Glass w/Septum (Headspace Free)	.25 ml HCl[A]	4 oz. jar	14 days until Analysis	14 days until Analysis
- Gasoline Range[F]	2X VOA (Headspace Free)	3 drops HCl[A]	4 oz. jar	14 days until Analysis	14 days until Analysis
- Diesel Range[G]	4 oz. Amber Glass w/Septum (Headspace Free)	.25 ml HCl[A]	4 oz. jar	14 days until Extraction 40 days after Extraction until Analysis	14 days until Extraction 40 days after Extraction until Analysis

Method	Container	Preservation	Container	Holding Time	Holding Time
8240	2X VOA (Headspace Free)	3 drops HCl[A]	4 oz. jar	14 days-Pres., 7 days-Unp.	14 days until Analysis
EDB	1 L Glass[B]	Unpreserved	8 oz. jar	28 days until Analysis	28 days until Analysis
8040	1 L Glass[B]	Unpreserved	4 oz. jar	7 days until Extraction; 40 days after Extraction until Analysis	14 days until Extraction; 40 days after Extraction until Analysis
8100/8310	1 L Amber Glass[B]	Unpreserved	4 oz. jar	7 days until Extraction; 40 days after Extraction until Analysis	14 days until Extraction; 40 days after Extraction until Analysis
8140	1 L Glass[B]	Unpreserved	4 oz. jar	7 days until Extraction; 40 days after Extraction until Analysis	14 days until Extraction; 40 days after Extraction until Analysis
8150	1 L Glass[B]	Unpreserved	4 oz. jar	7 days until Extraction; 40 days after Extraction until Analysis	14 days until Extraction; 40 days after Extraction until Analysis
Modified 619	1 L Glass[B]	Unpreserved	4 oz. jar	7 days until Extraction	14 days until Extraction

Appendix C Analysis Reference Chart (continued)

Analysis	Container For Water	Preservative For Water (Chill to 4°C)	Container For Soil (Chill to 4°C)[a]	Holding Time (From Sampling Date)	
				Water	Soil
8270	2×1 L Glass[B]	Unpreserved	4 oz. jar	40 days after Extraction until Analysis 7 days until Extraction 40 days after Extraction until Analysis	40 days after Extraction until Analysis 14 days until Extraction 40 days after Extraction until Analysis
Modified 632	1 L Glass[B]	Unpreserved	4 oz. jar	7 days until Extraction 40 days after Extraction until Analysis	14 days until Extraction 40 days after Extraction until Analysis
TCLP	N/A	N/A	4 oz. jar	N/A	14 days until Extraction
- Volatiles (zero headspace extraction)					14 days after Extraction until Analysis

Parameter	Container	Preservative	Holding Time
- Non-Volatiles	N/A	N/A	14 days until TCLP Leaching[H]
METALS (1 or more metals)			
• Total	6 oz. Plastic[B]	1 ml HNO₃[A]	6 mo. (28 days-Hg)
• Dissolved-Filtered in Field	16 oz. Plastic[B]	1 ml HNO₃[A]	6 mo. (28 days-Hg)
• Not Filtered	16 oz. Plastic[B] (Specify "To be lab filtered")	Unpreserved	6 mo. (28 days-Hg)
• Organic Lead	8 oz. Amber Glass (Glass Only) w/Septum (Headspace Free)	Unpreserved; Chill to 4°C	14 days until Analysis[H]
• Hexavalent Chromium (Cr + ^{6})	16 oz. Plastic	Unpreserved	24 hr.
• Total	16 oz. jar		
• Soluble	4 oz. jar		
- EP Toxicity	8 oz. jar		6 mo.
- WET	8 oz. jar		6 mo.
- TCLP (see also Organic)	8 oz. jar		
Chemistry			
• Hexavalent Chromium (Cr+^{6})	4 oz. jar		28 days
• Organic Lead	4 oz. jar		14 days until Analysis[H]

1. Fill all containers as much as possible. (Consult laboratory for minimum volume required.)
2. Holding time = the samples must be analyzed within the required time frame.
3. Most tests require samples to remain chilled @ 4°C after sampling.

A Typical volume needed to bring the pH < 2
B Additional volumes may be required - consult laboratory for recommendations
C Typical volume need the pH > 12
D Sample does not need to be chilled
E Typical volume needed to bring the pH > 9
F Applicable to all hydrocarbons in the Gasoline hydrocarbon range
G Applicable to all hydrocarbons in the Diesel hydrocarbon range
H Laboratory recommended holding time

4. VOA = 40 ml vial w/septum
5. Holding capacity: 4 oz. jar holds approx. 100 - 150g, 8 oz. jar holds approx. 300 - 400g, and 16 oz. jar holds approx. 600 - 800g
a Glass and teflon lined caps only

INDEX